PLANT MOLLUSCICIDES

PLANT MOLLUSCICIDES

Papers presented at a Meeting of the Scientific Working Group on Plant
Molluscicides, UNDP/WORLD BANK/WHO Special Programme for
Research and Training in Tropical Diseases, held in Geneva, Switzerland,
31 January to 2 February 1983

Edited by
KENNETH E. MOTT

Secretary, Steering Committee of the Scientific Working Group on
Schistosomiasis

A Wiley Medical Publication

Published on behalf of the UNDP/WORLD BANK/WHO Special
Programme for Research and Training in Tropical Diseases
JOHN WILEY & SONS LTD
Chichester · New York · Brisbane · Toronto · Singapore

Publications under copyright of the World Health Organization enjoy copyright protection in accordance with the provisions of Protocol 2 of the Universal Copyright Convention. For rights of reproduction or translation of such publications, in part or *in toto*, application should be made to the Office of Publications, World Health Organization, Geneva, Switzerland. The World Health Organization welcomes such applications.

This book contains the views and statements of an international group of scientists meeting under the auspices of the UNDP/WORLD BANK/WHO Special Programme for Research and Training in Tropical Diseases. Authors alone are responsible for views expressed in signed papers.

The geographical designations employed and the presentation of material in this book do not imply the expression of any opinion whatsoever on the part of the World Health Organization or of the Special Programme concerning the legal status of any country, territory, city, or area or of its authorities or concerning the delimitation of its frontiers or boundaries.

The mention of specific companies or of certain manufacturers' products does not imply that they are endorsed or recommended by the World Health Organization in preference to others of a similar nature that are not mentioned. No reference may be made to the World Health Organization or to the Special Programme in connection with any information contained in this book in statements or materials for promotional purposes. Errors or omissions excepted, the names of proprietary products are distinguished by initial capital letters.

Library of Congress Cataloging-in-Publication Data:

Plant molluscicides.

(A Wiley medical publication)
1. Plant molluscicides—Congresses.
2. Schistosomiasis—Prevention—Congresses.
I. Mott, Kenneth E. II. Special Programme for
Research and Training in Tropical Diseases.
Scientific Working Group on Plant Molluscicides.
III. Series.
RA641.M6P57   1987   614.5′53   86–19045
ISBN 0 471 91228 X

British Library Cataloguing in Publication Data:

Plant molluscicides: papers presented at a meeting of
   the Scientific Working Group on Plant Molluscicides,
   UNDP/WORLD BANK/WHO Special Programme for Research
   and Training in Tropical Diseases, held in Geneva,
   Switzerland, 31 January to 2 February 1983.
   1. Schistosomiasis—Prevention   2. Molluscicides
   I. UNDP/WORLD BANK/WHO Special Programme for
   Research and Training in Tropical Diseases,
   Scientific Working Group on Plant Molluscicides
   II. Mott, Kenneth E.
614.5′53   RA644.S3

ISBN 0 471 91228 X

Printed and bound in Great Britain.

CONTENTS

FOREWORD

The UNDP/WORLD BANK/WHO Special Programme for Research and Training in Tropical Diseases (TDR) is a goal-oriented research and training program with two interdependent objectives:

- research and development to obtain new and to improve existing tools for the control of major tropical diseases;
- strengthening of the research capabilities of the tropical countries.

The research is conducted on a global basis by multidisciplinary Scientific Working Groups; the training and institution-strengthening activities are limited to the tropical countries where the diseases are endemic.

The six diseases initially selected for attack are malaria, schistosomiasis, filariasis (including onchocerciasis), the trypanosomiases (both African sleeping sickness and the American form called Chagas' disease), the leishmaniases, and leprosy. Scientific Working Groups are also active in 'transdisease' areas: biological control of vectors, epidemiology, and social and economic research.

Scientists interested in participating in the Special Programme are invited to write for further information:

The Office of the Director
Special Programme for Research
and Training in Tropical Diseases
World Health Organization
1211 Geneva 27, Switzerland

PREFACE

New optimism about the control of schistosomiasis has a firm basis. The past decade has seen the development of safe and effective oral antischistosomal drugs and rapid, low-cost quantitative parasitological diagnostic techniques. It has also brought a broader understanding of the ecology and epidemiology of schistosomiasis. These advances will contribute directly to reducing the severity of the disease and the contamination of the environment by reducing the parasite burden in the affected human population; they will help control transmission as well.

In the midst of this remarkable progress, one important area – control of the snail intermediate hosts of schistosomiasis – has not benefited from the availability of new tools. It is generally agreed that chemotherapy and snail control should be coordinated to achieve maximum sustained reduction in prevalence and intensity of infection. But today, the chemical arsenal consists of only one molluscicide recommended for use in control programs. Although it is highly effective, it requires foreign exchange purchase, which may be restricted in many countries where the problem is worst.

Most of the 74 countries of the developing world where schistosomiasis is endemic have agriculturally based economies. The majority of the people at risk of becoming infected with schistosomiasis are poor farm workers and villagers who must derive their livelihood by seasonal, if not daily, exposure to contaminated waters. If a source of molluscicide could be derived from the agricultural activities of these countries, the effect would not only aid the national economy but also reduce the requirement for foreign exchange purchases.

Plant sources of pharmaceuticals are familiar to the biomedical community, and plants have been considered a logical source of molluscicides for many years. The rationale of the argument concerning their appropriateness in endemic countries is generally accepted. It will not be sufficient, however, to merely identify plants with molluscicidal activity. A number of scientific disciplines must contribute to development efforts if plants are ever to be used effectively.

The search for plants with molluscicidal activity must not be limited to the leads provided by folklore or anecdotal observations. The accumulated information on the botany of the world's plants goes far beyond mere identification and taxonomic classification (Balandrin et al., 1985). The growth characteristics, the agricultural potential, and even the chemical constituents of plants are becoming clearly defined. It should not be surprising that tannins and sesquiterpene compounds are molluscicidal. High levels of these compounds have been found in plants resistant to insects (Maugh, 1982). Perhaps other plants with molluscicidal properties can be identified by correlation of observations made in other areas of phytochemical research.

As molluscicidal plants are identified, their agricultural potential will be carefully evaluated. The adaptation of plants to new environments has been an area of intensive research (Boyer, 1982). Cultivation of plants taken from their natural environment may lead to unforeseen changes both in their chemical constituents and in their reproductive potential. In fact, it is estimated that only one quarter of all known plants have been or can be cultivated.

Our knowledge of plant genetics and plant biochemistry has steadily grown. Plant breeding now may proceed in a rational manner (Swaminathan, 1982). Genetic engineering is one of the specific tools that are aiding the adaptation of agricultural crops to human needs (Von Wettstein, 1983). Biochemical mutations can eliminate, increase, or decrease the content of specific constituents.

The Scientific Working Group (SWG) on Plant Molluscicides was convened by the Scientific Working Group on Schistosomiasis of the UNDP/ WORLD BANK/WHO Special Programme for Research and Training in Tropical Diseases (TDR) to examine the status of research in this area and to advise the SWG on Schistosomiasis on possible lines of future research. The working papers of this meeting collected in this volume will, it is hoped, stimulate basic research that will contribute to the development of new tools for the control of schistosomiasis. Professor Kurt Hostettmann and Dr. Andrew Marston have kindly provided Chapter 12, complementing the volume with a review of the literature on plant molluscicides since the time of the meeting.

In preparation of this volume the assistance of Miss Jane Carter and Mrs. Lynn Hollies, who typed the manuscripts, and Mrs. Marise Hostettmann, who reviewed the chemical and botanic nomenclature, is gratefully acknowledged.

K. E. Mott
Editor

REFERENCES

Balandrin, M. F., J. A. Klocke, E. S. Wurtele, and W. H. Bollinger. Natural plant chemicals: Sources of industrial and medicinal materials. *Science* 228:1154–1160 (1985).

Boyer, J. S. Plant productivity and environment. *Science* 218:443–448 (1982).

Maugh, T. H. Exploring plant resistance to insects. *Science* 218:722–723 (1982).

Swaminathan, M. S. Biotechnology research and Third World agriculture. *Science* 218:967–972 (1982).

Von Wettstein, D. Genetic engineering in the adaptation of plants to evolving human needs. *Experientia* 39:687–713 (1983).

1

MOLLUSCICIDES IN THE CONTROL OF SCHISTOSOMIASIS

G. Webbe

Department of Medical Helminthology
London School of Hygiene and Tropical Medicine
London, England

Molluscicides have 'a checkered history of success and failure' in the control of schistosomiasis (Olivier, 1971). This chapter puts that history in perspective with other methods of control. Particular attention is given to recent advances in epidemiological knowledge and techniques and to the new, safe, and highly effective schistosomicides (Webbe, 1981).

The future role of molluscicides may well be determined by economic considerations and the priority accorded the problem of schistosomiasis in relation to other public health issues. The future availability of the present generation of molluscicides is in doubt, and thus control prospects are compromised. No outstanding novel molluscicide has been developed in the past decade, and interest in such research by industry has diminished because of high development costs and lack of an assured market.

USE OF MOLLUSCICIDES

The early use of molluscicides in schistosomiasis control was reviewed and the factors affecting their performance were considered by Duncan (1974). He cited early attempts to kill snails with various chemicals and thus prevent infections, as reported by K. Fujinami and D. Nakamura in 1911, K. Fujinami and H. Narabayashi in 1913, Y. Miyagawa in 1913, and H. Narabayashi in 1915. Among the first compounds to be used as molluscicides and cercaricides were copper sulfate, calcium oxide, calcium cyanamide, chlorinated lime, calcium phosphate, and ammonium sulfate, a few of which were already used as fertilizers. Calcium arsenate and arsenite also have been used extensively in the Far East, including mainland China.

The Pacific campaigns of the Second World War helped draw attention to the disease (Ritchie and McMullen, 1961). Between 1946 and 1955, some 7000 compounds were screened as molluscicides, according to Ritchie (1973, quoted by Duncan, 1974). Pentachlorophenol (PCP), sodium pentachloro-phenate (NaPCP), and Dinex (dinitro-o-cyclohexyl phenol, DNCHP) emerged as potentially useful compounds and were extensively field tested. Later, Yurimin (3,5-dibromo-4-hydroxy-4' nitroazobenzene) was shown to be .16 to 18 times more active than NaPCP against *Oncomelania hupensis nosophora* and began to replace it in the Far East. However, Yurimin is no longer manufactured. Copper compounds of low or slow solubility have been evaluated in laboratory and field and may be of value where persistence and low toxicity to forms of aquatic life other than snails are desired. Copper carbonate has a proven residual effect in slow-flowing streams. Like the soluble copper sulfate, insoluble copper compounds act through their copper II^{++} ions. Because of their low solubility and high specific gravity, they are likely to be toxic only to organisms ingesting them on the bottom of water bodies. They are freely available throughout the world as commercial fungi-cides, formulated as small-sized particles and easily applied with conventional equipment (WHO, 1973; Cheng and Sullivan, 1974).

A review of the available and candidate molluscicides shows that only the nicotinanilide group of compounds, the organotins, and B-2 can be added to the list given in the report of the World Health Organization's Expert Committee on Schistosomiasis Control (WHO, 1973). B-2-sodium, 2–5 dich-loro-4-bromophenol (called B-2), in field evaluations against *O. nosophora*, achieved 95% snail mortality at a dosage of 10 g/m^2 using a 25% liquid formulation or 25 g/m^2 using a 10% granular formulation. The residual concentration of B-2 in the soil decreased rapidly, and the level in rice grains harvested from the treated paddy field did not exceed 0.03 mg/L (Kajihara et al., 1979).

The physical properties, toxicity, and formulations and field dosage of the compounds of chief interest today are listed in Tables 1 to 3. The two outstanding molluscicides developed during the past 20 years are niclosamide and trifenmorph (N-tritylmorpholine). Niclosamide, which is used in large-scale control operations, is available as a 70% wettable powder, a 25% emulsifiable concentrate, and a 60% wettable powder in two proprietary compounds, Bayluscide and Mollutox. Trifenmorph (called Frescon com-mercially) is produced as a 16.5% weight per volume emulsion concentrate but has not fulfilled its early promise and is no longer readily available.

Many tin and lead compounds have shown appreciable molluscicidal activity in the laboratory and in small-scale field trials. The high activity

Table 1. Effective Molluscicides – Physical Properties

Common Name and/or Active Ingredient	Form of Technical Material	Solubility in Water
Niclosamide; 2',5-dichloro-4'-nitrosalicylanilide	Crystalline solid	230 mg/L (pH dependent)
Trifenmorph; N-tritylmorpholine	Crystalline solid	0.02 mg/L
Sodium pentachlorophenate	Crystalline solid	33%
Copper sulfate	Crystalline solid	316 g/L
Yurimin; 3,5-dibromo-4-hydroxy-4' nitroazobenzene	Crystalline solid	Very slight
Bis (tri-n-butyltin) oxide	Liquid	20 mg/L
Nicotinanilide	Crystalline solid	Unknown

against *Biomphalaria glabrata* of a number of organotins, including bis (tri-n-butyltin) oxide (TBTO) (Hopf et al., 1967), has received extensive evaluation in slow-release formulations. Triphenyl lead acetate showed promise in field trials in Ethiopia.

The field use of these tin and lead compounds has been impeded by concern about their toxicity. In this regard, it is necessary to distinguish between inorganic tin and its salts on the one hand and organotin compounds on the other. The former have a low toxicological risk that is generally associated with poor absorption after ingestion. The latter vary: The monoalkyltins appear to be of low toxicity; the tetralkyltins are inactive *in vitro* but *in vivo* are converted to trialkyltin by the liver; and the trialkyltins are apparently more toxic than the dialkyltins. In view of the toxicological data available on organometals – TBTO, in particular – and the lack of information on their long-term cumulative effects in the aquatic environment, their use as molluscicides and larvicides is, as yet, not recommended (WHO, 1980; Duncan, 1980).

Nicotinanilide and its 3'- and 4'-chloro analogues reportedly are effective molluscicides in water at about 0.2 mg/L. Their ovicidal activity apparently varies, but many ova surviving treatments were said to be underdeveloped or abnormal. When these compounds were applied at dosages between 2 and 5 mg/L to fishponds, no obvious effects were noted on fish, frogs, tadpoles, or water weeds. The half-life in water of a solution of 4'-chloronicotinanilide was ten days (Dunlop, 1976).

These compounds seem to offer the possibility of selective control of snails; their development will require further laboratory and field trials of

Table 2. Effective Molluscicides – Toxicity

Common Name and/or Active Ingredient	Snail LC$_{90}$ (mg/L × h)	Snail Eggs LC$_{90}$ (mg/L × h)	Cercariae (mg/L)	Fish LC$_{90}$ (mg/L)	Rats Acute Oral LD$_{50}$ (mg/kg)	Herbicidal Activity
Niclosamide; 2',5'-dichloro-4'-nitrosalicylanilide	3–8	2–4	0.3	0.05–0.3 (LC$_{50}$)	>5000	None
Trifenmorph; N-tritylmorpholine	0.5–4	240	No effect	2–4	1400	None
Sodium pentachlorophenate	20–100	3–30	–	–	40–250	Phytotoxic
Copper sulfate	20–100	50–100	–	Toxic[a]	300	Phytotoxic
Yurimin; 3,5-dibromo-4-hydroxy-4'-nitroazobenzene	4–5	–	–	0.16–0.83 (LC$_{50}$)	168 (Mice)	None
Bis (tri-n-butyltin) oxide	3	0.01 (Prevented hatching)	0.001	0.3–16.2	194	Phytotoxic
Nicotinanilide	5	20–50	Unknown	>30	>2000 (Mice)	Unknown

[a] Toxicity depends very much on species of fish and water quality.

Table 3. Effective Molluscicides – Formulations and Field Dosages

Common Name and/or Active Ingredient	Formulations[a]	Field Dosage	
		Aquatic Snails (mg/L × h)	Amphibious Snails (g/m²)
Niclosamide; 2'5-dichloro-4'-nitrosalicylanilide	70% w.p. 25% e.c.	4–8	0.2
Trifenmorph; N-tritylmorpholine	16.5% e.c. 4% granules	1–2	–
Sodium pentachlorophenate	75% flakes 80% pellets 80% briquettes	50–80	0.4–10
Copper sulfate	98% granules (pentahydrate)	20–30	–
Yurimin; 3,5-dibromo-4-hydroxy-4'-nitroazobenzene	5% granules	–	5
Bis (tri-n-butyltin) oxide	6% in slow-release rubber pellets	20 g/m² (pellets)	–
Nicotinanilide	–	Unknown	Unknown

[a] e.c. = emulsifiable concentrate; w.p. = wettable powder.

slow-release formulations, investigation of field methods for analyzing low concentrations in water, and market evaluation and commercial production (J. Duncan, personal communication, 1978).

APPLICATION OF MOLLUSCICIDES

Two general strategies for snail control are in use: focal and areawide. The focal approach may be valuable where transmission is limited, but areawide control is more practical if transmission is spread over a watershed or an irrigation system. Focal control involves the identification and periodic mollusciciding of the transmission foci. To control an entire area or watershed unit, all snail habitats should be treated – a more difficult task initially than focal control but also likely to be longer lasting and therefore more economical.

Because of the diversity of snail habitats, both the chemical and the method of application must be carefully chosen. The final choice depends largely on the nature of the habitat, but cost-effectiveness must also be considered. Whichever chemical is chosen, it should be thoroughly dispersed in the given body of water. To do that, both natural and mechanical means must be relied on. Particular attention should be paid to impounded waters, where stratification due to vertical temperature differences may occur and the habitat may be so small that wave action and turbulence cannot help mix the chemical. In watercourses, a molluscicide may be carried along well in the main channel but fail to penetrate the lateral pools and marginal vegetation that harbor many snails. Attention should be given to possible attenuation of the chemical in flowing water; the farther it is carried the more diluted it becomes, and in long watercourses or channels it may be necessary to add more of the molluscicide at lower points (Jobin, 1968). Various colorimetric methods are available for determining chemical concentrations; bioassay methods also are used (Haskins, 1951; Strufe, 1961).

Certain molluscicides reportedly are degraded by bacterial action. This effect may be important in certain types of habitat (Etges et al., 1965; Bell et al., 1966), but many of the factors already discussed also cause loss of activity. It is important to know how long the molluscicide acts in all cases; in the case of flowing water it is essential, since dilution shortens the activity.

Little precise evidence exists that the snail intermediate hosts of schistosomes have acquired chemical resistance, although niclosamide resistance in *Bulinus truncatus* from Iran has been suggested (Jelnes, 1977). In the Sudan, *B. truncatus* from field populations exposed to trifenmorph for more than five years appeared to exhibit changes and take up the chemical more slowly than did snails from an untreated area (Daffalla and Duncan, 1979). On the other hand, *B. glabrata* from colonies exposed regularly for nine years to niclosamide in Saint Lucia showed no evidence of resistance when compared with snails from colonies from other parts of the island never subjected to chemical treatment (Barnish and Prentice, 1981). A test kit for molluscicide resistance has been assembled and is being evaluated (J. Duncan, personal communciation, 1980).

In comparing available molluscicides, the term milligrams per liter × hours (mg/L × h) or concentration × time (ct product) has been used to express the concentration of chemical applied and the duration of application. In general, concentration and duration are of equal importance in determining response to application, but there are theoretical and practical limits to both concepts; they are not fixed entities and may vary with circumstances. The variables include the chemical and physical properties of the molluscicide and the nature of its action, the chemical and physical constituents of the water, the species of snail, the ecological requirements, the microhabitat, and so forth. The ct product is normally calculated on the basis of a 24-hour exposure period, and because appreciable differences in ct values may exist over the range of times and concentrations used, the figures may not be realistic if applied to exposure periods of one to two hours (Ansari, 1973).

DELIVERY OF MOLLUSCICIDES

Available molluscicide formulations and application methods are limited in number. In flowing water habitats, both wettable powder suspensions and emulsion concentrates are usually applied with a constant head dispenser that delivers a specific concentration for a required period. Spraying with knapsacks and motor sprayers is usually required in impounded waters and delimited foci within them, the tail ends of irrigation ditches and drains, swampy areas, seepages, and paddies.

Aerial application has been carried out in some areas of endemic infection, with varying degrees of success. In the Sudan, this method apparently has been highly cost-effective (Barnish and Shiff, 1970; Sturrock and Barnish, 1973; Amin and Fenwick, 1977).

Recognition of focal areas of transmission has prompted attempts to devise cost-effective methods, including the application of slow-release formulations of a number of molluscicides (Cardarelli, 1974, 1977). Compounds that have been included in slow-release matrices (rubber sheets, baits, pellets) include copper salts, organotins, organoleads, niclosamide, and trifenmorph.

Field trials have been reported with TBTO, using Biomet SRM (slow-release matrix), the commercially available formulation containing 6% TBTO. The pellets were effective in controlling snails when applied at a dosage of 20 g/m^2, and no adverse effects were noted in the case of other biota in the treated area. The investigators believed that the molluscicide was confined to the depths of a body of water and exerted a selective action on snails (Shiff, 1974; Shiff and Evans, 1977). Other successful field trials of TBTO have been carried out in Saint Lucia (M. A. Prentice, personal communication, 1978).

Success in formulating either niclosamide or trifenmorph in slow-release matrices has been limited, and slow-release copper formulations appear to be of little value (Christie et al., 1978). The future of slow-release organometals, whether as larvicides or molluscicides, appears to be questionable unless more convincing and acceptable toxicologic data can be obtained.

The use of locally made formulations, including a gelatin granule formulation of niclosamide, has been reported in Saint Lucia (Upatham and Sturrock, 1977; Prentice and Barnish, 1980). In Ghana, field trials of a slow-release glass formulation showed that the material appears to have advantages over elastomeric matrices.

Better strategies and delivery sytems are needed to improve the cost-effectiveness of molluscicides. The possibilities of developing molluscicides from natural products of local origin in the affected countries need to be explored and necessary knowledge of feasible agronomy developed. Known plant molluscicides have been reviewed (Kloos and McCullough, 1981; see also Chapter 3) and will not be discussed further in this paper, but the need for adequate data on toxicology and methods of extraction and large-scale production is apparent.

CONTROL OF THE MOLLUSCAN INTERMEDIATE HOST

Planning and evaluating snail control measures demands a thorough knowledge of the molluscan host – its ecology, bionomics, population trends, and dynamics. Transmission of schistosomiasis is characterized by variability, and direct extrapolation of data, whether on snail populations and their infections from one area to another or on human ecology, may not always be valid, even for adjacent areas.

The failure to obtain control is usually due to a lack of such basic information. It is also often due to the dissipation of applied efforts through wrong emphasis or faulty timing. The abundance of snails and of concomitant cercarial infections fluctuates in response to local and seasonal climatic conditions. The number of infected snails that survive the winter or carry infections from one wet season to the next after aestivation, the rate of infection of new generations of snails, and the time required to develop cercariae are all pertinent to the choice and timing of control measures.

In many tropical areas, fluctuations in numbers of snails and in the production of cercariae are as great as in temperate conditions. Transmission may be quite limited, depending on the presence of water during certain periods of the year; in such a case, snail control measures may be necessary only at particular times and for limited periods. In other areas, where water and temperature remain at stable levels, snail populations and the production of cercariae are maximal throughout the year; chemical measures directed against snails must be applied continuously, or permanent measures involving habitat modification may be called for. In many areas, the focal nature of transmission is becoming increasingly recognized, resulting in the development of the more cost-effective transmission control methods. Furthermore, methods that are applicable for species of *Biomphalaria* and *Bulinus* in aquatic habitats differ in emphasis from those for the amphibious *Oncomelania* species (Jordan and Webbe, 1982).

In evaluating control measures, it is important to consider the period during which they are likely to remain effective. Few if any of the available methods of control that involve the use of chemicals are likely to completely eradicate the snails, and their high intrinsic rate of natural increase is likely to result in a rapid restoration of depleted populations. Prevention of breeding may therefore be more important than killing snails, and a permanent result may require an alteration first of the environment and then of the ecology of the population (Pesigan et al., 1958; Webbe, 1962). The intrinsic rate of natural increase will not be affected by such changes, but the actual, or realized, rate of increase will be influenced. If environmental resistance to snails is increased and the carrying capacity of the population lowered, population density will ultimately be decreased. Changes in the environment may bring this about by causing reproduction to slow down or by reducing survival rates. In either case, an unbalanced age structure will result.

No matter which factor is chosen, it should be density independent and lead to complete elimination of the snail population (Webbe, 1965). But even if complete eradication is not accomplished, reduction in transmission may still be achieved by decreasing the snail population and altering its age structure.

Furthermore, the technology involved in the use of molluscicides in natural habitats, including controlled water management, is now well established. Available data show that molluscicides are most cost-effective where the volume of water to be treated per capita at risk is small. Hence their use is well suited to arid areas where transmission is seasonal and confined to relatively small habitats – such places as Saudi Arabia, the Yemen Arab Republic, and parts of Zimbabwe and Tanzania. Their use may be unsuitable in large rivers and lakes unless transmission is focal as in Volta Lake, Ghana. Where the population density is high and the volume of water per person is low, molluscicides may be cost-effective even though the total volume of water is large (WHO, 1973). Irrigation units with controlled water management in densely populated areas (as in Egypt, the Sudan, China, and northeastern Brazil) are well suited to cost-effective control of transmission by the use of molluscicides.

Snail control by periodic, areawide application of molluscicides is being carried out successfully in several major control programs based on irrigation and controlled water management. Transmission control, based on the essential focality of transmission, is also being prosecuted successfully by means of molluscicides and surveillance. This latter approach, based on accurate knowledge of human water-contact patterns, may be highly cost-effective, and may result in a considerable saving of expensive chemicals. It must be realized, however, that focal transmission control may be totally impracticable in a flowing water system where the human population density is high and there is widespread diffuse domestic and occupational water contact. Further, it is labor intensive and requires considerable supervision and management, along with adequate logistic support. On a large scale, these requirements may prove too costly, offsetting the saving due to lower use of chemicals (Jordan and Webbe, 1982).

The efficacy of chemical control measures may be increased in natural habitats and irrigation systems by environmental modifications, along with improvements in water management and agricultural practices. In the Philippines and Japan, control of irrigation water, drainage works, and improved agricultural methods have been very successful. Habitats have been eliminated or are now amenable to control by molluscicides, and valuable areas of reclaimed land have been established (Hairston and Santos, 1961; Yokogawa, 1972). Ecological methods and habitat treatment in schistosomiasis control were reviewed by Bradley and Webbe (1978).

The costs and benefits of habitat modifications in any control process must be estimated. Coupled with improvements in sanitation and water supply, ecological methods may offer the long-term solution for schistosomiasis control – but in most cases they must be implemented in conjunction with short-term methods having predictable rapid effects, such as the use of molluscicides combined with chemotherapy.

The incidence and severity of schistosomiasis frequently increase as a result of multiple developmental activities and careless engineering practices. Many recent programs for developing water resources in affected areas have brought a pronounced concomitant increase in schistosomiasis. New man-made lakes and irrigation systems are important factors in the spread of infection (Hunter et al., 1980). Public health problems associated with the man-made lakes in Africa have become significant; these lakes all contain snail intermediate hosts of human schistosomiasis, and the need is urgent for suitable control measures.

CONTROL OF TRANSMISSION

Such specific measures as snail control and chemotherapy generally have little or no impact on health problems other than schistosomiasis and thus are likely to be charged directly to a schistosomiasis health budget. Non-specific control measures involving water supply, sanitation, and general improvement of the environment and living conditions are less dramatic, but they have wider social and medical benefits and their costs can be shared by public health and community development budgets. The effective use of nonspecific snail control measures brings about a slow reduction in the prevalence and intensity of infection and then a reduction in morbidity. The use of chemotherapy, however, has a direct impact on the disease and can quickly reduce its prevalence and intensity and then its transmission (Jordan and Webbe, 1982).

During the 1960s, the use of molluscicides was the only reliable approach to the control of schistosomiasis. During the next decade, it became apparent that integrated methods directed against different links in the life cycle were the best means of achieving rapid control. It is also clear now that any control program must take into account the particular approach determined by local conditions, the goals of each control effort, and the available resources.

SPECIFIC PROGRAMS

In the report of the Expert Committee on Schistosomiasis Control (WHO, 1973), it was noted that projects using molluscicides alone had made a clear impact on the incidence of infection in Ghana, Tanzania, Egypt, and Japan.

In other countries, where changes in the prevalence of infection after the use of molluscicides were estimated, pronounced reductions were observed in Brazil (Paulini et al., 1972) and Zimbabwe (Shiff et al., 1973).

Successful research programs and pilot and major control schemes employing both molluscicides and chemotherapy were carried out in Madagascar (Degremont, 1973), Tanzania (Fenwick, 1972; McCullough and Eyakuze, 1972), West Cameroon (Duke and Moore, 1976), Iran (Arfaa et al., 1970), the Fayum Governorate of Egypt (Mobarak, 1982), and Zimbabwe (Shiff et al., 1973).

Combined chemotherapy, molluscicide application, environmental control, sanitation, health education, and legislative action were applied in long-standing control programs in Japan (Yokogawa, 1972) and Venezuela (Faria, 1972). Both programs resulted in marked reductions in schistosomiasis as a public health problem.

The substantial and scientifically acceptable data available for many of these programs show the efficacy of both single and combined control measures. In particular, they demonstrate the valuable role molluscicides play in the control of schistosomiasis (Webbe and Duncan, 1978). Several notable research programs and pilot and national control programs are summarized in the following paragraphs.

Saint Lucia

An areawide mollusciciding campaign in Cul de Sac Valley in Saint Lucia reduced the incidence of *Schistosoma mansoni* from 22% to 4.3% between 1970 and 1979 (Jordan et al., 1978). The estimated annual cost was US$3.24 per person protected.

Afterward, a two-year focal surveillance and mollusciciding program was introduced. Sites of potential transmission were identified and routinely searched for *B. glabrata*. If any were found, the site was treated with Clonitralide (25% emulsifiable concentrate niclosamide). Two chemotherapy campaigns supplemented the snail control program.

As a result of these combined measures, the incidence of infection in two areas dropped from 4.3% to 1.0% and 2.2% to 0.6%. The total cost of protecting the population of 7000 was US$20362, of which labor absorbed 68%, transport 24%, equipment 4%, and molluscicide 4%. The annual cost per person protected was US$1.45, which compares favorably with the US$3.24 of the areawide campaign. Although this program was relatively inexpensive, its success depended on the provision of a high standard of management and supervision (Barnish et al., 1980).

From 1977 to 1981, routine focal mollusciciding was further evaluated. The annual cost per capita was US$0.46 (Barnish et al., 1982).

Ghana

Investigations in northwestern Ghana, followed by a pilot control project (Lyons, 1974), showed that selective, dry-season molluscicidal control of *Schistosoma haematobium* foci can reduce transmission. Ponds and small riverine habitats were treated with a 0.5 mg/L concentration of Bayluscide (70% wettable powder); large habitats received a 1 mg/L concentration. The annual cost per capita in the villages protected was US$2.34.

A research project on the epidemiology and methodology of schistosomiasis control in man-made lakes was conducted on Lake Volta by the Ghana government, UNDP, and WHO. Intervention measures, which began in 1976, involved snail control in focal lakeshore transmission sites by applying molluscicides and removing weeds. The project also included selective chemotherapy, health education, and provision of water supply in selected villages. A pronounced reduction in transmission ensued, with corresponding falls in the prevalence and intensity of infection in the population of the study villages (D. Scott, personal communication, 1978).

Klumpp and Chu (1977) established the essential focality of transmission, which was found to be seasonal and was correlated with lake-level drawdown and related vegetational changes. Transmission foci were sprayed monthly with a 0.5 mg/L concentration of Bayluscide (70% wettable powder) (Chu, 1978).

Annual per-capita costs were estimated in 1978 to be US$1.94 for chemotherapy and US$1.10 for snail control. For the combined measures, which protected a lakeside population of 7000 and an additional hinterland community of 8000, the per-capita cost was US$3.04 (WHO 1979).

Sudan

A pilot project area of 1,000 km^2 in the Gezira irrigated area of the Sudan received a series of applications of trifenmorph (Frescon, 16.5% emulsifiable concentrate). The results have been used since 1971 to formulate a regimen aimed at reducing the snail population in canals by at least 95% of the pretreated levels in order to control the transmission of *S. mansoni*.

Drip-feed application failed, except from September through December. Even then, when water use was maximal, the molluscicide did not cover the entire system, and each application had to be supplemented with knapsack spraying of canal tail ends. Subsequently, aerial spraying was established as the quickest, cheapest, and most effective application technique. In this regimen, all main, major, and minor canals were sprayed with Frescon (0.25 mg/L) five times a year, in early September, mid-November, late January, mid-March, and early June (Amin and Fenwick, 1977).

The total annual cost of aerial spraying in the 1,000 km^2 pilot area was estimated at US$0.73 per person protected. Epidemiological evaluation of the initial control measures indicated that results were equivocal in terms of reduced transmission, and a new regimen is now being evaluated. Alternative methods of control are also being considered, including the use of an alternative molluscicide (Bayluscide), focal rather than areawide control, provision of water supplies, and chemotherapy. Under the Blue Nile Health Project, engineering and biological control methods are also being evaluated.

Egypt

Several notable pilot control programs have been conducted in Egypt. In the Dakhla Oasis in the Western Desert and at Warraq El-Arab northwest of Cairo, copper sulfate and NaPCP were used. In a joint WHO/UNICEF/ Egyptian government project in Baheira Governorate, Bayluscide was employed (Gilles et al., 1973; Ayad, 1976).

In 1968, large-scale control operations began in Fayum Governorate, an irrigated area of 400,000 feddans with a population of 1,163,000 and an overall prevalence of *S. haematobium* of 45.7%. The areawide control strategy involved dispensing Bayluscide (70% wettable powder) in the spring, summer, and autumn, followed by surveillance and spraying as required. Specific population chemotherapy with tartar emetic and niridazole was also used. Epidemiological evaluation showed that the prevalence dropped to 9.1% after five years. In the later maintenance phase, areawide applications were made in the spring and autumn from a single dispensing point, followed by surveillance and spraying if required. Metrifonate (Bilarcil) has also been used to treat infected persons, at an annual cost of approximately US$0.50 per person protected.

In 1976, a major control program began in Middle Egypt, in an area of some 1,050,000 irrigated feddans with a population of 4,500,000. It is the same area in which a major tile drainage program was undertaken in collaboration with the World Bank. The area was selected not only because this large-scale reclamation operation was taking place there but also to reinforce the control operations in Fayum Governorate and to prevent transmission of *S. mansoni* south of Cairo. Overall prevalence of *S. haematobium* in the area was estimated to be 34.3%. The areawide control strategy involved dispensing of Bayluscide (70% wettable powder) in irrigation canals in the spring, summer, and autumn, with complementary spraying of all drains and static waters and population chemotherapy using Bilarcil. Snail control operations were based on synoptic data on irrigation rotations in the area. As in the Fayum control operations, the Bayluscide formulation proved to be highly efficient. It was carried from single dispensing points over long distances through secondary and tertiary canals and in main drains. Concentrations of 1 to 2 mg/L of active ingredient were applied, depending on the prevailing water conditions.

Epidemiological evaluation after three years in a sample of 58 villages showed that prevalence had been reduced from 29.7% to 16.3%. The annual per-capita cost of the program was estimated to be US$1.00 (Mobarak, 1982).

In 1979, control operations began in the irrigated area of 1 120 970 feddans between Aswan and Assiut. This area has a population of 5 099 000. The strategy used here was the same one that was applied in Middle Egypt. This project was to consolidate control operations in Middle Egypt and complete the plan to control *S. haematobium* transmission from Aswan to Giza, which should be done if potential settlement populations for new townships on the shores of Lake Nasser are to be drawn from areas in Upper Egypt. Perennial irrigation in Upper Egypt, which replaced basin irrigation when the Aswan High Dam became operational, resulted in a marked increase in the prevalence of infection.

Elsewhere in Egypt, limited snail control operations are using copper sulfate and niclosamide (as Mollutox, a formulation produced in Egypt). New control programs using chemotherapy and molluscicides are planned for the Delta.

Saudi Arabia

Because schistosomiasis transmission in Saudi Arabia is mostly focal – from wells, small canals and swamps, cisterns, temporary streams, residual pools, and ponds – the country offers the prospect of conducting successful snail control operations (Arfaa, 1976; Davis, 1977). Prevalence rates are high in many parts of the country, particularly in Gizan (Magzoub and Kasim, 1980). Bayluscide and Mollutox are being used in an attempt to control *Bulinus truncatus*, *B. beccarii*, *B. reticulatus*, *B. wrighti*, and *Biomphalaria 'arabica'* populations.

Iran

Control measures covering all of Khuzestan Province, an area of endemic infection, commenced in 1966. Bayluscide has been used, along with swamp drainage and filling operations, chemotherapy, improvements in water supply, sewage disposal, and health education. These measures brought a marked reduction in the prevalence of infections, at an annual estimated cost of US$0.40 per capita.

More recent information indicates, however, that effective control has not been maintained in all situations. The reasons are unknown to the author. Preliminary data suggest that *B. truncatus* from Dezful, where mollusciciding had been carried out for ten years, may have developed resistance to Bayluscide (Jelnes, 1977).

Brazil

Successful pilot control projects using Bayluscide as the sole means of control have been carried out in natural drainage and primitive irrigation areas in Belo Horizonte, Sao Laurenco, and Taquarendi.

A four-year national control program was launched in 1975 with a budget of US$195 million; it included population chemotherapy (using oxamniquine), mollusciciding, improved sanitation and water supplies, and health education. Encouraging results were obtained, but long-term evaluation was needed (Machado, 1982).

Venezuela

A 35-year control program in Venezuela reduced the prevalence of schistosomiasis infections from 14% to 1.8%. An estimated 250000 persons were at risk in a 15000 km² endemic area in the north-central part of the country. The control program included diagnosis of all cases and treatment with oxamniquine, which proved effective in 92% of the cases without important side effects; control of the snail host, *B. glabrata*, by the use of copper preparations, niclosamide, and trifenmorph; environmental and sanitary improvements; and health education (WHO, 1980).

Tunisia

Until a few years ago, urinary schistosomiasis was endemic in southern Tunisia. It occurred mainly in oases in the Governorates of Gabès and Gafsa, with an isolated focus in the Governorate of Kairouan. The population at risk, up to 200000 persons, was submitted to a parasitological survey that identified 11596 cases. The overall prevalence in endemic areas was 6.6%, but many localities had infection rates of 30% to 70%. The snail intermediate host was *B. truncatus*.

A WHO-assisted control project was established in June 1970, and control measures were started in December 1971. Niclosamide was applied to every water collection containing *Bulinus* in or around human settlements, and every person who had a positive urine examination was treated with niridazole (20 to 25 mg/kg daily for seven days), provided there were no contraindications.

In 1972, snail hosts were found in 42.6% of water sources – natural springs, artesian wells, and irrigation canals. The first niclosamide treatment was sufficient in 75% of the places treated; the others required two or more applications.

In addition to the malacological surveys, project evaluation was carried out each year. The prevalence of infection in ten localities with a total of 14000 inhabitants fell from 34.1% in 1972 to 0.6% in 1978. Every positive case now found is the object of an epidemiological inquiry to establish whether it is new and to assure identification of any active transmission site. No transmission has been registered in Tunisia during the past two years. These data suggest that the control measures, along with spontaneous cure of untreated cases, will result in the disappearance of human sources of schistosomiasis in the years to come (WHO, 1980).

Morocco

In Morocco, only *S. haematobium* is endemic; the total number of cases is probably fewer than 50000. *B. truncatus* may be the only snail host, but in certain foci the role of *Planorbarius metidjensis* merits investigation.

Because Morocco is almost at the extremity of *S. haematobium's* African distribution, seasonal climatic effects, particularly those of temperature, play a crucial role in transmission.

In the past, many foci of *S. haematobium* tended to be unstable; some even disappeared. In recent years, with the development of permanent water resources, mainly for irrigation, some more stable foci of *S. haematobium* created concern, and Moroccan health authorities initiated a nationwide control-and-surveillance program under the aegis of a national schisto-somiasis commission.

Manuals were published by the Ministry of Health providing standard practical procedures for the early phases of control commensurate with local resources of personnel, facilities, and equipment. The essential groundwork for a national control program is being laid, and it augurs well (WHO, 1980).

Philippines

An estimated 3.9 million persons in 22 Philippine provinces are exposed to transmission of *Schistosoma japonicum*. A five-year schistosomiasis control plan has been formulated, with a budget of US$188 million. Two mol-luscicides will be applied: Bayluscide and Biomet. Biomet, which contains TBTO, will be used in a rubber formulation and is intended to serve as a cercaricide in streams to which workers are exposed during reclamation work. The program also will include population chemotherapy; health education; environmental sanitation; and snail control by land reclamation, agro-engineering, and drainage and filling. The annual cost is expected to be US$9.64 per person protected.

China

In the multiple control approaches applied in endemic areas in China, extensive use is made of a variety of molluscicides. PCP, even though it is not soluble in water, is favored over NaPCP because it is cheaper. It is also much cheaper than niclosamide and other synthetics, which have been used in various areas. In some programs, an extract of *Camellia oleosa* (tea-cake seeds) and ethylene diamine have been used. Ethylene diamine (10 to 15 g/m^2) is sprayed on the banks of fishponds, and the saturated soil is then immersed in the water. While a concentration of less than 100 mg/L is lethal to snails, the lethal concentration for fish is 3200 mg/L; thus it is a most suitable compound for use in fishponds.

Apparently, the Chinese employ a combination of snail control techniques, including burying the snails, spraying molluscicides both on water surfaces and on snail habitats, and using PCP at 10 g/m^2. The mixture of bank weeds, soil, and snails is moved below water level near the shoreline; the compacted mass is reported to retain a high concentration of PCP, with subsequent high mortality of snails both above and below the water level. Rice fields are also treated: Before the young rice shoots are transplanted, the fields are leveled by hand and water and calcium cyanamide or PCP are added (A. Davis, personal communication, 1975).

Despite intensive integrated measures, control of schistosomiasis in marshlands, lake regions, and mountainous areas remains a difficult problem (Mao and Shao, 1982; see also Chapter 11).

Swaziland

Schistosomiasis is widespread in Swaziland. An estimated 150000 persons are infected, the more common form being *S. haematobium*. The intermediate snail hosts, *Bulinus globosus* and *Biomphalaria pfeifferi*, are widely distributed. Sugar and rice irrigation schemes favor their distribution, and the numerous small water reservoirs play an important role in transmission.

For the focal control of snails in limited areas, niclosamide is used during the transmission season (October to March). A strategy adopted in 1970 in connection with irrigation schemes was to treat all snail habitats with trifenmorph every seven weeks during the transmission season, with the aim of controlling transmission throughout the irrigation systems. These measures were a joint effort of the irrigation estates and the Ministry of Health. In 1974 and 1975, the schistosomicide hycanthone was given extensively. In 1976 a school health program started giving special attention to schistosomiasis. Four mobile teams, each with a public health nurse in charge, examined all new school entrants. These control programs have reduced the prevalence of schistosomiasis in Manzini and some of the sugar estates, but they appear to have been ineffective in the system of reservoirs (WHO, 1980).

Puerto Rico

Five pilot schistosomiasis control programs were initiated in Puerto Rico in 1953. At present, the primary aim is to control the snail intermediate host, *B. glabrata*. NaPCP was used first, together with drainage of swamps and seepages by hand ditching. In 1958, acrolein was introduced in the larger irrigation canals in the south for weed and snail control. In 1963, Bayluscide replaced NaPCP.

A biological control program began in 1958, with the use of *Marisa cornuarietis* against *B. glabrata* in irrigation systems, farm ponds, and big lakes (Ruiz-Tiben et al., 1969; Jobin et al., 1970, 1977). This program is estimated to be some two orders of magnitude cheaper than chemical control, with a high level of effectiveness. In ponds and lakes, its cost is 1% of the cost of chemical control.

Multiple control measures applied during the past 25 years (including snail control by environmental, biological, and chemical means; provision of improved public water supply; a latrine distribution program; and limited chemotherapy) have cost US$7 million, or approximately $1.00 per person protected (Webbe and Duncan, 1978; Cline, 1972).

ADVANCES IN EPIDEMIOLOGICAL KNOWLEDGE AND TECHNIQUES

Studies on epidemiology, pathology, and quantitative necropsy have established that egg output, measured in the field as eggs per gram of feces (*S. mansoni, S. japonicum*) or eggs per unit volume of urine (*S. haematobium*), accurately reflects intensity of infection (worm burden). A direct linear relationship appears to exist between intensity of infection and clinical disease. The linear relationship of hepatosplenic disease to intensity has been derived for *S. mansoni* from data obtained in Puerto Rico, Brazil, and Saint Lucia (Jordan, 1977). This correlation is, however, less positive for similar data obtained in Kenya and other African areas, and further studies are necessary.

Most infected individuals in a population harbor few parasites. Only a small proportion are heavily infected, and they are responsible for excreting the bulk of eggs and causing most of the contamination in the environment. In one instance, 6% of a population was found to excrete 50% of the eggs (Cook, 1974). Such data may be of direct relevance in developing new kinds of control strategies, including targeted or selective treatment of the heavily infected (Kloetzel, 1974). This approach may be highly effective in preventing disease in high-risk groups, but its value in controlling transmission is uncertain (Webbe, 1981).

Egg-counting techniques have been refined with the introduction of quality-control procedures (Bartholomew and Goddard, 1978; Goddard, 1980). Immunodiagnosis has also advanced, and recent studies have clarified the meaning of epidemiological indices and their implications in control programs.

Interest continues in the development of epidemiological models of schistosome transmission and their relevance to control strategy (Fine and Lehman, 1977). Progress will depend on greater use of recorded data and consideration of seasonal variation of transmission, immune phenomena, and differences in the epidemiological characteristics of endemic conditions.

Integrated methods can assure rapid and effective control of transmission. Where funds are limited, however, disease prevention may still be obtained by applying a single control approach through selective or targeted chemotherapy. The choice must be determined by local considerations.

Evidence is ample that areawide mollusciciding is now successfully controlling snails in major control programs, as in Egypt and China. Control of transmission based on the essential focality of transmission in many areas (Saint Lucia, Ghana, Yemen, and Saudi Arabia) is also being successfully prosecuted by killing snails and by surveillance. Although mollusciciding needs more adequate strategies and delivery systems to optimize its cost-effectiveness, it will continue to play a vital part in the integrated control of schistosomiasis. Nevertheless, the recent advances in epidemiological knowledge and techniques, together with the advent of new, safe, and highly efficacious schistosomicides, have appreciably increased the prospects for using population-based chemotherapy in controlling transmission and for more direct disease prevention.

Other available methods of control reduce transmission without any direct effects on the human worm load, but chemotherapy reduces egg output and thus reduces transmission. Clearly, much more information is required on the relative merits of such regimens as mass treatment (treatment of an entire community), selective population chemotherapy (examination of the community, followed by treatment of infected individuals), and targeted chemotherapy (treatment of high-risk groups only, after diagnosis). What are the effects on transmission of treating only high-risk groups? How does treatment affect acquired immunity, and should this effect be considered in formulating treatment policy? When is the best time to apply chemotherapy to stop transmission, and how frequently should it be used? The development of schistosome resistance to individual compounds and cross-resistance to others must be monitored along with long-term drug effects if periodic campaigns are to be used in a maintenance strategy (Webbe, 1981).

COSTS

The cost of schistosomiasis control measures is important because most countries where the disease is endemic are poor. Few pilot control projects or large-scale programs have developed cost-effectiveness data, and available figures are frequently not comparable because of the different measurements made and components included. Existing control programs have estimated annual recurrent costs ranging between US$0.40 and $12.00 per capita. Very poor countries with widespread transmission thus may be unable to attempt generalized control of transmission.

Where data are available, the per-capita cost has been a significant proportion of the national health budget (Hoffman et al., 1979). The annual cost of snail control to protect 6000 persons from exposure to *S. mansoni* in an irrigation system in Tanzania was estimated in 1970 at US$1.31 per capita (Fenwick, 1972); for an irrigation scheme in Kenya in 1973, it was approximately US$1.00 per hectare or US$0.19 per person protected. These costs were considered very favorable when compared with programs in Egypt and Zimbabwe (Choudhry, 1974). For measures in natural drainage systems, costs are substantially higher: US$7.80 to $9.70 per 100 m^3 of water treated in Puerto Rico, for example (Jobin et al., 1970).

Both labor and molluscicide costs have risen sharply in recent years, as reflected in programs already discussed. At Msungui in Tanzania, the cost per person protected was US$0.75 for mollusciciding and $3.37 for chemotherapy (McCullough and Eyakuze, 1972). In Saint Lucia the annual per-capita cost for the first two years was US$1.10 for chemotherapy, $3.24 for mollusciciding, and $4.59 for water supplies. Subsequent transmission control by surveillance and focal mollusciciding cost US$1.45 and then US$0.46 per capita annually (Barnish et al., 1982), while treatment with oxamniquine cost $0.94 per person protected (Jordan et al., 1980). In Ghana, the annual costs per capita were estimated in 1978 to be US$1.94 for population chemotherapy and US$1.10 for snail control. The relatively low cost per person protected (US$1.00) in the Egyptian control programs employing combined areawide mollusciciding and population-based chemotherapy shows the cost-effectiveness of measures where the volume of water being treated per capita is small. In this case, 80% of the budget was spent on the molluscicide, which cost US$17000 per metric ton.

The available data suggest that in many cases chemotherapy may be the most cost-effective short-term method of reducing prevalence, incidence, and intensity of schistosomiasis infection and disease, at least for *S. mansoni* and possibly for *S. haematobium* (Hoffman et al., 1979). An analysis of data from Iran has shown, however, that the minimum level of expenditure for chemotherapy was not effective in maintaining low levels of prevalence, and that it is important to distinguish between two different indications of the effectiveness of control measures: the prevalence level achieved at the end of a stated period and the number of case-years of infection prevented (Rosenfield et al., 1977). In general, the model results indicated that the integrated control program was more successful in reducing prevalence but that, at the end of the program, use of chemotherapy had prevented more case-years of infection. The authors concluded that control program planners need to look at a time horizon that is long enough to achieve effective control. They believed that mollusciciding could achieve and maintain low prevalence levels. Management strategies have also been examined in relation to data from Saint Lucia and Ghana through the application of transmission models for predictive epidemiology and cost-effectiveness analysis of schistosomiasis control measures (Rosenfield, 1979).

Control programs should aim to achieve a reduction in prevalence and intensity that is great enough to make the disease of minimal public health importance in relation to other endemic diseases. Such a reduction may be possible with current technology. Where schistosomiasis is not a clear priority, however, the high cost of drugs, molluscicides, and other methods may deter control operations. Thus it is imperative to develop less costly technology.

CONCLUSIONS

The availability of new efficient drugs and advances in epidemiological knowledge have led to the optimistic belief that various drug delivery systems can be used alone for control purposes in many situations. It is true that chemotherapy, by reducing the severity of disease and the number of eggs contaminating the environment, has a vital role to play in the control of transmission. It should be recognized, however, that experience in the use of drugs for large-scale population-based chemotherapy or targeted treatments is limited and the proposed delivery systems have not been adequately tested. It is difficult to treat large populations in a short period, particularly if there is substantial migration. The effect of targeted treatments on transmission is unknown; in large-scale programs, integrated measures may reduce prevalence and intensity and control transmission more cost-effectively in a short period. Other endemic disease control programs that relied exclusively on chemotherapy had difficulties after some years, either because the parasite developed drug resistance or because surveillance was relaxed after the incidence of infection was substantially reduced (WHO, 1980). The WHO Expert Committee in 1978 therefore considered it of paramount importance to encourage the use of environmental, behavioral, and other nonchemotherapeutic methods in programs using chemotherapy on a large scale. The committee distinguished three phases of control operations: first, definition of program goals in light of the priority accorded to the problem, epidemiologic evaluation, the design of a feasible strategy, and allocation of resources;

second, active intervention using the chosen strategy and achieving a substantial reduction in endemicity; and third, protracted maintenance of control measures with reduced application of the strategy. How much control activity is necessary following the intensive phase is not known because experience is limited; further evaluation of present programs in this phase of control is needed.

Costs must be related to the goal of any control initiative, and the cost-effectiveness of a particular method or an integrated strategy should be expressed in terms of the time scale involved in achieving a particular goal. The scale and periodicity of control inputs during a protracted consolidation phase will differ from those applied in the initial phase of intervention, and cost-effectiveness data should be carefully assessed in relation to their magnitude and timing and the possible duration involved.

Those who are responsible for control operations must ensure adequate provision for a long period of sustained effort and recurrent expenditure. They must also identify and apply measures that will result in effective control and achieve predictable goals in short-, medium- and long-term periods. Snail control by mollusciciding, coupled with chemotherapy, will undoubtedly continue to play an important role in the integrated control of transmission, particularly during an intensive phase of intervention. Moreover, control of snails will be an important factor during the consolidation phase of many operations, when surveillance and routine focal applications can be employed with considerable cost-effectiveness in well-defined situations.

Every effort must therefore be made to ensure the future availability of molluscicides, to improve their formulations if necessary, and to explore the possibility of using low-cost natural materials of local origin. Plant molluscicides could well be used in self-help projects, and more field research is required to couple effective habitat modifications with the appropriate use of chemical control of snails.

A great deal might also be done to restore the confidence of industry in the problem of molluscicide development if those responsible for making policy and shaping health delivery systems were more positive in recognizing the important role of snail control in the overall management of the schistosomiasis problem.

The present cost of control will make it difficult for poor countries to establish effective control programs within the constraints of their small health budgets. No such program will be seriously undertaken and no substantial progress can be achieved in these countries until costs are generally less than US$1.00 per person protected. It is, however, considered very unlikely that effective schistosomiasis control delivery systems can be designed within the primary health care concept, which itself is by no means a cheap option, and lacks the necessary framework for precise technology and time-related inputs.

REFERENCES

Amin, M., and A. Fenwick. The development of an annual regimen for blanket snail control on the Gezira irrigated area of the Sudan. *Ann. Trop. Med. Parasitol.* 71:205–212 (1977).

Ansari, N., ed. *Epidemiology and Control of Schistosomiasis*, 458–532. Basel: S. Karger (1973).

Arfaa, F. Studies on schistosomiasis in Saudi Arabia. *Am. J. Trop. Med. Hyg.* 25:295–298 (1976).

Arfaa, F., G. H. Sahba, and H. Bijan. Progress towards the control of bilharziasis in Iran. *Trans. R. Soc. Trop. Med. Hyg.* 64:912–917 (1970).

Ayad, N. Snail control and some significant control projects: Review. *Egypt. J. Bilharz.* 3:129–155 (1976).

Barnish, G., and M. A. Prentice. Lack of resistance of the snail *Biomphalaria glabrata* after nine years of exposure to Bayluscide. *Trans. R. Soc. Trop. Med. Hyg.* 75:106–107 (1981).

Barnish, G., and C. J. Shiff. Aerial application of the molluscicide Frescon at Lake McIlwaine. *Rhod. Agric.* 67:18–20 (1970).

Barnish, G., J. D. Christie, and M. A. Prentice. Schistosomiasis mansoni control in Cul de Sac Valley, Saint Lucia: 1. A two-year focal surveillance mollusciciding program for the control of *Biomphalaria glabrata*. *Trans. R. Soc. Trop. Med. Hyg.* 74(4):488–492 (1980).

Barnish, G., P. Jordan, R. K. Bartholomew, and E. Grist. Routine focal mollusciciding after chemotherapy to control *Schistosoma mansoni* in Cul de Sac Valley, Saint Lucia. *Trans. R. Soc. Trop. Med. Hyg.* 76:602–609 (1982).

Bartholomew, R. K., and M. J. Goddard. Quality control in laboratory investigations on *Schistosoma mansoni* on Saint Lucia, West Indies: A staff assessment scheme. *Bull. WHO* 56:309–312 (1978).

Bell, E. J., F. J. Etges, and L. J. Jenelle. The identity and source of a molluscicide-degrading bacterium. *Am. J. Trop. Med. Hyg.* 15(4):539–543 (1966).

Bradley, D. J., and G. Webbe. Ecological and habitat methods in schistosomiasis control. In *Proceedings of the International Conference on Schistosomiasis*, Vol. 2, 691–706. Cairo: Ministry of Health (1978).

Cardarelli, N. Slow-release molluscicides and related materials. In *Molluscicides in Schistosomiasis Control*, ed. T. C. Cheng, 177–240. London: Academic Press (1974).

———. *Controlled-Release Molluscicides*. Akron, Ohio: University of Akron (1977).

Cheng, T. C., and J. T. Sullivan. Mode of entry, action and toxicity of copper molluscicides. In *Molluscicides in Schistosomiasis Control*, ed. T. C. Cheng, 89–153. London: Academic Press (1974).

Choudhry, A. W. Seven years of snail control at Mwea Irrigation Settlement, Kenya: Results and costs. *East Afr. Med. J.* 51(8):600–609 (1974).

Christie, J. D., M. A. Prentice, E. S. Upatham, and G. Barnish. Laboratory and field trials of a slow-release copper molluscicide in Saint Lucia. *Am. J. Trop. Med. Hyg.* 27:616–622 (1978).

Chu, K. Y. Trials of ecological and chemical measures for the control of *Schistosoma haematobium* transmission in a Volta Lake village. *Bull. WHO* 56:313–322 (1978).

Cline, B. L. Control of schistosomiasis in Puerto Rico. In *Schistosomiasis: Proceedings of a Symposium on the Future of Schistosomiasis Control*, ed. M. J. Miller, 97–99. New Orleans: Tulane University Press (1972).

Cook, J. A. A controlled study of morbidity of schistosomiasis mansoni in Saint Lucian children, based on quantitative egg excretion. *Am J. Trop. Med. Hyg.* 23:625–633 (1974).

Daffalla, A. A., and J. Duncan. Relative susceptibility and uptake of Frescon applied to two field collections of B. truncatus. Pestic. Sci. 10:423–430 (1979).

Davis, A. Assignment report: Schistosomiasis control in the Kingdom of Saudi Arabia (with special reference to chemotherapy). WHO Unpublished Report EM/SCHIS/ 66, EM/SAA/MPD/022 (1977).

Degremont, A. A. Mangoky Project: Campaign Against Schistosomiasis in the Lower-Mangoky (Madagascar). Basel: Swiss Tropical Institute (1973).

Duke, B. O. L., and P. J. Moore. The use of a molluscicide in conjunction with chemotherapy to control Schistosoma haematobium at the Barombi Lake foci in Cameroun: I. The attack on the snail hosts, using N-tritylmorpholine, and its effects on transmission from man to snail. II. Urinary examination methods, the use of niridazole to attack the parasite in man, and the effect on transmission from man to snail. III. Conclusions and costs. Tropenmed. Parasitol. 27:297–313;489–504;505–507 (1976).

Duncan, J. A review of the development and application of molluscicides in schisto-somiasis control. In Molluscicides in Schistosomiasis Control, ed. T. C. Cheng, 9–40. London: Academic Press (1974).

——. The toxicology of molluscicides: The organotins. Pharmacol. Ther. 10:407–429 (1980).

Dunlop, R. W. Synthesis of isotopically labeled nicotinanilides and evaluation of their molluscicidal properties. Ph.D. thesis, University of London (1976).

Etges, F. J., E. J. Bell, and D. E. Gilbertson. Bacterial degradation of some molluscicidal chemicals. Am. J. Trop. Med. Hyg. 14(5):846–851 (1965).

Faria, H. F. Control of schistosomiasis in Venezuela. In Schistosomiasis: Proceedings of a Symposium on the Future of Schistosomiasis Control, ed. M. J. Miller, 100–103. New Orleans: Tulane University Press (1972).

Fenwick, A. Cost and cost-benefits of an S. mansoni control program on an irrigated estate in Tanzania. Bull. WHO 47:537–578 (1972).

Fine, P. E. M., and J. S. Lehman. Mathematical models of schistosomiasis: Report of a workshop. Am. J. Trop. Med. Hyg. 26:500–504 (1977).

Gilles, H. M., A. Abdel Aziz Zaki, M. H. Soussa, S. A. Samaan, S. S. Soliman, A. Hassan, and F. Barbosa. Results of a seven-year snail control project on the endemicity of Schistosoma haematobium in Egypt. Ann. Trop. Med. Parasitol. 67:45–65 (1973).

Goddard, M. J. A statistical procedure for quality control in diagnostic laboratories. Bull. WHO 58:313–320 (1980).

Hairston, N. G., and B. C. Santos. Ecological control of the snail host of Schistosoma japonicum in the Philippines. Bull. WHO 25:603–610 (1961).

Haskins, W. T. Colorimetric determination of microgram quantities of sodium and copper pentachlorophenate. Anal. Chem. 23:1672–1674 (1951).

Hoffman, D., K. S. Warren, V. Scott, J. S. Lehman, and G. Webbe. Report of a workshop on schistosomiasis control. Am. J. Trop. Med. Hyg. 28:249–259 (1979).

Hopf, H. S., J. Duncan, J. S. S. Beesley, D. J. Webley, and R. F. Sturrock. Molluscicidal properties of organotin and organolead compounds. Bull. WHO 36:955 (1967).

Hunter, J. M., L. Rey, and D. Scott. Disease prevention and control in water development schemes. WHO Unpublished Report PDP/80.1 (1980).

Jelnes, J. E. Evidence of possible molluscicide resistance to Schistosoma intermediate hosts from Iran? Trans. R. Soc. Trop. Med. Hyg. 71:451 (1977).

Jobin, W. R. Economics of the application of molluscicides to flowing water. Bull. WHO 38(2):322–328 (1968).

Jobin, W. R., F. F. Ferguson, and J. R. Palmer. Control of schistosomiasis in Guayana and Arroyo, Puerto Rico. *Bull. WHO* 42:151–156 (1970).

Jobin, W. R., R. A. Brown, F. F. Ferguson, and S. Velez. Biological control of *Marisa cornuarietis* in major reservoirs in Puerto Rico. *Am. J. Trop. Med. Hyg.* 26:1018–1024 (1977).

Jordan, P. Schistosomiasis – Research to control. *Am. J. Trop. Med. Hyg.* 26:877–886 (1977).

Jordan, P., and G. Webbe. *Schistosomiasis: Epidemiology, Treatment and Control.* London: Heinemann (1982).

Jordan, P., G. Barnish, R. K. Bartholomew, E. Grist, and J. D. Christie. Evaluation of an experimental mollusciciding program to control *Schistosoma mansoni* transmission in Saint Lucia. *Bull. WHO* 56:139–146 (1978).

Jordan, P., J. A. Cook, R. K. Bartholomew, E. Grist, and E. Augustie. *Schistosoma mansoni* control in Cul de Sac Valley, Saint Lucia: II. Chemotherapy as a supplement to a focal mollusciciding program. *Trans. R. Soc. Trop. Med. Hyg.* 74:493–500 (1980).

Kajihara, N., T. Horimi, M. Minai, and Y. Josaka. Field evaluation of B-2-sodium, 2–5 dichloro-4-bromophenol (B-2) against *Oncomelania nosophora*. *Japn. J. Med. Biol.* 32:225–228 (1979).

Kloetzel, K. Selective chemotherapy for schistosomiasis mansoni. *Trans. R. Soc. Trop. Med. Hyg.* 68:344 (1974).

Kloos, H., and F. S. McCullough. Plant molluscicides: A review. WHO Unpublished Report WHO/VBC/81.834; WHO/SCHISTO/81.59 (1981).

Klumpp, R. K., and K. Y. Chu. Ecological studies of *Bulinus rohlfsi*, the intermediate host of *Schistosoma haematobium* in the Volta Lake. *Bull. WHO* 55:715–730 (1977).

Lyons, G. R. L. Schistosomiasis in north-western Ghana. *Bull. WHO* 51:621–632 (1974).

Machado, P. A. The Brazilian program for schistosomiasis control, 1975–1979: Symposium on schistosomiasis control. *Am. J. Trop. Med. Hyg.* 31:76–86 (1982).

Magzoub, M., and A. A. Kasim. Schistosomiasis in Saudi Arabia. *Ann. Trop. Med. Parasitol.* 74:511–513 (1980).

Mao, S. P., and B. R. Shao. Schistosomiasis control in the People's Republic of China. *Am. J. Trop. Med. Hyg.* 31:92–99 (1982).

McCullough, F. S., and V. W. Eyakuze. Report of the schistosomiasis pilot control and training project, Mwanza District. WHO Unpublished Report WHO/AFR/SCHISTO/29 (1972).

Mobarak, A. B. The schistosomiasis problem in Egypt. *Am. J. Trop. Med. Hyg.* 31:87–91 (1982).

Olivier, L. J. The status of schistosomiasis control. In *Schistosomiasis: Proceedings of a Symposium on the Future of Schistosomiasis Control,* ed. M. J. Miller, 13–15. New Orleans: Tulane University Press (1972).

Paulini, E., C. A. de Freitas, and G. H. Aguirre. Control of schistosomiasis in Brazil. In *Schistosomiasis: Proceedings of a Symposium on the Future of Schistosomiasis Control,* ed. M. J. Miller, 104–110. New Orleans: Tulane University Press (1972).

Pesigan, T. P., N. G. Hairston, J. J. Jauregui, E. G. Garcia, A. T. Santos, B. C. Santos, and A. A. Besa. Studies on *Schistosoma japonicum* infection in the Philippines: 2. The molluscan host. *Bull. WHO* 18(4):481–578 (1958).

Prentice, M. A., and G. Barnish. Granule formulations of molluscicides for use in developing countries. *Ann. Trop. Med. Parasitol.* 74:45–51 (1980).

Ritchie, L. S. Chemical control of snails. In *Epidemiology and Control of Schisto-*

somiasis, ed. N. Ansari, 485–532. Basel: S. Karger (1973).

Ritchie, L. S., and D. B. McMullen. Review of molluscicide trials conducted in the Orient under the auspices of the U.S. Army. *Milit. Med.* 126:733–743 (1961).

Rosenfield, P. L. The management of schistosomiasis. R.F.F. Research Paper R-16. New York: Rockefeller Foundation (1979).

Rosenfield, P. L., R. A. Smith, and M. G. Wolman. Development and verification of a schistosomiasis transmission model. *Am. J. Trop. Med. Hyg.* 26(3):505–516 (1977).

Ruiz-Tiben, E., J. R. Palmer, and F. F. Ferguson. Biological control of *Biomphalaria glabrata* by *Marisa cornuarietis* in irrigation ponds in Puerto Rico. *Bull. WHO* 41:329–333 (1969).

Shiff, C. J. Focal control of schistosome-bearing snails using slow-release molluscicides. In *Molluscicides in Schistosomiasis Control*, ed. T. C. Cheng, 241–247. London: Academic Press (1974).

Shiff, C. J., and A. C. Evans. The role of slow-release molluscicides in snail control. *Cent. Afr. J. Med.* 23:6–11 (1977).

Shiff, C. J., V. de V. Clarke, A. C. Evans, and G. Barnish. Molluscicide for the control of schistosomiasis in irrigation schemes. *Bull. WHO* 48:299–307 (1973).

Strufe, R. Laboratory methods for determining Bayluscide in water samples. *Pflanzenshutz Nachrichten*, Leverkusen 18(3):130–139 (1961).

Sturrock, R. F., and G. Barnish. The aerial application of molluscicides with special reference to schistosomiasis control. *Bull. WHO* 49:283–285 (1973).

Upatham, E. S., and R. F. Sturrock. Preliminary trials against *B. glabrata* of a new molluscicide formulation: Gelatin granules containing Bayluscide wettable powder. *Ann. Trop. Med. Parasitol.* 71:85 (1977).

Webbe, G. The transmission of *Schistosoma haematobium* in an area of Lake Province, Tanganyika. *Bull. WHO* 27(1):59–85 (1962).

———. Transmission of bilharziasis: 1. Some essential aspects of snail population dynamics and their study. *Bull. WHO* 33(2):147–153 (1965).

———. Schistosomiasis: Some advances. *Br. Med. J.* 283:1104–1106 (1981).

Webbe, G., and J. Duncan. Molluscicides: Present and future roles in schistosomiasis control. WHO Unpublished Report SCHISTO/WP/78.9 (1978).

WHO. Schistosomiasis control. *WHO Tech. Rep. Series* 515:3–47 (1973).

WHO. Research on the epidemiology and methodology of schistosomiasis control in man-made lakes, Ghana and Egypt: Project findings and recommendations. WHO Unpublished Report PDP/79.2 (1979).

WHO. Epidemiology and control of schistosomiasis. *WHO Tech. Rep. Series* 643:3–63 (1980).

Yokogawa, M. Control of schistosomiasis in Japan. In *Schistosomiasis: Proceedings of a Symposium on the Future of Schistosomiasis Control*, ed. M. J. Miller, 129–132. New Orleans: Tulane University Press (1972).

2

THE BIOCHEMICAL AND PHYSIOLOGICAL BASIS OF THE MODE OF ACTION OF MOLLUSCICIDES

J. Duncan

Center for Overseas Pest Research
London, England

This chapter reviews the literature on the mechanisms by which molluscicidal compounds affect snails and inquires whether these studies offer any guidance for work on the mode of action of plant molluscicides. It also looks at studies of snail resistance to molluscicides.

Two main approaches have been used to discover new molluscicides. First, large-scale screening should isolate at least a few compounds with some molluscicidal action and so make it possible to examine related chemicals for activity. Structure-activity relationship studies may then help to explain the molluscicidal activity. All chemical molluscicides have been uncovered using screening methods.

Second, biochemical and physiological studies can be made to see whether there are any pathways, enzymes, or systems peculiar to snails that open them up to specific chemical control. An impressive amount of information already exists on molluscan biochemistry, physiology, and metabolism (for example, Wilbur and Yonge, 1964; Fretter, 1968; Florkin and Scheer, 1972; Fretter and Peake, 1975).

As a corollary, mode of action studies aim to discover which molluscan systems are affected by molluscicides. Mode of action is taken here to include not only activity of the molluscicide at the cellular level but also its uptake into the snail, distribution, metabolism, and excretion. Mode of action studies also seek information on the mechanism involved in molluscicide resistance. In recent years, with attempts being made to improve the cost-effectiveness

27

of molluscicide use through such means as new delivery systems and especially slow-release formulations, questions have been raised concerning the possible appearance of resistance.

MODE OF ACTION STUDIES

Copper Sulfate

The first use of copper sulfate in the field to control snails was recorded by Chandler (1920), and the compound has been used extensively in the Sudan Gezira (Sharaf el Din and Nagar, 1955).

In early screening tests for candidate molluscicides, Nolan et al. (1953) used complete retraction into the shell to denote poisoning. Harry et al. (1957) reported a 'distress syndrome' in snails exposed to levels of toxicants lower than those causing retraction, and Harry and Aldrich (1963) described this response more fully. It begins with extension of the cephalopedal mass as the snail lies immobile on the bottom of the container; attempts at this time to attach the foot to the side of the container usually fail. Later, the tentacles become swollen at the base and sloughing of cells occurs. The body stalk may contract spasmodically. Muscular action may eventually cease, but ciliary activity on the body surface continues. The heart continues to beat, though usually at a reduced rate.

A number of inorganic ions produced the distress syndrome, but silver, cadmium, and copper were the most active. Some organic compounds produced similar effects: urethane or nicotine sulfate (Michelson, 1957), certain antibiotics (Chernin, 1959), Nembutal (Van der Schalie, 1953), and menthol. Nembutal and menthol are commonly used to relax freshwater snails before fixing for anatomic or histologic study.

Azevedo et al. (1957, 1958) used ^{64}Cu in an autoradiographic study of the distribution of the metal in snail tissue; they noted deposition in the ovotestis and albumin gland. Cheng and Sullivan (1974) exposed adult *Biomphalaria glabrata* to copper (1 mgL^{-1}) containing ^{67}Cu and showed, also by autoradiography, that the copper tended to accumulate in the layer of mucus on the head-foot surface, in the sustentacular cells of the oviduct, in leukocytes in the lining of the alimentary canal, and on the surface of the rectal ridge. They thought that the rectal ridge surface was the most likely site for uptake of the molluscicide and that the chemical came into contact with the tissue during the entry of water currents circulating through the mantle cavity. Ryder and Bowen (1977) showed that copper is taken up through the spaces between epithelial cells of the foot of the slug *Agrolimax reticulatus*.

Yager and Harry (1964, 1966), after exposing snails to radioactive zinc, cadmium, and copper, measured the amounts taken up and the distribution among various organs. They concluded that distress depends on the concentration of the ion in the water, not on the amount of ion absorbed, and that it is more likely the result of a disruption of membrane permeability than of interference with any internal metabolic process. Sullivan and Cheng (1976) compared mortalities of *B. glabrata* when various concentrations of $CuSO_4$ were injected into the hemocoel and when snails were simply exposed to solutions of the same concentrations as those attained in the hemolymph of injected snails. Mortality was greater in the exposed snails, and the authors concluded that copper's molluscicidal action results from an effect on the external epithelia.

The structure and histology of the rectal ridge area were described by Sullivan and Cheng (1974) and Sullivan et al. (1974). Sullivan and Cheng (1975) also compared the damage sustained by the rectal ridge tissues when whole snails were exposed to high or low dosages (60 mgL^{-1} for 12 hours or 0.06 mgL^{-1} for 60 hours) of copper as $CuSO_4$. At the high dosage the snails retracted into their shells and produced copious mucus. At the low dosage they demonstrated the distress syndrome. The normal corrugated surface appearance, the digitiform projections at the base of epithelial cells, and the basal lamina all but disappeared; the loose vascular tissue below showed large spaces; and overall the tissue appeared swollen, distended, and disorganized, as if there had been an excessive uptake of water or a failure of osmoregulation. Snails receiving the high dosage may have been protected from such effects by the secretion of mucus. The authors suggested that copper may act on the membranes in some way that alters their permeability or may interfere with other metabolic or regulatory processes within the cell.

Cheng and Sullivan (1977) pursued further the idea that copper may affect osmoregulation. By measuring the wet and dry weights of soft tissue and the osmolality of hemolymph, they compared snails exposed to copper (0.06 mgL^{-1}) with unexposed snails. They found that the wet weight of the exposed snails increased with time, while that of the control snails decreased; the dry weights of both groups decreased. The ratio of wet to dry weight was significantly higher in the treated snails, and the hemolymph osmolality was significantly lower up to 36 hours of exposure. They concluded that water probably had accumulated in the tissue.

These investigators also noticed that the osmotic changes caused a lysis of pigment cells in the connective tissues underlying the rectal ridge. Since these cells are responsible for hemoglobin synthesis (Sminia et al., 1972), their loss during copper treatment may contribute to the snails' death. However, when Cheng (1975) compared the hemoglobin content of hemolymph serum from various treated groups with that from controls, no statistically significant difference was found. The conclusion was that cupric ion does not lower hemoglobin levels.

Von Brand et al. (1949) examined the effects of 72 candidate molluscicides on the respiration of whole snails. Ishak and Mohamed (1975) studied the survival rate and oxygen consumption of *Biomphalaria alexandrina* exposed for long periods to low levels of copper sulfate and found that respiration was depressed in relation to the concentration of chemical applied. In certain cases the depression may have been caused by retraction of the snail into its shell, which prevented measurement of the gaseous exchange by Warburg respirometry. Brown and Newell (1972) showed that copper sodium citrate reduced the respirations of the mussel *Mytilus edulis* and its gill tissue but not of tissue homogenates. Because gill ciliary activity was inhibited, they concluded that the copper salt affects an energy-consuming process (ciliary activity) rather than respiratory enzyme systems. El-Emam et al. (1981) showed that damsin, the sesquiterpene lactone from the composite plant *Ambrosia maritima*, reduces oxygen consumption of *B. alexandrina*. Ishak et al. (1970), in a study of the effects of copper on the oxidation of certain citric acid cycle intermediates added to *B. alexandrina* homogenates, found that such substrates as succinate and glutamate, when added alone, stimulated oxygen consumption, whereas copper sulfate inhibited their oxidation. The poisonous effect of copper on succinate oxidase is probably due to a reaction with a sulfhydryl group, since the effect lessened in the presence of cysteine (Passow et al., 1961; Ishak et al., 1972).

Cheng and Sullivan (1973a, 1974) also studied the effects of 26 copper-containing complexes on *B. glabrata* heart rate, respiration, and mortality. The degree of stereochemical encapsulation or biological 'exposure' of the copper determined whether these molecules would be effective. Copper also was shown to depress the heart rate of *M. edulis* (Scott and Major, 1972). Cheng and Sullivan (1973b) suggested that changes in heart rate and respiration might be used as molluscicidal bioassay techniques.

Trifenmorph

Trifenmorph is the common name for N-(triphenylmethyl)morpholine or N-tritylmorpholine; it is sold formulated under the trade name Frescon.

The investigation of trifenmorph's mode of action has centered on its supposed neurotoxicity, but evidence now indicates that its central nervous system effects are not the primary cause of death. Moreton and Gardner (1976) studied the electrophysiology of individual nerve cells from the visceral, or right parietal, ganglia of the isolated central nervous system of *Lymnaea stagnalis*. Although this snail is not an intermediate host for human schistosomiasis, it does provide a suitably large nervous system preparation.

When suspensions of trifenmorph (1 or 10 mgL^{-1}) were applied to the preparation, resting and action potentials were unaffected but the membrane potential was suddenly depolarized by a summating burst of synaptic potentials. These 'Frescon bursts' typically occurred about 10 to 20 minutes after the application and thereafter appeared spontaneously and at random. They occurred synchronously in different cells of the preparation and never independently in any one neuron. A synaptic origin for the bursts was suggested by the effect produced when trifenmorph was applied in low calcium/high magnesium Ringer's solutions, which are known to depress the release of transmitter substances. When synaptic potentials were reduced in amplitude, there was a corresponding decrease in burst activity.

Of two possible explanations for these observations, both interference with a presynaptic vesicular release mechanism and specific postsynaptic potentiation were discounted. In the first instance, massive release of transmitter would not be compatible with the observed discontinuous activity; in the latter instance, burst activity was not modified by topical application of a number of known transmitter substances. The authors suggested that trifenmorph may alter the delicate balance between excitatory and inhibitory inputs to the neurons, thus modifying synaptic networks. Such a 'synaptic network facilitation' might explain the involvement of the entire nervous system and the synchronous effect in various neurons.

Gardner and Moreton (1978) and Moreton and Gardner (1980) suggested that trifenmorph changes the neuronal intracellular chloride level, altering the chloride gradient across the cell membrane and leading to a state of 'disinhibition,' or a reversal of the normal chloride inhibitory postsynaptic potential. Chloride-free salines for bathing the ganglia were prepared by replacing this anion with acetate, propionate, sulfate, nitrate, or bromide. Replacement of external chloride with any of the anions did produce burst potentials, the response depending on the anion substituted. The relatively impermeable anions – acetate, propionate, and sulfate – produced bursts during the first 1.5 minutes of exposure. With the permeable anions, either the early surge of burst activity was less pronounced (nitrate) or no bursts at all were produced (bromide). Nitrate may have a permeability similar to that of chloride, although the observed persistence of bursts of both substitute anions indicates that neither can be effectively handled by a chloride pump. Moreton and Gardner (1981) were able to demonstrate an increase in intraneuronal chloride after trifenmorph treatment by using ion-selective microelectrodes. The rate of increase in chloride concentration was insensitive to changes in membrane potential. However, when the HEPES (N-2-hydroxyethylpiperazine-N'-2-ethanesulfonic acid) buffer solution used to bathe the ganglion preparation was replaced with bicarbonate saline, the chloride increase induced by trifenmorph was either wholly or partially counteracted, suggesting that the molluscicide modifies the normal exchange of HCO_3^-/Cl^- across the neuron membrane.

When Plummer and Banna (1979) applied trifenmorph to the isolated ganglia and a mantle collar preparation from the land snail *Archachatina marginata,* some nerve cells showed excitation, either in increased frequency of spikes or as Frescon bursts. Land snails are not susceptible to trifenmorph when it is presented in bait form and it seems that, to be effective, trifenmorph must enter via the surface epithelium. *Archachatina* was shown to be susceptible when the whole animal was immersed in trifenmorph solutions (Plummer and Banna, 1979). However, it has never been possible to demonstrate bursts in *Helix*, nor could this land snail be poisoned with trifenmorph by any method of application (R. B. Moreton, personal communication, 1980).

Brezden and Gardner (1980a) showed that trifenmorph and nine of its analogues were capable of producing burst activity. A tenth analogue, triphenylcarbinol (a hydrolytic product of trifenmorph), showed no effect. With the active compounds, varying lengths of time passed before burst activity occurred, and their relative effectiveness was assessed by means of an activity quotient that took into account the percentage of experiments in which bursts were induced and the mean time before burst activity occurred. Brezden and Gardner (1980b) compared the molluscicidal effectiveness of the analogues at three concentrations. A mortality quotient was used, based on the time taken to achieve 50% mortality. The rank order of effectiveness changed, depending on the concentration applied. A strong correlation was observed between activity and mortality quotients, suggesting that burst activity is responsible for the snail's death.

Poisoning symptoms – defined as early (mucus secretion and loss of foot muscle coordination, leading to slow muscular contractions or a tetanic contraction into the shell) or late (no response of the foot to light tactile stimuli) – occurred quicker in whole animals than did neurotoxic activity in the isolated central nervous system. No burst activity could be recorded in brain preparations excised from whole snails that had been exposed to lethal trifenmorph preparations and had shown early poisoning symptoms. Neither could burst activity be demonstrated after trifenmorph was injected into the hemocoel, although poisoning symptoms were observed. These results indicate that trifenmorph does not exert its lethal effect in the brain. Bursts could be recorded after continuous perfusion of the hemocoel with trifenmorph. Isolated brain preparations with intact perineural sheaths also showed burst activity after exposure to trifenmorph. These results together suggest that although trifenmorph takes some time getting to the central nervous system, the perineural sheath is not a barrier to the molluscicide.

Boyce and Milborrow (1965) suggested that the molluscicidal activity of triphenylmethyl compounds is associated with nucleophilic attack on the aliphatic carbon, displacing an inactive group and producing the molluscicidal agent. Brezden and Gardner (1980c) compared hydrolysis rates, particle size in suspension, and relative lipophilicity of trifenmorph and its analogues with their neurotoxic and molluscicidal effectiveness; they found that hydrolysis rate as an indicator of nucleophilic displacement could be correlated with neurotoxicity and, with the exception of one analogue, with molluscicidal effectiveness. Neither a retarded penetration rate nor an excessively large particle size could be invoked as an explanation for the aberrant behavior of this one analogue. However, particle size changes, related to the concentration of active ingredient in solution, could explain the change in order of the mortality quotients of the analogues at different concentration levels as described above.

It does seem, then, that molluscicidal action is the result of a nucleophilic displacement reaction but that it occurs at a site other than the central nervous system. Since contraction of the foot musculature is a symptom of poisoning, trifenmorph may act on cell membranes of muscle tissue.

Banna and Plummer (1978) investigated the effect of trifenmorph on molluscan heart. A sublethal dose (0.06 mgL^{-1}) caused a slowing of the heartbeat in intact *Bulinus truncatus*. A higher concentration (0.1 mgL^{-1}) caused not only a reduction in heart rate but also a diminution of the beat of an isolated heart preparation of *Archachatina*. At still higher concentrations, inhibition of the beat occurred, until at 4 mgL^{-1} contracture occurred and could not be reversed by the excitatory or restorative effect of 5-hydroxytryptamine. The authors suggested that derangement of cycling mechanisms of calcium and adenosine triphosphate (ATP) in the sarcoplasm was responsible for these effects. This suggestion was supported by histochemical studies in *B. truncatus* (Banna, 1977), which showed that trifenmorph could inhibit isocitrate dehydrogenase and NADPH diaphorase, leading to decreased ATP synthesis by reducing the transfer of hydrogen to respiratory chain enzymes. However, it has been shown that nerve resting potential is not affected by trifenmorph and therefore that there is no loss in electrogenic component; in short, energy stores are not affected, at least in this tissue (Brezden and Gardner, 1980b).

Brezden and Gardner (1983 a, b, c, d) examined *Lymnaea stagnalis* to determine the effect of trifenmorph on smooth (penis retractor and foot) and cross-striated (ventricular) muscle resting tensions. Calcium-free saline prevented the sustained contracture normally produced by trifenmorph. The heavy metal ions Mg^{2+}, Mn^{2+}, Ni^{2+}, and Co^{2+} and the rare earth La^{3+} – which inhibit calcium entry into cells – reversibly inhibit the trifenmorph-induced contracture if applied either before or after trifenmorph exposure. The calcium channel blocker D-600 (4×10^{-4} M) also inhibits trifenmorph action in the cross-striated muscle and appears to compete for the trifenmorph binding sites, since ^{3}H-trifenmorph no longer binds to a muscle preparation that has been pretreated with D-600. The action of trifenmorph is not mimicked by caffeine, which causes contractions in calcium-free saline, presumably by releasing calcium from intracellular stores such as the sarcoplasmic reticulum. Neither is the action of trifenmorph identical to that of the calcium ionophore A23187. This ionophore not only increases the sarcolemmal calcium permeability but also releases internal calcium stores.

There is no evidence that trifenmorph has any action on intracellular calcium compartments. It does not alter the resting potentials of cross-striated muscle fibers even after a contracture has been induced. Brezden and Gardner therefore concluded that trifenmorph produces its effect on *Lymnaea* muscle fibers by increasing the calcium permeability of the sarcolemma, thereby flooding the contractile apparatus with an irreversible influx of external calcium. Because the muscle fibers contract within 15 to 30 seconds of trifenmorph application, it seems that the main symptoms of poisoning can be accounted for by the direct action of trifenmorph on the snail's musculature.

Niclosamide and Sodium Pentachlorophenate

Niclosamide (available commercially as Bayluscide, the ethanolamine salt of 2',5-dichloro-4'-nitrosalicylanilide) is the molluscicide of choice in many parts of the world. Nevertheless, its mode of action has received relatively little attention.

Gonnert (1961) showed that low concentrations of niclosamide stimulate oxygen uptake by whole *B. glabrata*, whereas higher concentrations inhibit oxygen uptake. Ishak et al. (1970) and Ishak and Mohamed (1975) compared the effect of copper sulfate, sodium pentachlorophenate (NaPCP), and niclosamide on certain aspects of snail intermediary metabolism. The addition of tricarboxylic acid (TCA) cycle intermediates (citrate, succinate, glutamate) stimulated oxygen uptake of tissue homogenates. Copper sulfate inhibited the oxidation of all substrates and, although low concentrations of the other two molluscicides stimulated oxidation of substrates, higher concentrations inhibited the reaction. Weinbach (1954) suggested that NaPCP, when present in low concentrations, is an uncoupler of oxidative phosphorylation in snail tissues; but inhibition of the glycolytic pathways may occur at higher concentrations (Weinbach and Nolan, 1956). Ishak et al. (1972) indicated that inhibition of TCA cycle intermediates, due to NaPCP, could be attributed to accumulation of oxaloacetate.

Nicotinanilide

Dunlop et al. (1980) prepared a congeneric series of 22 substituted nicotinanilides and found a relationship between molluscicidal activity of para-substituted members of the series and the Hansch coefficient, which is a measure of the lipophilic character of the compounds. Daffalla (1978) and J. Duncan and N. Brown (unpublished data, 1976) showed that the length of time snails are acclimatized to the experimental conditions could influence their activity, their heart rate, and the amount of molluscicide they take up.

Duncan et al. (1977) constructed a flowing water exposure apparatus in which the snails were allowed to acclimatize before any experimentation. When the uptake rates of three para-substituted nicotinanilides of varying lipophilic character were then examined, no significant difference could be detected whether or not this apparatus was used. The uptake rates could, however, be explained in terms of the rate of water uptake. Thus it is possible that, although lipophilic character may be important in determining molluscicidal activity at the eventual site of action, carriage into the animal depends on water influx (Dunlop, 1976).

Organotins

The molluscicidal activity of organotin compounds has been known for some time (Hopf and Muller, 1962; Hopf et al., 1967). These chemicals appear to lend themselves particularly well to incorporation into slow-release formulations. This method of application offers the possibility of improved cost-effectiveness, very low dosage treatments, and reduced labor costs through less frequent applications. Tributyltin oxide (TBTO) is one organotin compound that has been so formulated and then investigated in the laboratory (Berrios-Duran and Ritchie, 1968) and in the field (Shiff and Evans, 1977).

The toxicity and mode of action of organotins have been investigated mainly in mammals (Duncan, 1980). Organotins fall into four classes, mono- (RSn X_3), di- (R_2Sn X_2), tri- (R_3Sn X), and tetraorganotins (R_4Sn), where R is an alkyl or aryl group linked directly to the tin atom and X is a simple or complex anion. Maximum biologic activity appears in the triorganotins. Increasing alkyl chain length leads to decreasing mammalian toxicity. The main site of action of trialkyltin compounds in mammals is the central nervous system, which is also the site of major pathologic change (Magee et al., 1957). Aldridge and Street (1964) showed that inhibition of oxidative phosphorylation could be demonstrated with a range of trialkyltins, but it is believed now that these compounds have a number of effects on mitochondrial functions, as reviewed by Aldridge (1976) and Aldridge et al. (1977). Triorganotins, especially the more hydrophobic triphenyltin and tributyltin, inhibit electron transport phosphorylation at the membrane-bound components of the coupling ATPase by inhibiting proton flow through these components. Triorganotins also inhibit the Na^+ to K^+ translocating ATPase of cell membranes and the Ca^{2+} translocating ATPases of sarcoplasmic reticulum. Triphenyltin and tributyltin are again the most effective compounds in this respect. Triorganotins are also ionophores mediating an exchange of Cl^-, Br^-, SCN^-, I^-, or F^- for OH^- across biologic and artificial phospholipid membranes. In media containing chloride, Cl^-/OH^- exchange causes pH changes across membranes, with a consequent shift away from the pH optima of some enzymes. The trioganotins, again particularly triphenyltin and tributyltin, cause gross swelling of mitochondria (Moore and Brody, 1961; Aldridge and Street, 1964). But, despite these many advances in understanding the action of the organotins at the cellular level, it is still not known, for example, whether the cerebral edema and the rapid fall in body temperature observed in triethyltin intoxication in mammals are initiated by an attack on mitochondrial function (Aldridge, 1976).

A common response in snails treated with organotin molluscicide is a swelling of the tissues, with the snail body relaxed in a way reminiscent of the distress syndrome reported by Harry et al. (1957). Patnode and Kanakkanatt (1971) described the response to lethal and sublethal doses of tributyltin in more detail: Yellowish inclusions appeared in connective tissue and black precipitates in epithelia; cell wall collapse was noted even after relatively short exposure periods, with a consequent invasion of tissues by hemolymph. It was thought that tributyltin had combined generally with structural protein, causing gross tissue damage. Aldridge and Rose (1969) pointed out that the trialkyltins combine a high biologic activity with very limited chemical reactivity. The activity on mitochondrial function has led to attempts to identify chemical binding sites. Smith et al. (1979) studied the reaction of these compounds with sulfhydryl-containing amino acids in polar media and found that air-stable derivatives were formed. Elliot et al. (1979) confirmed that reagents able to modify either cysteine alone or cysteine and histidine residues can destroy triethyltin binding sites. Thus both of these amino acids are involved in binding to protein, not only in the specific sense of activity against mitochondrial function but also possibly more generally.

Anacardic Acid

Sullivan et al. (1982) studied a molluscicide of plant origin. They reported that certain components of anacardic acid, obtained from shells of the cashew nut *Anacardium occidentale*, possess molluscicidal properties against *B. glabrata*. The triene component was the most active (LC$_{50}$, 0.35 mgL^{-1}), the diene and monoene components were less so (LC$_{50}$, 0.9 and 1.4 mgL^{-1}, respectively), and the saturated component was relatively inactive (LC$_{50}$, >5 mgL^{-1}). Since neither decarboxylated anacardic acid (cardanol) nor salicylic acid kills snails at concentrations up to 5 mgL^{-1}, it was proposed that both the carboxyl radical and the unsaturated side chain are required for molluscicidal activity. The same authors examined the ability of the unsaturated components of anacardic acid to inhibit prostacyclin action in human blood platelets by competing for the prostacyclin receptor site. The triene component was more active than the diene or monoene component in this regard. It remains to be demonstrated whether prostacyclin receptor sites occur in *B. glabrata*.

COMMON MODE STUDIES

De Villiers and MacKenzie (1963) suggested that molluscicides seem to fall into two classes: those that poison enzymes and rely on sulfhydryl groups for their activity and those that interfere with osmoregulation. These authors suggested that as molluscan tissue generally appears to have a low sulfhydryl content, snails may have little protection for enzymes utilizing -SH groups. Compounds with lachrymatory activity also inhibit -SH groups, and representatives of this group of compounds, the phenacyl halides, were shown to be molluscicidal. Zsolnai (1969) investigated a number of compounds that react with -SH groups and found some of them to be as active as niclosamide. As already noted, cupric ion and organotin compounds react with -SH groups.

The swollen tissue described in the distress syndrome by Harry et al. (1957) is suggestive of a failure in water balance control or osmoregulation. Similar signs have been described in organometal poisoning of snails (McMullen, 1952; Hopf et al., 1967). Cheng and Sullivan (1977) provided evidence of the accumulation of water in snail tissues after copper poisoning.

Evidence suggests that molluscan water balance is under neurosecretory control (Lever et al., 1961; Lever and Joosse, 1961). Hanumante et al. (1979) recorded aberrations in the neurosecretory cells of the freshwater pulmonate *Indoplanorbis exustus* after chronic exposure to sublethal concentrations of $BaCl_2$ and $CuSO_4$.

Molluscicidal action affecting water balance may take one of three forms. First, the snail may detect the presence of a chemical and react by producing excess mucus or reducing the permeability of the external membrane. Second, the molluscicide may react directly with the external membrane; Harry and Aldrich (1963) and Sullivan and Cheng (1976) believed that was the case with copper sulfate. Third, since the snail heart-kidney removes excess water from the body and Cheng and Sullivan (1974) demonstrated that heart rate declines under the influence of copper poisoning, accumulation of water in the tissue or increased osmolality of the hemolymph may be a secondary result of the uptake of copper.

RESISTANCE STUDIES

Jelnes (1977) reported that *B. truncatus* collected from an area in Iran that had been treated with niclosamide for ten years was more resistant than a laboratory stock taken from the same area before molluscicides were applied. But Barnish and Prentice (1981), comparing the susceptibility of *B. glabrata* from untreated areas with *B. glabrata* from areas in Saint Lucia that had been treated with niclosamide for nine years, found no statistically significant difference. These authors suggested that the susceptibility of laboratory snail colonies may differ from that of the field population from which they were derived, thus giving rise to spurious results in resistance studies. The rate at which any such change in susceptibility occurs merits further study.

Yasuraoka (1972) reviewed the findings of a number of Japanese workers and concluded that there is no evidence that resistance to NaPCP has occurred in *Oncomelania*.

A small but statistically significant difference in susceptibility to tri-fenmorph was noted between collections of *B. truncatus* from two areas of the Sudan Gezira, one untreated and one that had been treated for six consecutive years. The rate of uptake of the molluscicide was significantly less in snails from the treated area. It was suggested that the slow rate of uptake could be one way in which freshwater snails develop apparent resistance to molluscicides (Daffalla and Duncan, 1979).

There is no evidence that any freshwater snail has ever become resistant to any molluscicide to the degree that field populations can no longer be controlled. A molluscicide resistance test kit has been developed through the UNDP/WORLD BANK/WHO Special Programme for Research and Training in Tropical Diseases so that susceptibility of field populations of snails may be kept under surveillance.

EXPERIMENTAL CONDITIONS

The effects of experimental conditions on recorded responses in molluscicide studies have not been sufficiently investigated. Daffalla (1978) and Duncan et al. (1977) noted the importance of an acclimatization time before beginning any experimental recording. It could be shown, for example, that the heart rate and locomotor activity of untreated snails and the rate of uptake of molluscicide all declined during the first two hours after a snail was brought into new surroundings. Distilled water is commonly used in an attempt to reduce the number of variables in experiments involving snails; it does, however, cause a decrease in the water uptake of snails and affects molluscicide uptake rate also (J. Duncan, unpublished data, 1975).

CONCLUSIONS

It was noted at the beginning of this chapter that the study of molluscan physiology and biochemistry and the corollary study of the mode of action of molluscicides might lead to the discovery of novel compounds. Such studies might also lead to improvement in specificity of molluscicides. Some new compounds, such as the phenacyl halides and their action on sulfhydryl groups, have been uncovered by this approach, and it has been seen that copper and organotin compounds also have affinities with such groupings. At the same time, it must be accepted that all chemical molluscicides that have been developed to the stage of field testing and use have originated from screening procedures.

The question was also raised as to the contribution of mode of action studies to the problem of molluscicide resistance. It seems likely that such studies will only usefully be called into action if and when resistance actually appears under field conditions and not before.

Molluscan physiological, biochemical, and mode of action studies have nonetheless added to the overall body of information in this area. Certain techniques – those used, for example, in the study of the nervous system and in electrophysiology; neurosecretion; osmoregulation; Warburg respirometry applied to cellular respiration, metabolic pathways, and the respiration of whole animals; autoradiography; scintillation counting; and histology – have been applied and suitably adapted for snail studies. Such endeavors have grappled with the problems and pitfalls in interpreting and analyzing the results obtained with molluscan material. For example, apparent reductions in oxygen exchange of whole snails may be dependent on the covering of exposed surfaces with mucus or withdrawal into the shell rather than on any primary effect on respiratory chain enzymes. The release of hemolymph into the water commonly seen following molluscicide application may be due to rupture of external membranes; but another explanation could lie in the presence of a hemal pore, as described by Lever and Bekius (1965) in *Lymnaea*, that allows rapid loss of hemolymph and contraction of the snail into the shell when the animal is endangered. Alteration of water balance with swelling of tissues has been suggested as a primary response to molluscicides. But definitive proof of this explanation should be sought, employing the quantitative methods and neurosecretory investigations pioneered by Lever et al. (1961) and Van Aardt (1968).

Despite the complications, mode of action studies have yielded results. In particular, the investigations into trifenmorph toxicity to snails seem to have reached a point where the primary cause of death may have been identified. In addition to this specific activity, similarities of a general kind have been noted in reports on the effects of various compounds on, for example, sulfhydryl groups, reduction in heart rate, and osmoregulation. Whether any of these common effects indicate the common mode of action – a finding once hoped for that would have provided a justification for a biochemical and physiological approach to finding new molluscicides – is still debatable. A molluscicide does not, of course, need to affect only one system. Madrell and Reynolds (1972) suggested that insecticide poisoning may be a generalized disruptive effect consequent upon hormone release affecting several systems.

As far as the development of plant molluscicides is concerned, the screening approach is again likely to provide all or almost all of the new molluscicidal materials. The distribution among plant families and the chemistry and pharmacology of secondary plant products is already fairly well known, and plants showing activity in the screening process will therefore readily suggest chemotaxonomic relatives for testing. In a number of cases, it might also be possible to suggest what their molluscicidal mode of action might be. Duncan (1985) discussed the possibilities for the more promising candidate plant materials. Mode of action studies will not normally therefore be a preliminary in the research and development of a plant molluscicide but will more likely be of use at a later time in demonstrating the specificity or otherwise of material or in explaining its molluscicidal activity. In that case, all that has already been reported will naturally be a most useful reference for such work. Even if the diversity of chemical substances offering themselves as plant molluscicides will require novel modes of action to be studied, use can be made of the experience gained with the techniques and methods developed for investigation of the purely chemical molluscicides.

Acknowledgments

The author is grateful to Dr. T. C. Cheng, Medical University of South Carolina, and Dr. E. H. Michelson, Harvard School of Public Health, for their comments and contributions during the preparation of this chapter.

REFERENCES

Aldridge, W. N. The influence of organotin compounds on mitochondrial functions. *Adv. Chem. Series* 157:186–196 (1976).

Aldridge, W. N., and M. S. Rose. The mechanism of oxidative phosphorylation: A hypothesis derived from studies of trimethyltin and triethyltin compounds. *Fed. Eur. Biochem. Soc. Lett.* 4:61–69 (1969).

Aldridge, W. N., and B. W. Street. Biochemical effects and properties of trialkyltins. *Biochem. J.* 91:287–297 (1964).

Aldridge, W. N., B. W. Street, and D. N. Skilleter. Halide-dependent and halide-independent effects of triorganotin and triorganolead compounds on mitochondrial functions. *Biochem. J.* 168:353–364 (1977).

Azevedo, J. F. de, F. C. Gomes, A. M. Baptista, and F. B. Gil. Studies on the molluscicide action of copper sulfate using ^{64}Cu. *Tropenmed. Parasitol.* 8:458–464 (1957).

Azevedo, J. F. de, F. C. Gomes, F. B. Gil, A. M. Baptista, and E. M. de Magalhaes. Application of radioisotopes to the study of the metabolism of freshwater snails (Gastropoda: Pulmonata). *Am. J. Trop. Med. Hyg.* 7(1):84–89 (1958).

Banna, H. B. General histology and histopathology of *Bulinus truncatus* (Audouin) with special reference to the action of a molluscicide. Ph.D. thesis, University of London (1977).

Banna, H. B., and J. M. Plummer. The effects of Frescon on molluscan hearts. *Comp. Biochem. Physiol.* 62C:33–36 (1978).

Barnish, G., and M. A. Prentice. Lack of resistance of the snail *Biomphalaria glabrata* after nine years of exposure to Bayluscide. *Trans. R. Soc. Trop. Med. Hyg.* 75(1):106–107 (1981).

Berrios-Duran, L. A., and L. S. Ritchie. Molluscicidal activity of bis(tri-n-butyltin) oxide formulated in rubber. *Bull. WHO* 39:310–312 (1968).

Boyce, C. B. C., and B. V. Milborrow. A simple assessment of partition data for correlating structure and biological activity using thin-layer chromatography. *Nature* 208:537–539 (1965).

Brezden, B. L., and D. R. Gardner. The effect of some Frescon analogues on the aquatic snail *Lymnaea stagnalis*: 1. Effect on the isolated central nervous system. *Pestic. Biochem. Physiol.* 13:169–177 (1980a).

———. The effect of some Frescon analogues on the aquatic snail *Lymnaea stagnalis*: 2. Correlation of neurotoxicity to molluscicidal effectiveness. *Pestic. Biochem. Physiol.* 13:178–188 (1980b).

———. The effect of some Frescon analogues on the aquatic snail *Lymnaea stagnalis*: 3. Relation of some physical and chemical properties to neurotoxicity and molluscicidal effectiveness. *Pestic. Biochem. Physiol.* 13:189–197 (1980c).

———. The effect of the molluscicide Frescon on smooth and cross-striated muscles of *Lymnaea stagnalis* and *Helix aspersa*. *Pestic. Biochem. Physiol.* 20:259–268 (1983a).

———. Calcium entry blockers inhibit contractures induced by the molluscicide Frescon in *Lymnaea stagnalis* smooth and cross-striated muscles. *Pestic. Biochem. Physiol.* 20:269–277 (1983b).

———. Evidence that Frescon-induced contractures in *Lymnaea stagnalis* muscles do not depend on intracellular calcium stores: A comparison with caffeine action. *Pestic. Biochem. Physiol.* 20:278–285 (1983c).

———. A comparison of the action of Frescon with calcium ionophore A23187. *Pestic. Biochem. Physiol.* 20:286–293 (1983d).

Brown, B. E., and R. C. Newell. The effect of copper and zinc on the metabolism of the mussel *Mytilus edulis*. *Marine Biol.* 16(2):108–118 (1972).

Chandler, A. C. Control of liver fluke diseases by destruction of the intermediate host. *J. Agric. Res.* 20:193–208 (1920).

Cheng, T. C. Does copper cause anemia in *Biomphalaria glabrata*? *J. Invertebr. Pathol.* 26:421–422 (1975).

Cheng, T. C., and J. T. Sullivan. The effect of copper on the heart rate of *Biomphalaria glabrata* (Mollusca: Pulmonata). *Comp. Gen. Pharmacol.* 4:37–41 (1973a).

———. A new method for the preliminary testing of chemical molluscicides. WHO Unpublished Document WHO/SCHISTO/73.27 (1973b).

———. Mode of entry, action and toxicity of copper molluscicides. In *Molluscicides in Schistosomiasis Control*, ed. T. C. Cheng, 89–153. London: Academic Press (1974).

———. Alterations in the osmoregulation of the pulmonate gastropod *Biomphalaria glabrata* due to copper. *J. Invertebr. Pathol.* 29:101–104 (1977).

Chernin, E. Notes on the effects of various antibiotics on *Australorbis glabratus*. *J. Parasitol.* 45:268 (1959).

Daffalla, A. A.-R. The relative susceptibilities of *Bulinus truncatus* (Audouin) and *Sarotherodon mossambicus* (Peters) to certain molluscicides. Ph.D. thesis, University of London (1978).

Daffalla, A. A.-R., and J. Duncan. The relative susceptibility of two field collections of *Bulinus truncatus* (Audouin) to trifenmorph. *Pestic. Sci.* 10:423–428 (1979).

De Villiers, J. P., and J. G. MacKenzie. Structure and activity in molluscicides: The phenacyl halides, a group of potentially useful molluscicides. *Bull. WHO* 29:424–427 (1963).

Duncan, J. The toxicology of molluscicides: The organotins. *Pharmacol. Ther.* 10:407–429 (1980).

——. The toxicology of plant molluscicides. *Pharmacol. Ther.* 27:243–264 (1985).

Duncan, J., N. Brown, and R. W. Dunlop. The uptake of the molluscicide 4'-chloronicotinanilide into *Biomphalaria glabrata* (Say) in a flowing water system. *Pestic. Sci.* 8:345–353 (1977).

Dunlop, R. W. Synthesis of isotopically labeled nicotinanilides and evaluation of their molluscicidal properties. Ph.D. thesis, University of London (1976).

Dunlop, R. W., J. Duncan, and G. Ayrey. Quantitative structure-activity relationships for nicotinanilide molluscicides. *Pestic. Sci.* 11:53–60 (1980).

El-Emam, M. A., A. M. Mohamed, H. A. Shoeb, and M. F. El-Shafiee. The effect of damsin and Bayluscide homologue on the oxygen consumption of *Biomphalaria alexandrina* and *Bulinus truncatus*. *Helminthologia* 18:125–130 (1981).

Elliot, B. M., W. N. Aldridge, and J. W. Bridges. Triethyltin binding to cat hemoglobin: Evidence for chemically distinct sites and a role for both histidine and cysteine residues. *Biochem. J.* 177:461–470 (1979).

Florkin, M., and B. T. Scheer, eds. *Chemical Zoology, 7: Mollusca.* London: Academic Press (1972).

Fretter, V., ed. *Studies in the Structure, Physiology and Ecology of Molluscs.* Symposia of the Zoological Society of London, No. 22. London: Academic Press (1968).

Fretter, V., and J. Peake, eds. *Functional Anatomy and Physiology.* Pulmonates, Vol. 1. London: Academic Press (1975).

Gardner, D. R., and R. B. Moreton. Disinhibition caused by the molluscicide N-tritylmorpholine (Frescon) in the nervous system of a freshwater snail. *Pestic. Biochem. Physiol.* 8:225–240 (1978).

Gonnert, R. Results of laboratory and field trials with molluscicide Bayer 73. *Bull. WHO* 25:483–501 (1961).

Hanumante, M. M., R. Nagabhushanam, and D. P. Vaidya. Aberrations in the neurosecretory cells of a freshwater pulmonate, *Indoplanorbis exustus*, chronically exposed to sublethal concentrations of two molluscicides, $BaCl_2$ and $CuSO_4$. *Bull. Environ. Contam. Toxicol.* 23:70–72 (1979).

Harry, H. W., and D. V. Aldrich. The distress syndrome in *Taphius glabratus* (Say) as a reaction to toxic concentration of inorganic ions. *Malacologia* 1(2):283–287 (1963).

Harry, H. W., B. G. Cumbie, and J. Martinez de Jesus. Studies on the quality of freshwaters of Puerto Rico relative to the occurrence of *Australorbis glabratus*. *Am. J. Trop. Med. Hyg.* 6:313–322 (1957).

Hopf, H. S., and R. L. Muller. Laboratory breeding and testing of *Australorbis glabratus* for molluscicidal screening. *Bull. WHO* 27:783–789 (1962).

Hopf, H. S., J. Duncan, J. S. S. Beesley, D. J. Webley, and R. F. Sturrock. Molluscicidal properties of organotin and organolead compounds with particular reference to triphenyllead acetate. *Bull. WHO* 36:955–961 (1967).

Ishak, M. M., and A. M. Mohamed. Effect of sublethal doses of copper sulfate and Bayluscide on survival and oxygen consumption of the snail *Biomphalaria alexandrina*. *Hydrobiologia* 47:499–512 (1975).

Ishak, M. M., A. A. Sharaf, A. M. Mohamed, and A. H. Mousa. Studies on the mode of action of some molluscicides on the snail *Biomphalaria alexandrina*: I. Effect of Bayluscide, sodium pentachlorophenate and copper sulfate on succinate, glutamate and reduced TMPD oxidation. *Comp. Gen. Pharmacol.* 1(2):201–208 (1970).

Ishak, M. M., A. A. Sharaf, and A. M. Mohamed. Studies on the mode of action of some molluscicides on the snail *Biomphalaria alexandrina*: II. Inhibition of succinate oxidation by Bayluscide, sodium pentachlorophenate and copper sulfate. *Comp. Gen. Pharmacol.* 3(12):385–391 (1972).

Jelnes, J. E. Evidence of possible molluscicide resistance in *Schistosoma* intermediate hosts from Iran? *Trans. R. Soc. Trop. Med. Hyg.* 71(5):451 (1977).

Lever, J., and R. Bekius. On the presence of an external hemal pore in *Lymnaea stagnalis* L. *Experientia* 21:395–398 (1965).

Lever, J., and J. Joosse. On the influence of the salt content of the medium on some special neurosecretory cells in the lateral lobes of the cerebral ganglia of *Lymnaea stagnalis*. *Proc. Koninklijke Nederlandse Akad. Wetenschappen, Amsterdam* 64C:630–639 (1961).

Lever, J., J. Jansen, and T. A. de Vlieger. Pleural ganglia and water balance in the freshwater pulmonate *Lymnaea stagnalis*. *Proc. Koninklijke Nederlandse Akad. Wetenschappen, Amsterdam* 64C:531–542 (1961).

Madrell, S. H. P., and S. E. Reynolds. Release of hormones in insects after poisoning with insecticides. *Nature* 236:404–406 (1972).

Magee, P. N., H. G. Stoner, and J. M. Barnes. The experimental production of edema in the central nervous system of the rat by triethyltin compounds. *J. Pathol. Bacteriol.* 73:107–124 (1957).

McMullen, D. B. Schistosomiasis and molluscicides. *Am. J. Trop. Med. Hyg.* 1:671–679 (1952).

Michelson, E. H. Studies on the biological control of schistosome-bearing snails. *Parasitology* 47:413–426 (1957).

Moore, K. E., and T. M. Brody. The effect of triethyltin on mitochondrial swelling. *Biochem. Pharmacol.* 6:134–142 (1961).

Moreton, R. B., and D. R. Gardner. Frescon: Neurophysiological action of a molluscicide. *Experientia* 32:611–613 (1976).

—— . Disinhibition caused by the molluscicide N-tritylmorpholine (Frescon): A possible new mechanism for pesticide action. In *Insect Neurobiology and Pesticide Action*, 109–114, Neurotex 79, Conference Proceedings of the Society of Chemical Industry, University of York, England (1980).

—— . Increased intracellular chloride activity produced by the molluscicide, N-(triphenylmethyl)-morpholine (Frescon), in *Lymnaea stagnalis* neurons. *Pestic. Biochem. Physiol.* 15:1–9 (1981).

Nolan, M. O., H. W. Bond, and E. R. Mann. Results of laboratory screening tests of chemical compounds for molluscicidal activity. *Am. J. Trop. Med. Hyg.* 2:716–752 (1953).

Passow, H., A. Rothstein, and T. W. Clarkson. The general pharmacology of the heavy metals. *Pharmacol. Rev.* 13:185–202 (1961).

Patnode, P. P., and S. V. Kanakkanatt. Mechanisms of snail intoxication by organotin compounds. WHO Unpublished Document PD/MOL/71.15 (1971).

Plummer, J. M., and H. B. Banna. Activity changes in some large neurons of *Archachatina* caused by the molluscicide Frescon. *Comp. Biochem. Physiol.* 62C:9–18 (1979).

Ryder, T. A., and I. D. Bowen. The slug foot as a site of uptake of copper molluscicide. *J. Invertebr. Pathol.* 30:381–386 (1977).

Scott, D. M., and C. W. Major. The effect of copper (II) on survival, respiration and heart rate in the common blue mussel, *Mytilus edulis*. *Biol. Bull.* 143:679–688 (1972).

Sharaf el Din, H., and H. Nagar. Control of snails by copper sulfate in the canals of the Gezira irrigated area of the Sudan. *J. Trop. Med. Hyg.* 58:260–263 (1955).

Shiff, C. J., and A. C. Evans. The role of slow-release molluscicides in snail control. *Cent. Afr. J. Med.* 23:6–11 (1977).

Sminia, T., H. H. Boer, and A. Niemantsverdriet. Hemoglobin-producing cells in freshwater snails. *Z. Zellforsch. Mikrosk. Anat.* 135(4):563–569 (1972).

Smith, P. J., A. J. Crowe, V. G. Kumar Das, and J. Duncan. Structure-activity relationships for some organotin molluscicides. *Pestic. Sci.* 10:419–422 (1979).

Sullivan, J. T., and T. C. Cheng. Structure and function of the mantle cavity of *Biomphalaria glabrata* (Mollusca: Pulmonata). *Trans. Am. Microscop. Soc.* 93:416–420 (1974).

——. Heavy metal toxicity to *Biomphalaria glabrata* (Mollusca: Pulmonata). *Ann. N.Y. Acad. Sci.* 266:437–444 (1975).

——. Comparative mortality studies on *Biomphalaria glabrata* (Mollusca: Pulmonata) exposed to copper internally and externally. *J. Invertebr. Pathol.* 28:255–257 (1976).

Sullivan, J. T., G. E. Rodrick, and T. C. Cheng. A transmission and scanning electron microscopical study of the rectal ridge of *Biomphalaria glabrata* (Mollusca: Pulmonata). *Cell Tissue Res.* 154:29–38 (1974).

Sullivan, J. T., C. S. Richards, H. A. Lloyd, and G. Krishna. Anacardic acid: Molluscicide in cashew nut shell liquid. *J. Med. Plant Res.* 44:175–177 (1982).

Van Aardt, W. J. Quantitative aspects of the water balance in *Lymnaea stagnalis* (L.). *Neth. J. Zool.* 18(3):253–312 (1968).

Van der Schalie, H. Nembutal as a relaxing agent for mollusks. *Am. Midland Naturalist* 50:511–512 (1953).

Von Brand, T., B. Mehlman, and M. O. Nolan. Influence of some potential molluscicides on the oxygen consumption of *Australorbis glabratus*. *J. Parasitol.* 53:475–481 (1949).

Weinbach, E. C. The effect of pentachlorophenate on oxidative phosphorylation. *J. Biol. Chem.* 210:545–550 (1954).

Weinbach, E. C., and M. O. Nolan. The effect of pentachlorophenate on the metabolism of the snail *Australorbis glabratus*. *Exp. Parasitol.* 5:276–284 (1956).

Wilbur, K. M., and C. M. Yonge, eds. *Physiology of Mollusca*, Vols. 1 and 2. London: Academic Press (1964).

Yager, C. M., and H. W. Harry. The uptake of radioactive zinc, cadmium and copper by the freshwater snail, *Taphius glabratus*. *Malacologia* 1:339–353 (1964).

——. Uptake of heavy metal ions by *Taphius glabratus*, a snail host of *Schistosoma mansoni*. *Exp. Parasitol.* 19:174–182 (1966).

Yasuraoka, K. Studies on the resistance of *Oncomelania* snails to molluscicides. In *Research in Filariasis and Schistosomiasis*, Vol. 2, ed. M. Yokogawa, 103–111. Baltimore: University Park Press (1972).

Zsolnai, T. Die molluscizide Wirkung von Thiol-Reagentien. *Zentralbl. Bakteriol. Parasitenk. Infekt. Hyg.* 213(2):270–283 (1969).

3

PLANTS WITH RECOGNIZED MOLLUSCICIDAL ACTIVITY

H. Kloos

Department of Geography
Addis Ababa University
Addis Ababa, Ethiopia

F. S. McCullough

Division of Vector Biology and Control
World Health Organization
Geneva, Switzerland

Economic and ecologic considerations increasingly favor the use of molluscicides that are selectively active, biodegradable, inexpensive, and readily available in affected areas. The high cost of imported synthetic compounds, along with increasing concern over the possible buildup of snail resistance to these compounds and their toxicity in nontarget organisms, has given new impetus to the study of plant molluscicides (Al-Azzawil and Banna, 1980; Hostettmann, 1984; McCullough et al., 1980).

In the world's tropical and subtropical areas where schistosomiasis is endemic, many plants may be particularly suited for use in controlling snails, the intermediate hosts of the schistosomes. Such plants, as a result of constant exposure to pathogenic microorganisms, insects, and herbivores, have developed efficient defense mechanisms, as evidenced by their insecticidal, piscicidal, molluscicidal, medicinal, and other biologic activities (Whittaker and Feeny, 1971; Deverall, 1972). People living in developing countries are familiar with the properties and growing characteristics of these local plants and already use many of them as medicines, poisons, cosmetics, soap, food,

45

fuel, and building materials. Thus, by using local labor and simple technology, the development and production of inexpensive, efficient, ecologically sound, and culturally acceptable molluscicides may be feasible, although the basic framework and technical support may still need to be generated.

In a preliminary review of molluscicidal plants, Kloos and McCullough (1982) discussed 61 species. Recent acquisition of additional source material and isolation of numerous compounds now necessitate expansion of that review. The present chapter, incorporating information from the NAPRA-LERT computerized data-retrieval system (Farnsworth et al., 1979; see also Chapter 5), lists 571 species. Following a brief historical review, we focus on present uses, toxicology, and production of plant materials, with the aim of identifying suitable plants and making suggestions for their development and use as molluscicides.

HISTORICAL REVIEW

Since the 1930s more than 1100 plant species have been tested for molluscicidal activity, including nearly 600 in China (Maegraith, 1958; see also Chapter 11). Only a few of them are used regularly for snail control – all by Chinese farmers as part of a national program (Cheng, 1971). Little is known, however, about this program.

In other endemic areas, research has been carried out mostly in Africa and Central and South America. Archibald (1933) and Wagner (1936) found that the fruits of *Balanites aegyptiaca* and *Balanites maughamii*, used for medicine and as fish poison, also killed *Bulinus* snails and the miracidia and cercariae of schistosomes. Archibald (1933) and Anantaraman (1955) recommended that *B. aegyptiaca* and other trees be planted around schistosome transmission sites, where their fruit would fall into the water and suppress snail populations; these trees would also provide shade, fences, and food. Wagner (1936) rejected large-scale cultivation of *B. maughamii* because it is toxic to fish, tadpoles, and other nontarget organisms. A field trial in a pond in Puerto Rico showed that *Biomphalaria glabrata* could not be effectively controlled (Plank, 1945) by this approach. The only other attempt to grow molluscicidal plants along snail-infested watercourses was made by El-Sawy et al. (1981); their trials with the weed *Ambrosia maritima* on canal banks in the Nile Delta gave encouraging preliminary results, although further studies are needed. Integrated biological control programs involving molluscicidal plants and predatory organisms, such as fish and ducks, have been attempted at various times, but their results are not known (Anantaraman, 1955; Breuil et al., 1983).

Application of crude plant materials in snail habitats was, with few exceptions, unsuccessful. The berries of *Sapindus saponaria* temporarily reduced *Bulinus globosus* populations in several ponds in Zanzibar (Mozley, 1939;

Msangi and Zeller, 1965). Because of the need for large amounts of berries to achieve satisfactory snail kill, the difficulty of crushing the fruit, and the lack of residual effect, Msangi and Zeller (1965) recommended that they be used only in small bodies of water. Aqueous extract of berries from South American varieties was more active in laboratory trials than the crude African berries (Torrealba et al., 1953; Msangi and Zeller, 1965) and, when combined with pentachlorophenate, exhibited a synergistic effect (Barbosa et al., 1952; Morais et al., 1953).

The hard labor required to dig up, transport, and apply the tuberous roots of *Neorautanenia pseudopachyrhiza* and the difficulty of determining the correct concentration were considered major disadvantages, although the water extract was nontoxic to monkeys and fish at molluscicidal levels (500 ppm) (Teesdale, 1954). Mozley (1939) recommended the propagation and cultivation of *Derris elliptica* in small plantations near ponds in Tanzania using local labor. Ransford (1948) reported that *Tephrosia vogelii* leaves met criteria of acceptable plant molluscicides in Malawi: molluscicidal and ovicidal activity, absence of toxic effects in man and livestock, and easy cultivation in areas where needed. However, Mozley (1944), Cowper (1946), and Kloos et al. (in press) found the leaves active only at very high concentrations.

A suggestion that *Ammi majus* plants, which have snail-killing properties and dermal photosensitizers (bergapten, pimpinellin, and xanthotoxin), be thrown into snail-infested waters (Soine, 1964) was not pursued. Several East African saponin fish poisons were either weakly molluscicidal or too scarce for practical use (Mozley, 1939, 1944). Encouraged by the finding that snails were rare in lakes in Finland where sawmills discarded sawdust, Mozley (1944) applied aqueous extracts of pine trees and wood ash in high concentrations in Rhodesian waters, with varying results.

PRESENT USES

In recent searches for more effective and selective molluscicides, four trends can be identified: Plants are used as extracts or are placed directly into snail habitats; active components are identified and isolated; plant materials are subjected to secondary screening to determine their stability under field-simulated physicochemical conditions and their toxicity for both snails and nontarget organisms; and attempts are being made to cultivate candidate plants for large-scale use.

The advantages of applying plant substances as extracts, preferably water extracts, are several. They include more efficient extraction of active substances, greater miscibility in water (Lugt, 1981), and greater dosing accuracy; extracts are also easier to apply than unprepared materials. More than 95% of all plants tested in the laboratory and field (Tables 1 to 4) were used as extracts. A large proportion of them were in organic solvents, which

Table 1. Toxicity of Plant Molluscicide Extracts

Species (Family)	Part (Solvent) Used[a]	Snail LC90 (LC50) ppm	Snail Eggs LC90 ppm	Mammals Acute Oral LD50; Rats, Mice mg/kg	Fish Conc. ppm (% Mort.)	Cercaricidal Effect[b]
1. *Croton macrostachys* (Euphorbiaceae)	s(W)	0.1–20	20+	No effect at 3000 g/rat/ 28 days	20–45 (50)	*
2. *Croton tiglium* (Euphorbiaceae)	s(W)	0.7	?	1.6 mg/kg	0.007	?
3. *Croton tiglium* (Euphorbiaceae)	s(E)	0.1	?	1.6 mg/kg	?	?
4. *Euphorbia cotonifolia* (Euphorbiaceae)	l(H)	1.2–8	48	No effect at 600 mg/kg	2.5 (100)	+
5. *Jatropha curcas* (Euphorbiaceae)	s(M)	(7)	?	100% mort. at 10 g/kg/ day × 3	500 (40)	?
6. *Anacardium occidentale* (Anacardiaceae)	SH(H)	3(1)	18	No effect at 100 ppm/day x 3 months	10 (100)	?
7. *Ambrosia maritima* (Compositae)	lf(W)	200– 1000	1000	?	No effect at 1000 ppm	+

8. Tetrapleura tetraptera (Mimosoideae)	SB(M)	(2)	?	1.49	0.35(50)	+
9. Entada phaseoloides (Mimosoideae)	b(S)	(3–6)	?	?	1(50)	?
10. Pithecellobium multiflorum (Mimosoideae)	l(S)	5(3)	21	35 mg	5 (50–100)	?
11. Phytolacca dodecandra (Phytolaccaceae)	BE(W)	6–29	?	2.6 g (mice) 1 g (sheep)	3–90(90)	+
12. Phytolacca dodecandra (Phytolaccaceae)	BE(B)	3–6	100±	1.2 g (mice)	2–4(90)	?
13. Gnidia kraussiana (Thymelaeaceae)	rlST (W)	(1.5–3)	?	?	?	?
14. Solanum nodiflorum (Solanaceae)	l(W)	79	?	?	?	?

[a] b = bark; BE = berries; fl = flowers; l = leaves; r = root; s = seeds; SB = stem bark; SH = shell; ST = stem; (B) = butanol; (E) = ethanol; (H) = hexane; (M) = methanol; (S) = successive extractions with different organic solvents; (W) = water. Italic letter denotes hot solvent.

[b] * = No effect observed after eight hours; + = effect observed at concentrations higher than molluscicidal levels; ? = lack of data.

Table 2. Stability of Plant Molluscicide Extracts

Species (Family)[a]	Stability Affected by:					
	UV Light	Temperature	Mud, Silt	pH	Minerals	Storage
1. *Croton macrostachys* (Euphorbiaceae)	No data	No	Yes	No	?	No
2. *Croton tiglium* (Euphorbiaceae)	?	?	?	?	?	?
3. *Croton tiglium* (Euphorbiaceae)	?	?	?	?	?	?
4. *Euphorbia cotonifolia* (Euphorbiaceae)	No	?	?	Yes	?	?
5. *Jatropha curcas* (Euphorbiaceae)	?	?	?	?	?	?
6. *Anacardium occidentale* (Anacardiaceae)	?	?	?	No	?	?
7. *Ambrosia maritima* (Compositae)	Yes	Yes	Yes	Yes	?	No
8. *Tetrapleura tetraptera* (Mimosoideae)	No	Yes	Yes	Yes	No	?
9. *Entada phaseoloides* (Mimosoideae)	No	?	No	Yes	No	?
10. *Pithecellobium multiflorum* (Mimosoideae)	?	?	?	?	?	?
11. *Phytolacca dodecandra* (Phytolaccaceae)	No	Yes	Yes	No	No	No
12. *Phytolacca dodecandra* (Phytolaccaceae)	No	Yes	No	No	No	No
13. *Gnidia kraussiana* (Thymelaeaceae)	?	?	?	No	?	?
14. *Solanum nodiflorum* (Solanaceae)	?	Yes	?	No	?	No

[a]Same plant part and solvent used as in Table 1.

Table 3. Field Data on Plant Molluscicides

Species (Family)[a]	Aquatic Snails ppm (% Mort.)	Amphibious Snails on Soil
1. *Croton macrostachys* (Euphorbiaceae)	2(100)	No data
2. *Croton tiglium* (Euphorbiaceae)	?	4 g/m^2
3. *Croton tiglium* (Euphorbiaceae)	?	?
4. *Euphorbia cotonifolia* (Euphorbiaceae)	20(100)	?
5. *Jatropha curcas* (Euphorbiaceae)	?	4 g/m^2
6. *Anacardium occidentale* (Anacardiaceae)	20(92)	?
7. *Ambrosia maritima* (Compositae)	70± (100)	?
8. *Tetrapleura tetraptera* (Mimosoideae)	?	10(?)
9. *Entada phaseoloides* (Mimosoideae)	?	40 g/m^2
10. *Pithecellobium multiflorum* (Mimosoideae)	?	?
11. *Phytolacca dodecandra* (Phytolaccaceae)	35–100 (80+)	?
12. *Phytolacca dodecandra* (Phytolaccaceae)	?	5
13. *Gnidia kraussiana* (Thymelaeaceae)	?	?
14. *Solanum nodiflorum* (Solanaceae)	?	?

[a]Same plant part and solvent used as in Table 1.

Table 4. Summary of Snail Toxicity Studies and Phytochemical Screening[a]

Plant Species by Family, Order,[b] Subphylum, and Phylum	Parts Tested[c]	Extract[d]	Conc. Tested ppm[e]	% Mort.	Snail Species[f]	Inactive Part[c]	Author	Compound[g]
1. ANGIOSPERMS (FLOWERING PLANTS)								
MONOCOTYLEDONAE:								
ARACEAE								
Pistia stratiotes	?	W	?	active	?	?	Manson & Bahr (1954)	
Another 6 species were inactive: Adewunmi & Sofowora (1980), Medina & Woodbury (1979)								
DIOSCOREACEAE								
Dioscorea cayenensis subsp. rotundata (= D. rotundata)	l	W	1000	100	LcLm	rf	Medina & Woodbury (1979)	S,ST
AGAVACEAE								
Agave sisalana	w	W	5000	90	Bg		Otieno (1966)	S
Agave spp. (2)	f	W	1000	0	LcLm		Medina & Woodbury (1979)	
Sanseviera guineensis	f	W	1000	0	LcLm		Medina & Woodbury (1979)	
S. guineensis	lr	M	100	0	BG		Adewunmi & Sofowora (1980)	
S. liberica	lr	M	100	5	BG		Adewunmi & Sofowora (1980)	S,T
S. trifascata	l	M	100	0	BG		Adewunmi & Sofowora (1980)	S,T
Yucca aloifolia	f	W	1000	0	LcLm		Medina & Woodbury (1979)	S,ST

Cont'd.

Y. pallida	l	W	800	20	Bg		Ritchie et al. (1963)
Y. schidigera	l	W	700	100	Bg	A	Ritchie et al. (1963)
AMARYLLIDACEAE							
Amaryllis vittata	w	W	650	50	BaBT	ST	El-Kheir & El-Tohami (1979)
—							
LILIACEAE							
Aloe secundiflora	l	W	1000	20	Bp		Kloos et al. (in press)
Asparagus currilus	See Table 5						
A. plumosa	See Table 5						
Dipcadi fesoghlense	b	W	40	100	BT		El-Kheir & El-Tohami (1979)
Eriospermum abyssinicum	r	M	100	100	Bg		Adewunmi & Sofowora (1980)
—							
ORCHIDACEAE							
Eulophia guineensis	ST	M	100	10	BG		Adewunmi & Sofowora (1980)
—							
BROMELIACEAE							
'gravata-acu'	l	W	1000	40	Bg		Amorin & Pessoa (1962)
Bromelia spp. (2)	lfr	W	1000	0	LcLm		Medina & Woodbury (1979)
—							
PONTEDERIACEAE							
Eichhornia natans	r	M	100	30	BG	F,ST	Adewunmi & Sofowora (1980)
—							
ZINGIBERACEAE							
Alpinia speciosa	l	W	10000	0	Bg		Silva et al. (1971)
Alpinia spp.		W	1000	0	LcLm		Medina & Woodbury (1979)

Table 4. cont'd..

Plant Species by Family, Order,[b] Subphylum, and Phylum	Parts Tested[c]	Extract[d]	Conc. Tested ppm[e]	% Mort.	Snail Species[f]	Inactive Part[c]	Author	Compound[g]
Hedychium coronarium	s	W	25	100	LcLm		Medina & Woodbury (1979)	
H. gardnerianum	?	S	?	?	Ba		Saleh et al. (1982)	
Zingiber officinale	RH	M	100	20	Bg		Adewunmi & Sofowora (1980)	
CANNACEAE								
Canna indica	lr	M	100	5–10	Bg	ST	Adewunmi & Sofowora (1980)	T
C. indica	w	W	650	100	Ba		Mahran et al. (1977)	
XYRIDACEAE								
Xyris anceps	l	M	100	100	Bg		Adewunmi & Sofowora (1980)	
DICOTYLEDONAE:								
ANNONACEAE								
Annona senegalensis	ST	M	100	85	BG	f	Adewunmi & Sofowora (1980)	T,F
A. senegalensis	l	M	100	20	BG		Adewunmi & Sofowora (1980)	A,G,ST. T
A. squamosa	s	E,W	10000	40;100	Bs	l	Silva et al. (1971)	
A. squamosa	RB	E,W	10000	80;100	Bg		Silva et al. (1971)	
Cleistopholis patens	r	M	100	5	Bg		Adewunmi & Sofowora (1980)	

Another 5 species were inactive: Adewunmi & Sofowora (1980), Medina & Woodbury (1979)

CANELLACEAE

Warburgia ugandensis   See Table 5

LAURACEAE

Species							Reference	
Persea americana (= *P. gratissima*)	s	W.A	10000	100	Bs		Silva et al. (1971)	SS

PIPERACEAE

Species							Reference
Piper marginatum	l	eo	?	60+	Bg		Rouquayrol et al. (1980)
P. tuberculatum	SBRB	W	200	40–80	BgBs		Sousa & Rouquayrol (1974)

Another 4 species of *Piper* were inactive: Medina & Woodbury (1979). Adewunmi & Sofowora (1980)

PHYTOLACCACEAE

Species							Reference
Phytolacca americana	BE	B	150	100	Bg		Johnson (1974)
P. dodecandra	BE	W	6–29	90	LnBT		Lemma (1970). Lugt (1981)
P. dodecandra	fl	W	100	80+	BT	r	Lemma (1970)
P. dodecandra	bST	W	500	0–60	BT		Lemma (1970)
P. dodecandra	BE	WF	2	100	Bg		Yohannes et al. (1979)
P. dodecandra	BE	B	3–6	90+	BgBc BpBN		Baalawy (1972) Lemma et al. (1972)
P. icosandra	BE	W	200	100	LcLm	lr	Medina & Woodbury (1979)
P. octandra	BE	W	50	90	Bg		Lugt (1981)
P. octandra	BE	W	100	100	Bg		Lugt (1981)
P. octandra	BE	B	26–	47	Bg		Ruffino (1975)
P. rivinoides	BE	W	200	100	LcLm	lr	Medina & Woodbury (1979)

CHENOPODIACEAE

Species							Reference	
Chenopodium ambrosioides	l	W	1000+	?	BgBs		Sousa & Rouquayrol (1974)	S.ST

Cont'd.

Table 4. cont'd..

Plant Species by Family, Order,[b] Subphylum, and Phylum	Parts Tested[c]	Extract[d]	Conc. Tested ppm[e]	% Mort.	Snail Species[f]	Inactive Part[e]	Author	Compound[g]
POLYGONACEAE								
Polygonum meisnerianum	w	E	30	19	Bp		Brossat et al. (1979)	
P. mite	w	E	30	14	Bp		Brossat et al. (1979)	
P. pulchrum (= *P. tomentosum*)	w	E	90	96	Bp		Brossat et al. (1979)	H
P. senegalense forma *senegalense*	1	W	5000	60	BpLn		Dossaji et al. (1977)	H
P. senegalense forma *senegalense*	1	W	1000	60	Bp		Kloos et al. (in press)	
P. senegalense forma *senegalense*	ST	W	1000	20	Bp		Kloos et al. (in press)	
P. senegalense forma *senegalense*	s	W	1000	40	Bp		Kloos et al. (in press)	
P. senegalense (= *P. glabrum, P. sambesicum*)	w	E	69,90	50,11	Bp		Brossat et al. (1979)	H

Another 2 species were active at 1000 ppm+; 2 were inactive: Medina & Woodbury (1979), Sousa & Rouquayrol (1974)

Plant Species by Family, Order,[b] Subphylum, and Phylum	Parts Tested[c]	Extract[d]	Conc. Tested ppm[e]	% Mort.	Snail Species[f]	Inactive Part[e]	Author	Compound[g]
PAPAVERACEAE								
Argemone mexicana	1	W	1000+	?	BgBs		Sousa & Rouquayrol (1974)	
DILLENIACEAE								
Curatella americana	ST	W	10000	80	Bs		Silva et al. (1971)	

Cont'd.

STERCULIACEAE						
Guazuma ulmifolia	SB	E	10000	100	Bg	Sousa et al. (1970)
COCHLOSPERM-ACEAE						
Cochlospermum insigne	SB	W	10000	80	BgBs	Sousa & Rouquayrol (1974)
C. regium	1	eo	?	0	Bg	Rouquayrol et al. (1980)
BOMBACEAE						
Bombax costatum	sr	M	100	100	BG	Adewunmi & Sofowora (1980)
CANNABACEAE						
Cannabis sativa (= C. indica)	fl	E	1000	100	Bg	Sousa et al. (1970)
ULMACEAE						
Trema guineensis	1	M	100	0	BG	Adewunmi & Sofowora (1980)
MORACEAE						
Chlorophora tinctoria	SB	E	10000	40	Bg	Sousa et al. (1970)
Dorstenia cayapia	r	W	1000+	?	BgBs	Sousa & Rouquayrol (1974)
Ficus glumosa	ST	M	100	40	BG	Adewunmi & Sofowora (1980)
F. sycomorus	1	W	1000	20	Bp	Kloos et al. (in press)
F. sycomorus	ST	W	1000	73	Bp	Kloos et al. (in press)
F. thoningii	1	W	1000	7	Bp	Kloos et al. (in press)
F. thoningii	ST	W	1000	7	Bp	Kloos et al. (in press)

Table 4. cont'd..

Plant Species by Family, Order, Subphylum, and Phylum	Parts Tested[e]	Extract[d]	Conc. Tested ppm[c]	% Mort.	Snail Species[f]	Inactive Part[c]	Author	Compound[g]
RHAMNACEAE								
Maesopsis eminii	r	M	100	100	BG	1	Adewunmi & Sofowora (1980)	T
Ziziphus joazeiro	STsf	W	10?	20–30	Bg		Barbosa & Mello (1969)	
Z. undulata	SB	W	10000	0	BgBs		Sousa & Rouquayrol (1974)	
THYMELAEACEAE								
Gnidia kraussiana	STrl	W	14–120	100	BG	1	El-Emam et al. (1981)	
G. kraussiana	STrl	W	2–10	100	BG		El-Kheir & El-Tohami (1979)	
G. kraussiana	STrl	Pe	0.1–5	100	BG		El-Kheir & El-Tohami (1979)	
G. kraussiana	STrl	Be	0.1–20	100	BG		El-Kheir & El-Tohami (1979)	
G. latifolia	l	W	1000	10	Bp	STfl	Kloos et al. (in press)	
EUPHORBIACEAE								
Acalypha ornata	lr	M	100	10+	BG	ST	Adewunmi & Sofowora (1980)	T
Bridelia atroviridis	ST	M	100	100	BG		Adewunmi & Sofowora (1980)	
Croton macrostachys	s	W	0.1	90	BT		Daffalla & Amin (1976)	A,ST
C. macrostachys	s	W	20	90	Bp		Daffalla & Amin (1976)	
C. macrostachys	s	W	200	0	Bg		Lugt (1981)	
C. megalocarpus	l	W	1000	47	Bp		Kloos et al. (in press)	
C. megalocarpus	ST	W	1000	60	Bp	S	Kloos et al. (in press)	
C. tiglium	s	W	0.7*	50	Oq		Yasuraoka et al. (1979a)	A

Cont'd.

Species							Reference	
C. tiglium	s	Er	0.1*	50	Oq		Yasuraoka et al. (1979a)	
Cyrtogonone argentea	r	M	100	100	BG	—	Adewunmi & Sofowora (1980)	
Euphorbia aegyptiaca	w	Pe	50	100	BTBp		Ahmed et al. (1984)	ST
E. candelabrum	la	W	8	100	Bg		Tesfaigzi (1978)	
E. cotinifolia	l	H	2–8	90	Bg		Pereira et al. (1980)	
E. lactea	?	E	2.4?	90	Ba		Abou El-Hassan et al. (1980)	
E. lactea	w	B(S)	22.0	90	Ba		El-Emam et al. (1982)	
E. lactea	w	Pe(S)	10.2	90	Ba		El-Emam et al. (1982)	
E. lactea	w	E(S)	14.5+	90	BTBa		El-Emam et al. (1982)	
E. mauritania	?	E	10	90	?		Abdel-Alim & Kamel (1984)	SS,A.F. G,T T
E. pulcherrima	r	M	100	10	BG		Adewunmi & Sofowora (1980)	
E. royleana	la	W	1.9	90	La		Singh & Agarwal (1984)	
Jatropha aceroides	s	W	40	80	BT		El-Kheir & El-Tohami (1979)	
J. aceroides	s	S	40	60	BT		El-Kheir & El-Tohami (1979)	
J. aethiopica	rST	W	250	10,20	BT		El-Kheir & El-Tohami (1979)	
J. curcas	s	W	27–48*	90	Oq		Yasuraoka et al. (1979b)	
J. curcas	Ss	B	45*	50	Oq		Yasuraoka et al. (1979b)	
J. curcas	s	M	7*	50	Oq		Yasuraoka et al. (1979b)	
J. curcas	sST	M	100	20	BG	lfST	Adewunmi & Marquis (1980, 1981a)	
J. curcas	r	W	160	50	BT		El-Kheir & El-Tohami (1979)	
J. gossypiifolia	f	M	100	100	BG	lpSB	Adewunmi & Marquis (1980)	

Table 4. cont'd..

Plant Species by Family, Order,[b] Subphylum, and Phylum	Parts Tested[c]	Extract[d]	Conc. Tested ppm[e]	% Mort.	Snail Species[f]	Inactive Part[c]	Author	Compound[g]
J. podagrica	sST	M	100	20	BG	1	Adewunmi & Marquis (1980)	A.F
Manihot esculenta	ST	W	1000	13	Bp	1	Kloos et al. (in press)	
M. glaziovii	ST	M	100	15	BG		Adewunmi & Sofowora (1980)	A.S
Phyllanthus niruri	r	Pe	25	100	BTBp		Ahmed et al. (1984)	ST,T
Ricinus communis	1	W	1000	20	Bp	ST	Kloos et al. (in press)	
Synadenium glaucescens	1	W	1000	80	Bp		Kloos et al. (in press)	

Another 8 species were active at 1000 ppm+; 26 were inactive: Adewunmi & Sofowora (1980), Medina & Woodbury (1979), Rouquayrol et al. (1980), Silva et al. (1971), Sousa & Rouquayrol (1974), Sousa et al. (1970)

FLACOURTIACEAE

Casearia guianensis	RB	W	1000+	?	Bg		Sousa & Rouquayrol (1974)	

CUCURBITACEAE

Cayaponia america	w	W	1000	0	LcLm		Medina & Woodbury (1979)	
Cucumis abyssinicus	1	W	1000	27	Bp		Kloos et al. (in press)	
C. abyssinicus	ST	W	1000	13	Bp		Kloos et al. (in press)	
C. abyssinicus	sf	W	1000	40	Bp		Kloos et al. (in press)	

Cucurbita foetidissima	r		W	450	100	Bg	Ritchie et al. (1963)
C. pepo	s		W	10000	0	Bs	Silva et al. (1971)
Luffa operculata	f		W	1000	60	Bs	Silva et al. (1971)
Momordica charantia	f	S.ST.T	W	1000+	?	BgBs	Sousa & Rouquayrol (1974)
M. tuberosa	l		Pe	25	100	BT	Ahmed et al. (1984)
M. tuberosa	l		Pe	50	100	Bp	Ahmed et al. (1984)
Wildebrandia sp.	ST		W,E	10000	10	Bg	Sousa et al. (1970)
TURNERACEAE							
Turnera sp.	r		W	1000	40	Bs	Silva et al. (1971)
Turnera sp.	r		W	10000	0	BgBs	Sousa & Rouquayrol (1974)
OCHNACEAE							
Lophira alata	l		M	100	100	BG	Adewunmi & Sofowora (1980)
Ouratea fieldingiana	RB		W	10000	0	Bg	Sousa et al. (1970)
THEACEAE							
Camellia sinensis (=*Thea oleosa*)	w	T	W	5000	100	Oh	Cheng (1971)
Schima wallichii subsp. *noronhae* var. *superba* (=*Schima argentea*)	l		W	500	100	Oh	Fang (1959)
SAPOTACEAE							
Butyrospermum paradoxum	RB	f	M	100	70	BG	Adewunmi & Sofowora (1980)
STYRAXACEAE							
Styrax officinalis	h		W	100	100	BT	Saliternik & Witenberg (1959)

Cont'd.

Table 4. cont'd..

Plant Species by Family, Order,[b] Subphylum, and Phylum	Parts Tested[c]	Extract[d]	Conc. Tested ppm[e]	% Mort.	Snail Species[f]	Inactive Part[c]	Author	Compound[g]
MYRSINACEAE								
Embelia schimperi	f	W	100	40	Bg		Kloos et al. (1982)	
Myrsine africana	f	W	1000	93	Bg		Kloos et al. (1982)	T
CASUARINACEAE								
Casuarina equisetifolia (= C. equinaceae)	lf	W	100	100	LcLm		Medina & Woodbury (1979)	
C. equisetifolia	b	W	500	50	Pc		Cowper (1946)	
ROSACEAE								
Quillaja saponaria	l	W	2000	100	BgPc		Cowper (1946)	
Q. saponaria	l	W	2000	100	BgPc		Cowper (1946)	
Q. saponaria	b	W	7500	100	Pc		Cowper (1946)	
Q. saponaria	ST	W	10000	50	Pc		Cowper (1946)	
Rosa sp.	w	W	1000	0	LcLm		Medina & Woodbury (1979)	
CHRYSOBALANACEAE								
Acioa barteri	ST	M	100	100	BG		Adewunmi & Sofowora (1980)	
A. rudatisii	ST	M	100	100	BG		Adewunmi & Sofowora (1980)	
(=A. lehmbachii) Hirtella racemosa, var. hexandra (=H. americana)	SB	W	1000	?	Bg		Sousa and Rouquayrol (1974)	

Cont'd.

LEGUMINOSAE-CAESALPINIOIDEAE

Species								Reference
Caesalpinia coriaria	f	W	?	?	L			Anantaraman (1955)
Cassia singueana	ST	M	100	5	BG		S.T	Adewunmi & Sofowora (1980)
C. singueana	l	W	1000	7	Bp			Kloos et al. (in press)
C. singueana	ST	W	1000	7	Bp			Kloos et al. (in press)
Delonix regia	ST	M	100	20	BG		T	Adewunmi & Sofowora (1980)
Dialium guineense	f	M	100	100	BG		ST	Adewunmi & Sofowora (1980)
Swartzia madagascariensis	lST	W	5000	100	BG			Mozley (1944)
S. madagascariensis	ps	W	200	100	BG			Mozley (1944)
Tamarindus indica	s	M	10	10	BG			Adewunmi & Sofowora (1980)

Another 6 species were active at 1000 ppm: Mahran et al. (1977), Sousa & Rouquayrol (1974); 14 species were inactive: Adewunmi & Sofowora (1980), Medina & Woodbury (1979), Sousa and Rouquayrol (1974), Sousa et al. (1970)

LEGUMINOSAE-MIMOSOIDEAE

Species								Reference
Acacia dudgeoni	lSB	M	100	60	BG		T	Adewunmi & Sofowora (1980)
A. nilotica	f(p)	W	120	100	BG	t	T	Adewunmi & Sofowora (1980)
A. nilotica	f(p)	E	100	100	BT			El-Kheir & El-Tohami (1979)
A. nilotica	f(p)	W(sp)	75	100	BT			Hussein Ayoub (1982)
A. nilotica	ST	W	500	100	Bg			Kloos et al. (1982)
A. nilotica	l	W	1000	20	Bp			Kloos et al. (in press)
A. nilotica	s	W	1000	100	Bp			Kloos et al. (in press)
A. nilotica	p	W	1000	40	Bp			Kloos et al. (in press)

Table 4. cont'd..

Plant Species by Family, Order,[b] Subphylum, and Phylum	Parts Tested[c]	Extract[d]	Conc. Tested ppm[e]	% Mort.	Snail Species[f]	Inactive Part[c]	Author	Compound[g]
Calliandra portoricensis	r	E	20	?	BG		Adewunmi & Marquis (1981a)	
Distrostachys cinerea	l	M	100	100	BG		Adewunmi & Marquis (1981a)	
(=*D. glomerata*)								
Entada phaseoloides	b	S	3.6	100	Oq		Yasuraoka et al. (1977)	
E. phaseoloides	b	WEB	500+	50	Oq		Yasuraoka et al. (1977)	
Piptadenia biuncifera	RB	W	500	60	BgBs		Silva et al. (1971)	
P. macrocarpa	RB	E	500	100	BgBs		Silva et al. (1971)	
Samanea saman	f	W	500	100	BgBs		Silva et al. (1971)	
(=*Albizia saman*)								
S. saman	s	W	100	100	Bs		Rouquayrol et al. (1973)	A
S. saman	s	S	4.9	90	Bs		Rouquayrol et al. (1973)	
Stryphnodendron coriaceum	SB	W	1000+	?	BgBs		Sousa et al. (1970)	
Tetrapleura tetraptera	f	W	100	100	BG		Adewunmi et al. (1982)	S,G
T. tetraptera	f	M	10	100	BG		Adesina et al. (1980)	
T. tetraptera	SB	M	2	50	BG		Adewunmi & Marquis (1981b)	
T. tetraptera	SB	M	2	50	Ln		Adewunmi & Marquis (1981b)	
T. tetraptera	SB	M	5.2	50	BP		Adewunmi & Marquis (1981b)	

Another 9 species were active at 1000 ppm+: Alzérreca et al. (1981), Amorin & Pessoa (1962), Medina & Woodbury (1979), Mozley (1939), Sousa & Rouquayrol (1974); 2 species were inactive: Adewunmi & Sofowora (1980), Sousa & Rouquayrol (1974), Sousa et al. (1970)

LEGUMINOSAE-PAPILIONOIDEAE

Species								Reference
Calopogonium velutinum (=Stenolobium velutinum)	bfl	W	1000	100	Bg			Amorin & Pessoa (1962)
Calpurnia aurea	?	M	?	active	Bg		A	Kubo et al. (1984b)
Crotalaria senegalensis	l	Pe	250	100	BT			Ahmed et al. (1984)
Derris elliptica	r	W	20	100	BG			Mozley (1939)
Dioclea reflexa	r	M	100	70	BG	IST		Adewunmi & Sofowora (1980)
Erythrina abyssinica	ST	W	1000	30	Bp	l		Kloos et al. (in press)
Indigofera kerstingii	lr	M	100	20	BG	ST		Adewunmi & Sofowora (1980)
I. secundiflora	rST	M	100	20	BG	fl		Adewunmi & Sofowora (1980)
I. spicata	ST	M	100	30	BG	r		Adewunmi & Sofowora (1980)
I. suffruticosa	s	W	100	100	LcLm			Medina & Woodbury (1979)
Indigofera sp.	l	W	1000	10	Bp			Kloos et al. (in press)
Indigofera sp.	RB	W	1000	20	Bp			Kloos et al. (in press)
Neorautanenia mitis (=N. pseudopachyrhiza)	r	W	500	100	BG	ST		Teesdale (1954)
Requienia obcordata	l	A.Pe	150–200	100	BTBp			Ahmed et al. (1984)
Sesbania sesban	l	W	10	100	Bg		S	Lugt (1981)
S. sesban	l	W	50–100	100	Bg			Teesdale (1954)
S. sesban	l	Pe	100–200	100	BTBp			Ahmed et al. (1984)
S. sesban	l	A	100	100	BTBp			Ahmed et al. (1984)
Stylosanthes viscosa	ST	M	100	35	BG		S.T	Adewunmi & Sofowora (1980)

Cont'd.

Table 4. cont'd...

Plant Species by Family, Order, Subphylum, and Phylum	Parts Tested[c]	Extract[d]	Conc. Tested ppm[e]	% Mort.	Snail Species[f]	Inactive Part[e]	Author	Compound[g]
Tephrosia pseudolongipes	l	Pe	10	100	BT		Ahmed et al. (1984)	F,ST,T
T. pseudolongipes	l	Pe	25	100	Bp		Ahmed et al. (1984)	
T. sinapou (= *T. toxicara*)	r	W	1000	?	Bs		Sousa & Rouquayrol (1974)	
T. vogelii	l	W	250*	100	BG		Ransford (1948)	A(stem)
T. vogelii	lfST	W	?	28–100	BG		Mozley (1944)	
T. vogelii	l	W	1000	20	Bp		Kloos et al. (in press)	
T. vogelii	ST	W	1000	10	Bp		Kloos et al. (in press)	
T. vogelii	p	W	1000	13	Bp		Kloos et al. (in press)	
T. vogelii	s	W	1000	10	Bp		Kloos et al. (in press)	
Tephrosia sp.	l	W	10000*	80	Pc		Cowper (1946)	
Vigna coerulea	l	Pe	50–75	100	BTBp		Ahmed et al. (1984)	
Zornia setosa	l	W	1000	100	Bp		Kloos et al. (in press)	
subsp. *obvata*	ST	W	1000	20	Bp		Kloos et al. (in press)	

Another 10 species were active at 1000 ppm[+]:Amorin & Pessoa (1962), Medina & Woodbury (1979), Mozley (1939), Alzérreca et al. (1981), Sousa & Rouquayrol (1974), Sousa et al. (1970); 11 species were inactive: Adewunmi & Sofowora (1980), Silva et al. (1971), Sousa & Rouquayrol (1974)

MYRTACEAE *Eucalyptus acmenoides*								
(= *E. triantha*)	l?	W	500–1000*	100	L		Broberg (1982)	A,T
E. citriodora	l	eo	?	100	Bg		Rouquayrol et al. (1980)	T
E. largiflorens (= *E. bicolor*)	l?	W	500–1000	100	L		Broberg (1982)	

Cont'd.

Species							Reference	
E. saligna	l	eo	?	0	Bg		Broberg (1982)	T
Eucalyptus sp.	ST	W	2000	75–100	BG		Mozley (1944)	T
Eucalyptus sp.	l	W	1000	20	Bp	ST	Kloos et al. (in press)	
Eucalyptus sp.	s	W	1000	20	Bp		Kloos et al. (in press)	
Eucalyptus spp.	l	W	200–500	0	Pc		Cowper (1946)	T
Eucalyptus spp. (6 species)	ll	W	1000*	100	L		Broberg (1982)	T
Eugenia sp.	l	eo	?	100	Bg		Rouquayrol et al. (1980)	
Psidium guajava	l	W	1000	73	Bp		Kloos et al. (in press)	
Syzygium cordatum	l	W	1000	7	Bp	fl	Kloos et al. (in press)	
S. cordatum	ST	W	1000	7	Bp		Kloos et al. (in press)	
S. cumini (=S. jambolana)	SB	WE	1000	10	Bg		Sousa et al. (1970)	
S. guineense	l	W	1000	87	Bp		Kloos et al. (in press)	
S. guineense	ST	W	1000	33	Bp		Kloos et al. (in press)	

Another 7 species were inactive: Broberg (1982), Medina & Woodbury (1979), Rouquayrol et al. (1980)

PUNICACEAE								
Punica granatum	r	W	1000+	?	BgBs		Sousa & Rouquayrol (1974)	
COMBRETACEAE								
Combretum fragrans (=C. ghasalense)	rST	M	100	100	Bg	f	Adewunmi & Sofowora (1980)	
C. molle	l	W	1000	80	Bp		Kloos et al. (in press)	
C. molle	ST	W	1000	20	Bp		Kloos et al. (in press)	
Terminalia brownii	l	Pe	200	100	Bp		Ahmed et al. (1984)	
T. kilimandscharica	l	W	1000	100	Bp		Kloos et al. (in press)	
T. kilimandscharica	ST	W	1000	40	Bp		Kloos et al. (in press)	
T. kilimandscharica	p	W	1000	40	Bp		Kloos et al. (in press)	
T. mollis	r	M	100	100	BG	l	Adewunmi & Sofowora (1980)	
T. mollis	SB	M	100	50	BG		Adewunmi & Sofowora (1980)	

Table 4. cont'd..

Plant Species by Family, Order,[b] Subphylum, and Phylum	Parts Tested[c]	Extract[d]	Conc. Tested ppm[e]	% Mort.	Snail Species[f]	Inactive Part[c]	Author	Compound[g]
Terminalia spp. (2)	ST	M	100	10–15	BG		Adewunmi & Sofowora (1980)	
Another 2 species were active at 1000 ppm; 4 species were inactive: Adewunmi & Sofowora (1980), Sherif & El-Sawy (1962). Silva et al. (1971)								
ONAGRACEAE								
Ludwigia leptocarpa	lf	W	1000	0	LcLm		Medina & Woodbury (1979)	
L. octonervis subsp. *brevisepala* (=*L. angustifolia*)	l	W	1000	100	LcLm	fr	Medina & Woodbury (1979)	
RUTACEAE								
Clausena anisata	r	M	6–10?	?	BG		Adesina & Adewunmi (1981)	
C. anisata	l	W	1000	53	Bp		Kloos et al. (in press)	
C. anisata	ST	W	1000	7	Bp		Kloos et al. (in press)	
C. anisata	b	W	1000	40	Bp		Kloos et al. (in press)	
Fagaropsis hildebrandtii	l	W	1000	27	Bp		Kloos et al. (in press)	
F. hildebrandtii	ST	W	1000	33	Bp		Kloos et al. (in press)	
Pilocarpus sp.	l	eo	?	60	Bg		Rouquayrol et al. (1980)	
Ruta chalepensis	See Table 5							
Zanthoxylum chalybeum	l	W	1000	0	Bp		Kloos et al. (in press)	
Z. fagara	SB	E	10000	40	Bg		Sousa & Rouquayrol (1974)	

Plant							Reference
Z. macrophylla (=Fagara macrophylla)	See Table 5						
Zanthoxylum sp. (Fagara sp.)	l	eo	?	20	Bg		Rouquayrol et al. (1980)
SIMAROUBACEAE							
Simarouba versicolor	SB	W	10000	100	Bg		Sousa & Rouquayrol (1974)
S. versicolor	RB	WE	10000	100	Bg		Sousa & Rouquayrol (1974)
BURSERACEAE							
Bursera simaruba (=Elaphrium simaruba)	s	W	1000	1000	LcLm	fr	Medina & Woodbury (1979)
Dacryodes edulis	rs	M	100	0	BG		Adewunmi & Sofowora (1980)
Protium heptaphyllum	b	eo	?	60	Bg		Rouquayrol et al. (1980)
Protium sp.	b	eo	?	0	Bg		Rouquayrol et al. (1980)
MELIACEAE							
Azadirachta indica (='neem margosa')	f	W	5000–100	100	Ms	ST	Muley (1978)
Ekebergia senegalensis	l	M	100	20	BG		Adewunmi & Sofowora (1980)
Guarea trichilioides	lfr	W	1000	0	LcLm		Medina & Woodbury (1979)
ANACARDIACEAE							
Anacardium occidentale	SH	S	0.4	50	Bg		Sullivan et al. (1982)
A. occidentale	SH	H	0.6	50	Bg		Pereira & Pereira de Souza (1974)
A. occidentale	SH	W	1000	?	BgBs		Sousa & Rouquayrol (1974)
Astronium fraxinifolium	SB	AW	1000	20,40	Bg		Sousa et al. (1970)

Cont'd.

Table 4. *cont'd..*

Plant Species by Family, Order,[b] Subphylum, and Phylum	Parts Tested[c]	Extract[d]	Conc. Tested ppm[e]	% Mort.	Snail Species[f]	Inactive Part[c]	Author	Compound[g]
Mangifera indica	l	W	1000	40	Bp		Kloos et al. (in press)	
Spondias mombin (=*S. lutea*)	SB	W	10000	100	Bg		Sousa et al. (1970)	
Tapirira guianensis	SB	W	10000	100	Bs		Silva et al. (1971)	
Another 4 species were inactive: Medina & Woodbury (1979), Sousa et al. (1970)								
SAPINDACEAE								
Magonia pubescens	s	?	9	80–90	BgBs		Barbosa & Mello (1969)	
Pappea capensis	l	W	1000	100	Bp		Kloos et al. (in press)	
P. capensis	ST	W	1000	40	Bp		Kloos et al. (in press)	
P. capensis	fl	W	1000	20	Bp		Kloos et al. (in press)	
Paullinia pinnata	lb	W	1000	100	Bg		Amorin & Pessoa (1962)	
Sapindus emarginatus	f	WA	20000	100	Ms		Muley (1978)	
S. saponaria	f	WA	25	94	BgLc		Torrealba et al. (1953)	S
S. saponaria	f	B	3	83	Bg		Ruffino (1975)	
S. saponaria	l	W	250	0	Pc		Cowper (1946)	
S. saponaria	lb	W	500	100	BA		Msangi & Zeller (1965)	
S. saponaria	SB	W	10000	40	Bg		Sousa et al. (1970)	
Another 4 species were inactive: Medina & Woodbury (1979), Sousa & Rouquayrol et al. (1974)								
MALPIGHIACEAE								
Brysonima sericea	SB	W,E	200	80	BgBs		Silva et al. (1971)	
POLYGALACEAE								
Polygala erioptera	w	Pe	50	100	BTBp		Ahmed et al. (1984)	F.S.ST
P. paniculata	w	W	1000	0	LcLm		Medina & Woodbury (1979)	

Cont'd.

Name						S	Reference
P. paniculata	?	Pe					See Table 5
P. paniculata			400	active	Bg		Hamburger et al. (1985)
Securidaca longepedunculata	r	W	350	100	BG		Azevedo & Medeiros (1963)
BALANITACEAE							
Balanites aegyptiaca	f	W	1000+*	?	Bg	S	Plank (1945)
B. aegyptiaca	f	M	100	100	BG		Adewunmi & Sofowora (1980)
B. maughamii	f	W	10	?	BA,Ln		Wagner (1936)
B. roxburghii	f	W	500	100	BgPc		Cowper (1946)
B. roxburghii	f	W	1000?	50	BgPc		Cowper (1946)
OLACACEAE							
Ximenia americana	l	M	100	100	BG	S	Adewunmi & Sofowora (1980)
OPILIACEAE							
Agonandra brasiliensis	SB	W	1000	80	Bs		Silva et al. (1971)
UMBELLIFERAE (APIACEAE)							
Ammi majus							See Table 5
Eryngium foetidum	rf	M	100	0	BG		Adewunmi & Sofowora (1980)
Eryngium sp.	a	W	1000+*	0	Bg		Ruffino (1975)
Steganotaenia araliaceae	l	W	1000	100	Bp	ST	Kloos et al. (in press)
S. araliaceae	b	W	1000	40	Bp		Kloos et al. (in press)
ARALIACEAE							
Aralia balfouriana	rlf	W	1000	0	LcLm		Medina & Woodbury (1979)

Table 4. cont'd..

Plant Species by Family, Order,[b] Subphylum, and Phylum	Parts Tested[c]	Extract[d]	Conc. Tested ppm[e]	% Mort.	Snail Species[f]	Inactive Part[c]	Author	Compound[g]
Hedera helix	BE	M	40	?	Bg		Hostettmann (1980)	ST
Polyscias guilfoylei	lST	W	100	100	LcLm		Medina & Woodbury (1979)	
COMPOSITAE (ASTERACEAE)								
Ambrosia maritima (*A. senegalensis*)	lST	W	375?	100	LnBG		Vassiliades & Diaw (1980)	S,ST
A. maritima	lfl	W	1000	30	BaBT		Sherif & El-Sawy (1962)	
A. maritima	w	W	200	50	Bg		Lugt (1981)	
A. maritima	w	A	2000	0	BaBT		Sherif & El-Sawy (1962)	
A. maritima	ls	W	1000	100	Bg		Kloos et al. (1982)	
Artemisia maritima	l	eo	800?	20	La		Khand & Qadri (1974)	
A. kurramensis	l	eo	800?	20	La		Khand & Qadri (1974)	
Aspilia mossambicensis	l	W	1000	53	Bp		Kloos et al. (in press)	
A. mossambicensis	ST	W	1000	53	Bp		Kloos et al. (in press)	
A. mossambicensis	fl	W	1000	60	Bp		Kloos et al. (in press)	
Baccharis genistelloides, var. trimera (=*B. trimera*)	sffl	E	?	100	Bg	S	Frischkorn (1978)	
Bidens pilosa	l	W	1000	27	Bp		Kloos et al. (in press)	
B. pilosa	ST	W	1000	27	Bp		Kloos et al. (in press)	
B. pilosa	fl	W	1000	7	Bp		Kloos et al. (in press)	
Chrysanthemum cinerariafolium	l	W	50	8	Bg		Tekle (1977)	

Cont'd.

						S,ST,F	
Eremanthus glomerulatus	w	H	100	65	Bg		Barros et al. (1985)
Ethulia conyzoides	l	Pe	200	100	Bp		Ahmed et al. (1984)
Gundelia tournefortii	See Table 5						
Heliopsis longipes	r	W	3000	100	L		Johns et al. (1982)
Lactuca capensis	l	W	1000	7	Bp	fl	Kloos et al. (in press)
L. capensis	ST	W	1000	7	Bp		Kloos et al. (in press)
Launaea leynaena	w	Pe	250	100	BT		Ahmed et al. (1984)
L. leynaena	w	Pe	100	100	Bp		Ahmed et al. (1984)
Podachaeminum eminens	See Table 5						
Pulicaria crispa	l	Pe	100	100	Bp		Ahmed et al. (1984)
P. crispa	l	Pe	50	100	BT		Ahmed et al. (1984)
P. cristata	lST	Pe	200	100	Bp		Ahmed et al. (1984)
Sphaeranthus gomphrenoides	l	W	1000	60	Bp		Kloos et al. (in press)
S. gomphrenoides	ST	W	1000	10	Bp		Kloos et al. (in press)
S. gomphrenoides	fl	W	1000	60	Bp		Kloos et al. (in press)
Tithonia diversifolia	l	W	1000	60	Bp		Kloos et al. (in press)
T. diversifolia	fl	W	1000	33	Bp	ST	Kloos et al. (in press)
Spilanthes maritima	l	W	1000	7	Bp		Kloos et al. (in press)
S. maritima	ST	W	1000	7	Bp		Kloos et al. (in press)
S. maritima	fl	W	1000	7	Bp		Kloos et al. (in press)
Vernonia brachycalyx	l	W	1000	33	Bp	ST	Kloos et al. (in press)
V. brachycalyx	fl	W	1000	13	Bp		Kloos et al. (in press)
V. lapiopus	l	W	1000	40	Bp		Kloos et al. (in press)
V. lapiopus	ST	W	1000	10	Bp		Kloos et al. (in press)
V. lapiopus	fl	W	1000	10	Bp		Kloos et al. (in press)
Wedelia caracasana (=*W. scaberrima*)	See Table 5						
W. parviceps	See Table 5						

Another 3 species were active at 1000 ppm+: Rouquayrol et al. (1980), Silva et al. (1971); 12 species were inactive: Adewunmi & Sofowora (1980), Medina & Woodbury (1979), Rouquayrol et al. (1980).

Table 4. cont'd..

Plant Species by Family, Order,[b] Subphylum, and Phylum	Parts Tested[c]	Extract[d]	Conc. Tested ppm[e]	% Mort.	Snail Species[f]	Inactive Part[c]	Author	Compound[g]
SOLANACEAE								
Capsicum frutescens	lf	W	100	100	LcLm	r	Medina & Woodbury (1979)	
Cestrum laurifolium	lf	W	100	100	LcLm	r	Medina & Woodbury (1979)	
C. macrophyllum	l	W	100	100	LcLm		Medina & Woodbury (1979)	
Lycopersicon esculentum	l	W	1000	100	LcLm		Medina & Woodbury (1979)	
Nicotiana tabacum	l	W	168?	?	L		Rao et al. (1980)	
Solanum americanum (=S. nodiflorum)	lr	W	50	100	LcLm		Medina & Ritchie (1980)	A
S. americanum	?-l	W	100	85	Pc		Medina & Ritchie (1980)	
S. americanum	l	W	100	100	Bg		Medina & Ritchie (1980)	
S. americanum		W	100	0	TM		Medina & Ritchie (1980)	
S. mammosum	fp	W	100	100	LcLm	lr	Medina & Woodbury(1979)	
S. nigrum	l	W	1000	73	Bp		Kloos et al. (in press)	
S. nigrum	ST	W	1000	33	Bp		Kloos et al. (in press)	
S. nigrum	f	W	1000	100	Bp		Kloos et al. (in press)	
S. nigrum	r	W	1000	100	Bp		Kloos et al. (in press)	
Withania obtusifolia	l	A.Pe	75	100	BTBp		Ahmed et al. (1984)	

Another 2 species were active at 1000 ppm; 7 species were inactive: Medina & Woodbury (1979). Silva et al. (1971). Sousa & Rouquayrol (1974)

Cont'd.

Species								Reference
HYDROPHYLLACEAE								
Hydrolea spinosa	a	W	1000+	?	BgBs			Sousa et al. (1970)
CORNACEAE								
Cornus florida	b	M	100	?	Bg		S	Hostettmann et al. (1978)
LOGANIACEAE								
Antonia ovata	l	eo	?	50	Bg			Rouquayrol et al. (1980)
Strychnos parviflora	SB	W	1000+	?	BgBs			Sousa et al. (1970)
RUBIACEAE								
Canthium subcordatum (=Randia nilotica)	rST	M	100	20–25	BG			Adewunmi & Sofowora (1980)
Catunaregam nilotica (=Randia nilotica)	RB	S	100	0	BT			El-Kheir & El-Tohami (1979)
	RB	W	40	100	BpBT			El-Kheir & El-Tohami (1979)
C. nilotica	f	A	25–50	100	BTBp		S,ST,T	Ahmed et al. (1984)
Coffea arabica	h(fresh)	W	1000	0	Bp			Kloos et al. (in press)
C. arabica	h(dried)	W	1000	0	Bp			Kloos et al. (in press)
C. spathicalyx	l	M	100	20	BG		A.T	Adewunmi & Sofowora (1980)
Gonzalagunia spicata (=Duggena hirsuta)	f	W	100	100	LcLm	l		Medina & Woodbury (1979)
Gardenia lutea	fFP	A	100	100	BTBp		A.S,ST	Ahmed et al. (1984)
G. terrifolia	ST	M	100	15	BG	r	A.T	Adewunmi & Sofowora (1980)
G. vogelii	FP	W	40	100	BpBT			El-Kheir & El-Tohami (1979)
G. vogelii	FP	S	40	60	BT			El-Kheir & El-Tohami (1979)

Table 4. cont'd..

Plant Species by Family, Order,[b] Subphylum, and Phylum	Parts Tested[c]	Extract[d]	Conc. Tested ppm[e]	% Mort.	Snail Species[f]	Inactive Part[c]	Author	Compound[g]
G. vogelii	SB fl	W	200	80+	BpBT		El-Kheir & El-Tohami (1979)	
Morinda lucida	l	M	100	100	BG		Adewunmi & Sofowora (1980)	
M. lucida	See Table 5							
Oldenlandia affinis	r	M	100	20	BG	1ST	Adewunmi & Sofowora (1980)	
Rothmania urcelliformis	l	M	100	20	BG		Adewunmi & Sofowora (1980)	
R. whitfieldii	ST	M	100	100	BG	1	Adewunmi & Sofowora (1980)	

Another 4 species were active at 1000 ppm+; 15 species were inactive: Medina & Woodbury (1979), Silva et al. (1971). Sousa & Rouquayrol (1974), Sousa et al. (1970)

APOCYNACEAE

Himatanthus bracteata (=Plumeria bracteata)	ST	E	10000	80	Bg		Sousa et al. (1970)	
Hunteria caffra	l	M	100	5	BG		Adewunmi & Sofowora (1980)	
Peschiera affinis	RB	E	10000	100	Bg		Sousa et al. (1970)	
Rauvolfia caffra	r	M	100	100	BG	1	Adewunmi & Sofowora (1980)	A,S
R. caffra	ST	M	100	10	BG		Adewunmi & Sofowora (1980)	S
R. ternifolia	RB	W	1000+	?	BgBs		Sousa & Rouquayrol (1974)	

Cont'd.

Species							Reference	
Voacanga africana	ST	M	100	45	BG		Adewunmi & Sofowora (1980)	A.S

Another 6 species were inactive: Adewunmi & Sofowora (1980), Medina & Woodbury (1979)

ASCLEPIADACEAE

Asclepias curassavica	r	W	100	100	LcLm	rl	Medina & Woodbury (1979)	
Cryptostegia grandiflora	ST	M	100	100	BG		Adewunmi & Sofowora (1980)	F,ST

BIGNONIACEAE

Crescentia cujete	w	W	1000	0	LcLm		Medina & Woodbury (1979)	
Kigelia africana	f	M	100	100	Bg		Adewunmi & Sofowora (1980)	
Tabebuia caraiba	SB	W	10000	100	Bg		Sousa et al. (1970)	

SCROPHULARIACEAE

Capraria biflora	r	W	10000	20	Bs		Silva et al. (1971)	
Scoparia dulcis	SB	E	10000	80	Bg		Sousa et al. (1970)	

ACANTHACEAE

Brillantaisa vogeliana	ST	M	100	40	BG	lr	Adewunmi & Sofowora (1980)	
Crossandra flava	f	M	100	20	BG	ST	Adewunmi & Sofowora (1980)	
Lankesteria elegans	l	M	100	2	BG	?	Adewunmi & Sofowora (1980)	A,ST

Another 5 species were inactive: Adewunmi & Sofowora (1980), Medina & Woodbury (1979)

Table 4. cont'd..

Plant Species by Family, Order,[b] Subphylum, and Phylum	Parts Tested[c]	Extract[d]	Conc. Tested ppm[e]	% Mort.	Snail Species[f]	Inactive Part[c]	Author	Compound[g]
VERBENACEAE								
Lantana trifolia	l	W	1000	7	Bp	ST	Kloos et al. (in press)	
L. trifolia	f	W	1000	7	Bp		Kloos et al. (in press)	
L. trifolia	fl	W	1000	7	Bp		Kloos et al. (in press)	
Lippia javanica	l	W	1000	40	Bp		Kloos et al. (in press)	
L. javanica	ST	W	1000	10	Bp		Kloos et al. (in press)	
L. javanica	fl	W	1000	47	Bp		Kloos et al. (in press)	
Vitex oxycuspis	ST	M	100	100	BG	l	Adewunmi & Sofowora (1980)	ST
V. oxycuspis	rl	M	100	20	BG	l	Adewunmi & Sofowora (1980)	S,T

Another 6 species were active at 1000 ppm+; 7 species were inactive: Medina & Woodbury (1979). Rouquayrol et al. (1980), Sousa & Rouquayrol (1974)

Plant Species by Family, Order,[b] Subphylum, and Phylum	Parts Tested[c]	Extract[d]	Conc. Tested ppm[e]	% Mort.	Snail Species[f]	Inactive Part[c]	Author	Compound[g]
LABIATAE (LAMIACEAE)								
Fuerstia africana	l	W	1000	40	Bp		Kloos et al. (in press)	
F. africana	ST	W	1000	60	Bp		Kloos et al. (in press)	
Hyptis pectinata	l	W	1000	60	Bp	ST	Kloos et al. (in press)	
H. pectinata	fl	W	1000	100	Bp		Kloos et al. (in press)	
H. pectinata	l	M	100	25	BG		Adewunmi & Sofowora (1980)	S,T
Ocimum basilicum (=O. canum)	l	M	100	40	BG		Adewunmi & Sofowora (1980)	ST

O. basilicum	ST	W	1000	7	Bp	lfl	Kloos et al. (in press)
O. basilicum	l	W	1000	60	Bp	ftt	Kloos et al. (in press)
O. basilicum	ST	W	1000	33	Bp		Kloos et al. (in press)
O. gratissimum	a	eo	?	80	Bg		Rouquayrol et al. (1980)
Plectranthus barbatus	ST	W	1000	20	Bp	l	Kloos et al. (in press)
P. barbatus	fl	W	1000	60	Bp		Kloos et al. (in press)
Another 6 species were inactive: Medina & Woodbury (1979), Rouquayrol et al. (1980), Sousa & Rouquayrol (1974)							
AMARANTHCEAE							
Achyranthes aspera	l	Pe	200	100	BTBp		Ahmed et al. (1984)
Aerva javanica	w	A, Pe	150–200	100	BTBp		Ahmed et al. (1984)
Alternanthra nodiflora	w	Pe	150	100	Bp		Ahmed et al. (1984)
BORAGINACEAE							
Cordia ruthii	r	Pe	100	100	Bp		Ahmed et al. (1984)
CAPPARIDACEAE							
Cadaba glandulosa	r	A	250	100	BT		Ahmed et al. (1984)
Cleome brachycarpa	lST	Pe	75	100	Bp		Ahmed et al. (1984)
CAPRIFOLIACEAE							
Lonicera nigra	See Table 5						
CARICACEAE							
Carica papaya	s	W	1000	20	Bp	f	Kloos et al. (in press)
CYPERACEAE							
Cyperus rotundus	r	Pe	250	100	Bp		Ahmed et al. (1984)
Scirpus praelongatus	w	Pe	100	100	Bp		Ahmed et al. (1984)
DROSERACEAE							
Drosera rotundiflora	See Table 5						
EBENACEAE							
Diospyros usambarensis	See Table 5						
ELATINACEAE							
Bergia suffriticosa	l	Pe	100	100	BT		Ahmed et al. (1984)
B. suffriticosa	l	Pe	150	100	Bp		Ahmed et al. (1984)
B. suffriticosa	w	A.Pe	150–200	100	BTBp		Ahmed et al. (1984)

Cont'd.

Table 4. *cont'd.*.

Plant Species by Family, Order,[b] Subphylum, and Phylum	Parts Tested[c]	Extract[d]	Conc. Tested ppm[e]	% Mort.	Snail Species[f]	Inactive Part[c]	Author	Compound[g]
FICOIDACEAE								
Trianthema pentadra	l	A	250	100	BTBp		Ahmed et al. (1984)	
JUGLANDACEAE								
Juglans regia	See Table 5							
LYTHRACEAE								
Lawsonia inermis	See Table 5							
MALVACEAE								
Pavonia burchelli	lST	Pe	250	100	BT		Ahmed et al. (1984)	
PEDALIACEAE								
Rogeria adenophylla	r	A,Pe	150–200	100	Bp		Ahmed et al. (1984)	
PORTULACACEAE								
Portulaca quadrifida	fl	W	1000	7	Bp	ST	Kloos et al. (in press)	
Talinum tenuissimum	See Table 5							
ZYGOPHYLLACEAE								
Fagonia cretica	r	Pe	50–75	100	BTBp		Ahmed et al. (1984)	
Peganum harmala	s	W	600	100	Bg		Cowper (1946)	S
2. ALGAE								
CYANOPHYCEAE								
Microcystis farlowiana/ Pseudanaboena franquetii	w	W	10000*	100	Lc		Gevrey et al. (1976)	

CHARACEAE						
Chara vulgaris	Plants placed in aquaria		?	100	Bg	Renno (1972)
3. MYCOPHYTAE (FUNGI)						
BASIDOMYCETAE						
Unidentified basidiomycete	*w*	W	1000	100	Lc	Medina & Woodbury (1979)
Aspergillus parasiticus	See Table 5					

[a] Source: Mostly NAPRALERT data.

[b] – at left margin=order division.

[c] a=aerial parts; b=bark; BE=berries; f=fruit; fl=flowers; FP=fruit pulp; h=husk; l=leaves; la=latex; p=pods; r=root; RB=rootbark; RH=rhizome; s=seeds; SB=stembark; SH=shell; ST=stem; t=thorns; w=whole plant. Italic letters denote fresh plant material.

[d] A="alcohol'; B=butanol; BE=benzene; E=ethanol; eo=essential oils; Er=ether; H=hexane; M=methanol; Pe=petroleum ether; S=success-ive extractions with different organic solvents; sp=spray-dried powder; W=water; WF=water/fermentation. Italic letters denote hot solvents.

[e] – =Exposures of less than 24 hours; * =exposures of more than 24 hours; ? =lack of data.

[f] Ba, Bc, Bg, Bp, and Bs = *Biomphalaria alexandrina, B. choanomphala, B. glabrata, B. pfeifferi,* and *B. stramina;* BA, BG, BT, and BN = *Bulinus africanus, B. globosus, B. truncatus,* and *B. (P.) nasatus;* L, La, Lc, Lm, and Ln = *Lymnaea* sp., *L. acuminata, L. cubensis, L. columella,* and *L. natalensis;* M=*Marisa cornuarietis;* Ms=*Melania scabra;* Oh and Oq = *Oncomelania hubensis* and *O. quadrasi;* Pc=*Planorbia corneus;* T=*Tarebia granifera.*

[g] Major compound group identified: A=alkaloids; F=flavonoids; G=glycosides; H=HCN; S=saponins; SS=steroidal saponins; ST=sterols and/or triterpenes; T=tannins.

require elaborate extraction apparatus and skilled technicians, thus sharply increasing costs and reducing their suitability for use in most endemic areas. It seems surprising, therefore, that many investigators nevertheless used only organic solvents in primary screens (Table 4). Notable exceptions are the water/fermentation procedure for endod (*Phytolacca dodecandra*) (Yohannes et al., 1979), boiled water extracts (Shoeb and El-Sayed, 1984), and the continuous aqueous extraction (Hussein Ayoub, no date) and spray-drying procedures (Hussein Ayoub, 1982) for *Acacia nilotica* seed pods.

While the use of plant extracts may give more effective snail control on a long-term basis than do present methods, it also requires adequate plant supplies and proper management, including good coordination of many activities during all phases of a program. These requirements apparently are met only in China, where a long-term control plan, an infrastructure, and financial and technical support permit the regular use of water-extracted camellia (*Thea oleosa*). Multiple use of the camellia, which is nontoxic to man, domestic animals, and crops, was made possible by extracting its oil; the aqueous extract of tea-cake is used as a molluscicide and plant fertilizer and the powdered residue as a cercaricide in the footwear of farmers (Cheng, 1971; Teh-Lung, 1958). The leaves of the tree *Schima argentea,* which Chinese farmers traditionally used to kill insects, leeches, and eels by spraying rice seedlings before transplanting them in paddy fields, have also been used as a molluscicide. In field trials, 100% snail kill was obtained with water-extracted leaves at 500 ppm (Fang, 1959). In Mozambique, a major producer of cashew nuts (*Anacardium occidentale*), a snail control program used the highly potent nutshells with encouraging results (L. Rey, personal communication, 1982). The effect of these molluscicides on the incidence of schistosomiasis and other snail-transmitted infections must still be determined.

CANDIDATE MOLLUSCICIDAL PLANTS

Several plants with promising molluscicidal activity have been tested in secondary screens, some in the field (Tables 1 to 3). They were selected on the basis of their biologic activity and the physicochemical criteria used by WHO (1965) in the evaluation of synthetic molluscicides. The list is tentative and will need to be amended after the species are more fully evaluated and data on additional plants become available. Species of Euphorbiaceae and Thymelaeaceae, particularly the genera *Euphorbia*, *Croton*, and *Gnidia*, contain tumor-producing and irritant principles that may exclude them from use as molluscicides. Results of toxicological studies of these and other plants are presented in Chapter 5 and by Duncan (1985).

The berries of endod (*P. dodecandra*) are the most thoroughly studied plant molluscicide. Because Lemma et al. (1979) have compiled a literature review on endod and the International Scientific Workshop held in Lusaka reported on ongoing research (Lemma et al., 1984), only some key features of its use and recent findings are discussed here. Epidemiologic studies of the effect of an aqueous extract of endod on schistosome transmission during a five-year snail control program in an Ethiopian town showed a significant reduction in *Schistosoma mansoni* infection rates in children and in the general population (Lemma et al., 1978). This program encountered problems of inadequate berry supplies and insect damage, which were largely solved by selecting and cultivating high-potency and pest-resistant strains (Lugt, 1981).

Endod's lack of ovicidal activity, also noted for other molluscicidal plants, is apparently due to its high molecular weight, which prevents penetration through the tough outer membrane of *Bulinus* and *Biomphalaria* eggs (Lemma and Yau, 1974). The necessary repeated applications can be detrimental to fish populations, although fish eggs may be resistant (Lugt, 1981). The problem of broad-spectrum faunal toxicity, which is not peculiar to endod use and is inherent particularly in the use of synthetic compounds, is believed to have been lessened by the now common practice of focal and periodic rather than areawide blanket application (McCullough et al., 1980). Nevertheless, piscicidal activity may be a problem locally where fish are an important source of protein or are used as agents in the control of snails, mosquito larvae, and aquatic weeds (Coates and Redding-Coates, 1981; McCullough, 1981). Like *P. dodecandra*, the American, Asian, and Australian species of *Phytolacca* studied thus far possess piscicidal triterpenoids as their major active principle (Parkhurst et al., 1974; and see Table 5).

Endod, as butanol and water extracts, was found not to be mutagenic (Lemma and Ames, 1975; Pezzuto et al., 1984; see also Chapter 5). Also, the berries of *P. americana* showed no mutagenic activity (Pezzuto et al., 1984). It is clear, however, that a full evaluation of any potential molluscicide must consider all findings on toxicity with the aim of eliminating products that are acutely or chronically toxic or both. Even plants like endod that have been used for many years by people and livestock in endemic areas must be investigated for mammalian toxicity in well-designed epidemiologic and ethnobotanic studies to supplement laboratory findings, evaluate their long-term effects, and establish minimum toxicity levels and guidelines for their development as molluscicides. Highland Ethiopians have washed their laundry with endod berry soap for centuries in streams and lakes that were also the only source of drinking water for most of the rural population. Although endod was lethal to sheep in oral doses of 1 g/kg of body weight, they tolerated it well in doses of 200 mg/kg, which is equivalent to 10 to 20 times the molluscicidal concentration (Mamo et al., 1979).

A water extract of the seeds of *Croton macrostachys* was nearly as effective as the major synthetic molluscicides against *Bulinus truncatus* in laboratory and field trials. Little damage resulted to other aquatic fauna. The extract was

Table 5. Molluscicidal Compounds and Their Biological Activities

Compound or Class of Compounds and Plant Source	Molluscicidal Activity				Other Activities[b]
	ppm	% Mort.	Snails Tested[a]	Author	
Lemmatoxin (triterpenoid saponin) *Phytolacca dodecandra*	1.5	90	Bg	Parkhurst et al. (1974)	Fungicide, larvicide (mosquito), anti-implantation, anti-fertility, and spermicidal effects
Lemmatoxin-C (triterpenoid saponin) *Phytolacca dodecandra*	3	90	Bg	Parkhurst et al. (1973a)	Spermicidal effect
Oleanoglyco-toxin-A (triterpenoid saponin) *Phytolacca dodecandra*	3	90	Bg	Parkhurst et al. (1973b)	Antiimplantation, antifertility, embryotoxic, and spermicidal effects
Phytolacca-dodecandra glycoside (triterpenoid saponin) *Phytolacca dodecandra*	1–5	50	Bg	Jewers & King (1972)	No data
Phytolaccagenin, glycosyl-xylosyl (triterpenoid saponin) *Phytolacca americana*	80	100	Bg	Johnson (1974), Johnson & Shimizu (1974)	Mitogenic, hemaglutination, antiviral, antibacterial
Phytolaccagenin, xylosyl (triterpenoid saponin) *Phytolacca americana*	60	100	Bg	Johnson (1974), Johnson & Shimizu (1974)	Mitogenic, hemaglutination, antiviral, antibacterial
4 compounds of hederagenin (triterpenoid saponins) *Hedera helix*	15 12 8 3	100 100 100 100	Bg Bg Bg Bg	Hostettmann (1980), Hostettmann et al. (1982)	Antibacterial, antifungal effects

Table 5 *cont'd.*

Compound or Class of Compounds and Plant Source	Molluscicidal Activity				
	ppm	% Mort.	Snails Tested[a]	Author	Other Activities[b]
4 compounds of monodesmosidic saponins (triterpenoid saponins) plant sources unspecified	2–32	100	Bg	Hostettmann et al. (1982), Hostettmann (1980)	?
Primulic acid (triterpenoid saponin) source unspecified	8	100	Bg	Hostettmann et al. (1982)	?
(triterpenoid saponin) *Lonicerca nigra*	2	100	Bg	Marston & Hostettmann (1985), Hostettmann et al. (1982)	Antibacterial, antifungal, antineoplasmic, termiticide
Balanitin 1,2,3 (spirostanol saponins) *Balanites aegyptiaca*	5–10	100	?	Nakanishi (1982), Liu & Nakanishi (1982)	Insect antifeedant, cercaricide, antimicrobial
2 compounds of sarsapogenin (spirostanol saponins) *Cornus florida*	6 12	? ?	Bg Bg	Hostettmann et al. (1978)	Astringent, probably antipyretic
7 spirostanol saponins *Asparagus curillus* *A. plumosus*	5–20 20–25	100 100	Bg Bg	Marston & Hostettmann (1985)	No data
Digitonin *Digitalis pupurea*	10 25	100 100	Bg Lc	Alzérreca et al. (1981)	

Cont'd.

Table 5 *cont'd.*

Compound or Class of Compounds and Plant Source	Molluscicidal Activity			Author	Other Activities[b]
	ppm	% Mort.	Snails Tested[a]		
Solasonine Solamargine (steroid glycoalkaloids)				Marston & Hostettmann (1985),	Antipyretic, WBC stimulant, photo-sensitizer, general toxicity, cardiac antiaccelerator effect
Solanum mammosum	10	100	Lc		
S. nodiflorum	25	100	Bg		
Tomatine (steroid glycoalkaloid)	4	100	Bg	Hostettmann et al. (1982), Marston &	?
Lycopersicum esculentum	10	100	Lc	Hostettmann (1985)	
Compound I (diterpene) *Euphorbia lactea*	0.18	90	Ba	El-Emam et al. (1982)	Cocarcinogenic irritant
Ent-kaur-16-en-19-oic acid (diterpene) *Wedelia carasana* (=*W. scaberrima*)	10	100	Bg	Tomassini, personal communi-cation (1982), Tomassini & Oliveira Matos (1979)	?
Muzigadial (sesquiterpene) *Warburgia ugandensis*	5	50	Bp,Ln	Nakanishi & Kubo (1977)	Antibacterial, insect antifeedant, cytotoxic, hemolytic action
Warburganal (sesquiterpene) *Warburgia stuhlmanni*	10	50	Bp,Ln	Nakanishi & Kubo (1977)	Antibacterial, insect antifeedant, cytotoxic, hemolytic action
Mukaadial (sesquiterpene) *Warburgia stuhlmanni*	20	50	Bg	Marston & Hostettmann (1985)	No data
(sesquiterpene) *Podachaeminum eminens*	1	100	Bg	Marston & Hostettmann (1985)	No data
Ambrosin	10–14	90	Ba,BT	Shoeb & El-Emam (1978)	Spasmogenic, antitumor, cytotoxic, hypotensive effects
Damsin	9–12	90	Ba,BT		
Tribromodamsin	8–15	90	Ba,BT		

Table 5 *cont'd.*

Compound or Class of Compounds and Plant Source	Molluscicidal Activity				Other Activities[b]
	ppm	% Mort.	Snails Tested[a]	Author	
(sesquiterpene lactones) *Ambrosia maritima*					
Helenalin (sesquiterpene lactone)	10	100	Bh	Marchant et al. (1984)	No data
Another 10 sesquiterpene lactones plant sources not specified	10	active	Bh	Marchant et al. (1984)	No data
Oleanolic acid (Saponin I) *Talinum tenuissimum*	?	active	Bg	Gafner et al. (1985)	No data
7 oleanic acids Saponin A Saponin B Mixture of saponins	40	100	Bg	Wagner et al. (1984)	No data
3 compounds of anacardic acid (alkenyl phenols) *Anacardium occidentalis*	0.35 0.9	50 50	Bg Bg	Sullivan et al. (1982)	Antibacterial
Epigallo- catechin-7- gallate (I) (phenol) *Acacia nilotica*	75	10	Bp	Hussein Ayoub (1984a, b)	No data
Epigallo- catechin-5,7- digallate (phenol) *Acacia nilotica*	100	10	Bp	Hussein Ayoub (1984a, b)	No data
(coumarin) *Polygala paniculata*	25	0	Bg	Hamburger et al. (1985)	No data
Affinin (isobutylamide) *Spilanthes*	50+	100	Po	Johns et al. (1982)	Cercaricide, insecticide, analgesic

Cont'd.

Table 5 *cont'd.*

Compound or Class of Compounds and Plant Source	Molluscicidal Activity				Other Activities[b]
	ppm	% Mort.	Snails Tested[a]	Author	
oleractea *Wedelia* *parviceps* *Heliopsis* *longipes*					
2 isobutylamide compounds *Fagara* *macrophylla*	200	50	Bg	Kubo et al. (1984b)	Insecticide
Quercetin 3-(2'-galloylgluco-side) (flavonoid) *Polygonum* *senegalense*	10	100	Bp,Bg, Ln	Dossaji & Kubo (1980)	Quercetin (has high mutagenic activity)
Lactone I Flavone II (flavonoid) *Baccharis* *trimera*	100 100	100 100	Bg Bg	Dos Santos Filho et al. (1980)	Cytotoxic effect
Chalepensin (furanocoumarin) *Ruta* *chalepensis*	2	100	Bg	Brooker (1967)	Barbiturate potentiation, antitumor and respiration on inhibition effects
Bergapten Isopimpinellin Xanthotoxin (furanocoumarins) *Ammi majus*	5 5 50	100 69 100	? ? ?	Soine (1964)	Antischistosomal, antitumor effect, photosensitizer, antispasmodic, plant growth inhibitor
Aflatoxin (furocoumarin) *Aspergillus* *parasiticus*	0.5	100	Bte	Purchio & Campos (1970)	Carcinogenic in fish and mammals, phytotoxicity
Oruwacin (anthraquinone?) *Morinda lucida*	1.3-5.3	90	Bp,BR, BG	Adewunmi & Adesoga (1984)	No data
2,'4'-dihydroxy-3', 6'-dimethoxy-chalcone	40	100	Bp,Bsu	Maradufu & Ouma (1977)	No data

Table 5 *cont'd.*

| Compound or Class of Compounds and Plant Source | Molluscicidal Activity | | | | |
	ppm	% Mort.	Snails Tested[a]	Author	Other Activities[b]
(chalcone) *Polygonum senegalense*					
Rotenone[c] (rotenoid) *Derris elliptica*	2–5	95	BG	Mozley (1939)	Insecticide
Isopiper-longumina (?) *Piper tuberculatum*	10	?	?	Sousa & Rouquayrol (1974)	No data
(quinolizine alkaloid) *Calpurnia aurea*	?	active	Bg	Kubo et al. (1984b)	No data
Wedelin (I) (galactopyrano-siduronic acid)	13	active	Bh	Oliveira Matos & Tomassini (1984)	No data
6 naphthoquinone compounds				Marston et al. (1984)	No data
Juglone *Juglans regia*	10	100	Bg		
Isojuglone *Lawsonia inermis*	50+	100	Bg		
7-methyljuglone *Diospyros usambarensis*	50+	100	Bg		
Plumbagin *Drosera rotundiflora*	50+	100	Bg		
Isodiospyrin *Diospyros usambarensis*	50+	100	Bg		
Mamegakinone *Diospyros usambarensis*	50+	100	Bg		

[a] See key of Table 4 for snail identification. Other species listed here are: Bh = *Biomphalaria havanensis*; Bsu = *B. sudanica*; Bte = *B. tenagophila*; Po = *Physa occidentalis*; BR = *Bulinus rohlfsii*.

[b] Most sources are listed in this table; some others, unlisted, were obtained through the NAPRALERT system (see Farnsworth et al., 1979).

[c] Rotenone from an unspecified plant source was 100% lethal to *Bulinus truncatus* at 20 ppm (Saliternik & Witenberg, 1959).

ovicidal against the late stages of *Bromphalaria pfeifferi* eggs, but it was nontoxic to *Tilapia nilotica* at the lethal concentration, 90% (LC$_{90}$) for *B. pfeifferi* (20 ppm). Below the LC$_{90}$ for *B. truncatus* (0.1 ppm), several other fish species apparently were unaffected. The seeds' reportedly high toxicity to humans (Daffalla and Amin, 1976) and the failure of Lugt (1981) to obtain any molluscicidal effect require further study. The seeds of *Croton tiglium*, the most potent plant tested (LC$_{50}$, 0.1 ppm), are highly toxic to fish and mammals and are carcinogenic (Yasuraoka et al., 1979a).

Of the two other well-researched species in the family Euphorbiaceae, *Jatropha curcas* (Yasuraoka et al., 1979b) and *Euphorbia cotinifolia* (Rouquayrol et al., 1973), *J. curcas* in an aqueous extract was harmless to fish at molluscicidal levels (LC$_{50}$, 7 ppm). Its other advantages are relatively low mammalian toxicity, production of active fruit almost year-round (Yasuraoka et al., 1979b), and widespread cultivation, in the tropics of Asia, Africa, and the Americas (see Chapter 6).

Ambrosia maritima (or *A. senegalensis;* see Chapter 6) was found to be molluscicidal and ovicidal but harmless to fish at 1000 ppm. At a lower concentration, snails stopped feeding (Sherif and El-Sawy, 1962), and field trials indicate that control may be achieved by this effect. Submergence of *A. maritima* in canals after monthly applications of the dry plant material and planting of the weed along canal banks within the drawdown area caused most *B. truncatus* and *B. alexandrina* to stop feeding, drift downstream, and die at concentrations of about 70 ppm. Fish and other aquatic fauna were not affected. However, Kloos et al. (1982) noted no antifeedant action.

Although complete destruction of snail populations (and presumably of snail eggs) could not be achieved and *A. maritima* grows only poorly along irrigation drains (El-Sawy et al., 1981), the possibility of planting suitable species near water level or submerged in snail habitats, using a minimum of labor and technology inputs, should be studied. An important advantage of *A. maritima* in Egypt is that its seasonal production of active flowering parts and leaves coincides with the summer peak of schistosome transmission. Another advantage of *Ambrosia*, which is essentially an American genus, is that ten species besides *A. maritima* contain ambrosin or damsin or both, the two sesquiterpene lactones isolated from *A. maritima* (Shoeb and El-Emam, 1978; NAPRALERT data).

Among the disadvantages of *A. maritima* are that in dried, ground form it has an irritant effect on handlers (Abu-Shady and Soine, 1953) and that, being a weed, it may discourage farmers from cooperating fully in its production and use. In addition, the gradual disappearance of the plant in Egypt (Kloos and McCullough, 1982) would have to be reversed.

Yasuraoka et al. (1977) questioned the practical use of the bark of *Entada phaseoloides* because of its moderate activity against *Oncomelania quadrasi* and its high fish toxicity.

The leaves of the weed *Solanum nodiflorum* killed *Lymnaea* species but were harmless against the effective competitor snails *Marisa cornuarietis* and *Tarebia granifera*. This selective molluscicidal effect of the water extract may encourage combined snail control strategies involving the use of molluscicides and intramolluscan snail competition (Medina and Ritchie, 1980).

The stability of the shells of the cashew nut (*A. occidentale*), the seeds of *Pithecellobium multiflorum*, the leaves of *Polygonum senegalense*, and the various parts of *Gnidia kraussiana* remains to be studied. All but the last were highly active in organic solvents and weakly so as water extracts. Moreover, the possible carcinogenicity of quercetin, one of the two active principles isolated from *P. senegalense* (Dossaji and Kubo, 1980) and present in the five other active species of *Polygonum* (Brossat et al., 1979; NAPRALERT data), should be studied. The leaves, roots, and stem of *G. kraussiana* could be more efficiently extracted with hot water (LC_{100}, 2 to 10 ppm) than with cold (LC_{100}, 14 to 120 ppm); the same is true of several other molluscicidal plants (Table 4; Shoeb and El-Sayed, 1984). Although the roots and stem were more potent, it will probably be necessary to use the regenerating leaves so as to maintain living plants; the leaves are also easier to harvest.

The stembark and fruits of the Nigerian medicinal plant *Tetrapleura tetraptera* were studied in secondary screens; the LC_{100} of the methanolic extract of the fruits was 1.5 ppm against *B. globosus*, whereas that of the aqueous extract was 50 ppm (Adewumni et al., 1982). The plant's disadvantages include loss of molluscicidal activity in dry heat, pH above 7.5, and organic matter (Adewunmi and Marquis, 1981b). Adesina et al. (1980) identified oleanolic glycoside as the major glycoside of the saponin fraction in *T. tetraptera*.

Most plants screened for cercaricidal and miracidicidal effects were active, some of them above and others below molluscicidal levels. They include leaves of *E. cotinifolia* (Rouquayrol et al., 1973), berries of *P. dodecandra* (Lemma, 1970), leaves of *T. oleosa* (Teh-Lung, 1958), fruits of *Balanites* species (Archibald, 1933; Wagner, 1936), leaves of *Euphorbia maideni* (Broberg, 1982), various parts of *Amaryllis vittata* (Galal et al., 1976), and fruits of *Solanum mammosum* (Medina, 1984).

For practical reasons and greater cost-effectiveness, it is preferable to give priority to plants that are both molluscicidal and larvicidal, although slow-release preparations need to be developed to achieve prolonged effects. The combined use of plants with specific molluscicidal and cercaricidal activity (Graham et al., 1980) or miracidicidal activity (Broberg, 1980) should thus be discouraged, except for some powerful agents that may be used in special situations on a short-term basis. Extracts of the leaves of *Bidens pilosa* and the roots of *Tagetes patula,* for example, when activated under ultraviolet radiation, lower their cercaricidal LC_{100} thirtyfold, to 0.01 ppm (Graham et al., 1980), making them highly effective on the sunlit surface of waters where cercariae may be concentrated.

Cercaricidal oils and extracts, which are found in more than 50 plants (Farnsworth and Cordell, 1976; Frischkorn, 1978; Graham et al., 1980; Ritchie et al., 1963; Vichnewski et al., 1976; Warren and Peters, 1969), may be cost-effective as prophylactic preparations applied to human skin, but their effect on the skin and the time/concentration relationships of applied materials remain to be studied. Any such preparation should give protection for at least several hours and preferably for a day or more.

Some 80 molluscicidal compounds have been isolated from plants (see chapters 4 and 12). Marston and Hostettmann (1985) summarized structure/activity relationships of the natural products responsible for the molluscicidal activity of these plants. They properly noted that isolation and identification of the active principles is also essential for the study of their toxicity levels and stabilities under field conditions, for dosage determinations, and for mode of action investigations. In addition to saponins – which were earlier considered the major if not the only molluscicidal compound in plants (Cowper, 1971) – several groups have given encouraging results. They include spirostanol saponins, sesquiterpene lactones, alkaloids, steroid glycoalkaloids, diterpenes, monoterpenes, iridoids, naphthoquinones, alkenyl phenols, chalcones, flavonoids, furanocoumarins, isobutylamides, and tannins (Table 5, see chapters 4 and 12).

Most compounds have a wide range of biological effects (Table 5; NAPRA-LERT data), further suggesting that plants may be developed for multiple purposes economically. However, time/concentration relationships and physicochemical influences, so far studied only for the three compounds isolated from *A. maritima* (Shoeb and El-Emam, 1978), still must be evaluated.

The metabolite aflatoxin, from the fungus *Aspergillus parasiticus* (Purchio and Campos, 1970), is evidently far too dangerous for use because of its toxicity and carcinogenicity in fish and mammals and its phytotoxicity in crop plants (Moss, 1972). The findings by Hostettmann et al. (1982) that bidesmosidic saponins and some dammarane glycosides and alkaloid saponins are inactive and that monodesmosidic saponins are active may help explain variations in results obtained by investigators and should encourage studies of other structure/activity relationships of compounds. Several bidesmosidic and monodesmosidic saponins have also been isolated from methanolic and water extracts of *P. dodecandra* berries (Domon and Hostettmann, 1984). It is hoped that such studies will result in the identification of plants with more selective biocidal effects.

With the accumulation of information from phytochemical screening and from the cultivation, harvesting, storing, and application of plant materials in both community and centralized snail control programs, additional criteria will be needed to evaluate the suitability and cost-effectiveness of candidate plants.

BOTANIC AND TOXICOLOGIC RELATIONSHIPS AND PLANT SELECTION

Phytogenetic relationships among groups of plants are expressed not only in floristic, physiological, and genetic terms but also by chemical characters (Dahlgren et al., 1981). Increasing recognition of this fact further facilitates the search for suitable plant molluscicides (Kloos and McCullough, 1985; see also Chapter 5) and makes it possible to arrange the plant species by families in taxonomic sequence (Table 4), using the system of Dahlgren et al. (1981).

Molluscicidal compounds, like other phytotoxins, may be present and synthetized more frequently in some plant groups than in others, although many compounds do occur independently (Young and Seigler, 1981). Whereas sesquiterpene lactones, for example, are most common in the Compositae, triterpenoid saponins are widespread in nature (Seigler, 1981). Because of the small number of species and compounds tested, the differences in selection and collection of plant materials, and the diverse methods of extraction and bioassay used by various investigators (see Kloos and McCullough, 1982), the data in Tables 4 and 5 cannot be compared. In addition, they are of limited predictive value. Nevertheless, several broad patterns can be discerned.

Of the 571 species (106 families) tested, a relatively high proportion (62%) were active, apparently because most investigators selected medicinal plants for primary screening. Particularly high proportions (97%) of active medicinal plants were reported by Kloos et al. (in press). Similarly, Ahmed et al. (1984) found 28 of 50 medicinal plants (56%) to be 100% lethal against *B. pfeifferi*, but they failed to report the negative species. No significant difference in activity is found between the monocotyledons and dicotyledons. The highest proportions of molluscicidal species were in the families Phytolaccaceae (all five species tested), Mimosaceae (18 of 20), Rutaceae (all nine species), Papilionaceae (28 of 39), and Polygonaceae (seven of nine). Highly active plants were also most commonly found in these families, but an unusually large proportion (25%) of the 57 Euphorbiaceae tested at concentrations of 100 ppm or lower killed 90% to 100% of the snails.

Of particular interest is the finding that the three families of the order Leguminales (Mimosaceae, Papilionaceae, and Caesalpiniaceae) – all of which contain many drought-resistant trees that are easily cultivated and produce high-quality seed protein (Felker and Bandursky, 1979) – also contained many active species. Plants that can grow beside transmission sites may be of special value in the semiarid parts of the world that have a high prevalence of schistosomiasis (Iarotski and Davis, 1981).

Information on the distribution of molluscicidal principles in various plant parts that might have predictive value is lacking, owing to the failure of most investigators to study whole plants systematically. In the *Phytolacca* species, the berries were most active, the leaves less so, and the roots not at all, a finding that is consonant with the distribution of saponins and inactive alkaloids. In most species in the family Euphorbiaceae, the highest activity levels were in the seeds; and in the Solanaceae, in the leaves and fruits. Localization of toxins in general in various plant tissues is also inadequately known (Rothschild, 1972). Thus efforts should be made to test all plant parts for molluscicidal activity; but for sustained use in snail control programs, preference in perennial plants should be given to regenerating leaves, fruits, and flowers rather than roots, stems, and nonregenerating bark.

In 29 families (23 dicotyledonae and six monocotyledonae), all 56 species tested were inactive (Adewunmi and Sofowora, 1980; Amorin and Pessoa, 1962; Medina and Woodbury, 1979; Rouquayrol et al., 1980; Sousa and Rouquayrol, 1974). Nevertheless, the small number of species tested renders most of these data of limited usefulness. Absence of activity in the eight species of Graminaceae is noteworthy, however.

PLANT PRODUCTION AND EXTRACTION: ALTERNATIVE METHODS

While cultivation of plants exclusively for snail control has been successful on a small scale in Egypt (El-Sawy et al., 1981) and Ethiopia (Lugt, 1981), it remains to be demonstrated on a large scale. Also, the cost constraints of using organic solvents and the failure of water to efficiently extract many potent compounds warrant a search for alternatives.

In view of the increasing scarcity of arable land in many developing countries and the agronomic difficulties and socioeconomic constraints associated with the cultivation of nonfood, nonprofit speciality crops, greater emphasis should be placed on plants (and their unused products) already cultivated by local farmers and on pharmaceutical, pesticidal, and molluscicidal aquatic plants. Some indigenous plants may be unsuitable for large-scale agricultural development and more appropriately harvested in wild systems (Bollinger, 1980). The extensive knowledge farmers in indigenous tropical agricultural systems have of both cultivated and wild plants in relation to soil requirements, cropping techniques, and crop protection (Glass and Thurston, 1978) favors the use of these plants in community-based snail control programs.

Certain trees are particularly suited for soil-erosion, land-reclamation, shelter-belt, or beautification and reforestation projects. Legumes, the pioneers in nature's plant succession, are most appropriate for revegetation efforts. *Sesbania* species, for example, are not only molluscicidal but are also among the world's fastest-growing trees (Tekle, 1977; Vietmeyer, 1979).

Some other wild plants, including species of *Balanites*, *Yucca*, and *Agave*, grow too slowly to meet molluscicidal requirements of snail control programs.

Plant tissue culturing for propagation of high-potency varieties and the direct extraction of active compounds in greenhouse systems (Bollinger, 1980) should be studied further. Encouraging results have been obtained in tissue culturing of *P. dodecandra* and several other saponin plants (Adams and Balandrin, 1984). Production of uniform cell and organ biomass economically remains a challenge (Staba, 1985).

In addition to cashew nutshells and camellia leaves, plant wastes with snail-killing properties include the leaves of tomatoes, peppers (*Capsicum frutescens*), and tobacco, all in the Solanaceae family; the seeds of avocado (*Persea gratissima*, Lauraceae) and tamarind (*Tamarindus indica*, Caesalpiniaceae); waste of sisal (*Agave sisalana*, Agavaceae); the leaves of eucalyptus trees (*Eucalyptus* spp., Myrtaceae), olives (*Olea europaea*), and yams (*Dioscorea rotundata*, Punicaceae); and the essential oils of several garden flowers and herbs (Table 4; Kubo and Matsumoto, 1984). The leaves of eucalyptus, which are among the most prolific trees around settlements in many parts of the tropics and subtropics, where their wood is widely used for fuel, lumber, and poles, may be useful if their active ingredients can be more efficiently extracted. Highly active compounds have been isolated from common ivy (*Hedera helix*) and rue (*Ruta chalepensis*), the medicinal and culinary herb (Table 5). The rapid growth and small size of herbs may be advantageous in certain situations. Products of many wild plants (Jardin, 1970), some of them toxic to snails (Table 4), are commonly gathered during seasonal food shortages and may constitute another source of mollusicicides if their supply can be assured.

Several plants now used as insecticides, larvicides, and herbicides may prove suitable for snail control as well (see Chapter 6). Multiple uses of toxic plants for agricultural pest and vector control purposes favor an ecologic approach to integrated pest management programs. *D. elliptica* contains the insecticidal and highly molluscicidal compound rotenone and can be easily cultivated (Mozley, 1939). The dried leaves of pyrethrum (*Chrysanthemum cinerariafolium*) were mildly molluscicidal in a water extract at 50 ppm (Tesfaigzi, 1978).

Increased emphasis on using plants in the control of root-knot nematodes (Roy, 1979) and other plant diseases may encourage collaborative research by agricultural and public health workers. In comparative studies, Khand and Qadri (1974) found that the insecticidal and molluscicidal oils from *Artemisia maritima*, *A. kurramensis*, and *Ereua sativa* were significantly less harmful to various aquatic insects than DDT and lindane.

Another promising plant is the African, Central American, and Indian neem or margosa tree (*Azadirachta indica*). Its seeds, at 0.1 ppm, prevent damage to agricultural crops and stored products by at least 25 species of economic pests in the United States, yet its mammalian toxicity is low and no mutagenic activity has been observed. Programs are under way to mass cultivate neem in Cameroon, Gambia, Puerto Rico, and the United States. The present trend among funding organizations to reduce the dependency of developing countries on imported pesticides (Jacobson, 1980) may facilitate the use of this and other indigenous plants. The minimum molluscicidal concentration of neem seeds is lower than the 5000 ppm reported by Muley (1978) for the flowers after 20 minutes' exposure; at the University of California a butanol extract of the seeds was 100% lethal at 500 ppm (D. Heyneman, personal communication, 1982).

The frequent association between the absence of aquatic snails and the presence of certain plants may offer possibilities of introducing or encouraging ecologically suitable plants as another form of biologic control. Oxygen depletion (Beadle, 1974; Ferguson, 1978) and emission of plant metabolites that may act as attractants or repellents through the chemical sensory mechanisms of snails (Bousfield, 1979) have been implicated as causative factors. Dense mats of the alga *Chara vulgaris* coincided with stretches of *B. glabrata*-free waters in Brazil (Ferguson, 1978) and caused high snail mortalities in aquaria (Renno, 1972). The algal complex *Microcystis farlowiana/Pseudanaboena franquetti* was mildly active (Gevrey et al., 1976). No snails were observed in parts of canals fringed by *Canna indica*, which was moderately molluscicidal (Mahran et al., 1974). Dense mats of aquatic plants may inhibit the development of snail host populations, but in some habitats they may be inimical, for other diverse reasons, to the aquatic environment.

Use of the water/fermentation extraction process increased the molluscicidal potency of endod sevenfold over water-extracted crude berries. This extraction method and the chemical process that frees the active glycoside from the inhibitive oils and lipids were described by Lemma et al. (1979) and Yohannes et al. (1979). The spray-drying procedure developed by Hussein Ayoub (1982, 1984a) increased the LC_{100} of a water extract of the seed pods of *A. nilotica* from 150 to 50 ppm; the continuous water extraction process gave 56.4% molluscicidal tannins (Hussein Ayoub, no date). Possibilities of using these methods in the cost-effective development of other plant molluscicides should be explored. Use of hot water to extract molluscicidal principles should also be considered, particularly in areas where solar heating technology is being developed. Shoeb and El-Sayed (1984) found the boiled water extracts of several Agavaceae highly potent and economical for field use. The finding that water extracts yield larger quantities of monodesmosidic saponins than organic solvents (Marston and Hostettmann, 1985) should encourage the development of cost-effective molluscicides.

Cost-effectiveness studies of plant molluscicides are still in their infancy, but the economic feasibility of developing and using these natural products will have to be considered. In the only such analysis made to date, the cost of the Ethiopian schistosomiasis control program in the town of Adwa, where crude endod berries were used, was US$0.03 per capita. Although that figure is higher than the cost of using Bayluscide (Lemma et al., 1978), it may be reduced by the use of cultivated high-potency strains or water/fermentation extracts or both.

FUTURE ACTION

This review shows that many plants from a large number of families are toxic to snail hosts and that a few of them are as effective and stable as the major synthetic compounds. Reports of ovicidal action and low fish toxicity of several plants with molluscicidal activity are highly significant, suggesting that other species with these desirable features are likely to occur in nature. The recent emphasis on plant compounds can facilitate the search for fairly predictable chemical characters and effects. Their long-term biologic effects (Domon et al., 1984; Marston and Hostettmann, 1985) and their safety to man and other nontarget biota remain to be demonstrated. The cost of developing and using plant molluscicides must be acceptable to developing countries; the availability of plant materials must be assured for use on a sustained basis; and social, cultural, and economic conditions must be amenable to the use of appropriate technologies. So far, none of the candidate plants has satisfactorily met all these criteria.

Possible ecologic effects on crop plants and aquatic organisms need further study. Graham et al. (1980) found that sublethal doses of a cercaricidal extract of *Tagetes patula* protected *Physa occidentalis* from bacterially complicated death without reducing cercarial production. Plant extracts containing steroidal hormones may also enhance snail survival (Pereira de Souza et al., 1978). Adewunmi et al. (in press), on the other hand, reported significant reductions in oviposition and hatchability of *B. globosus* eggs after prolonged application of sublethal doses of a methanolic extract of *Bridelia atroviridis*.

More information is needed on organisms that are commensals, competitors, and predators of snail hosts. Torrealba et al. (1953) noted markedly varying degrees of susceptibility of some 65 species of crustaceans, rotifers, free-living protozoa, fish, snails, and insect larvae to extracts from *Sapindus saponaria* fruits. The large number of nontarget aquatic organisms affecting snail survival and the potential use of biologic agents in snail control programs (McCullough, 1981) point out the need for ecological studies.

Longitudinal phytotoxicity studies are needed, particularly in areas where treated water is being used for irrigation. The brief studies done so far did not show any effects on seed germination, physiological structure, or plant growth (Adewunmi and Marquis, 1981a; Daffalla and Amin, 1976; Yohannes, 1970; Adewunmi and Adesogan, 1984).

Integrated use of plants in insect vector control needs to be studied. Larvicidal or repellent action of some plants against *Anopheles, Simulium,* and houseflies (Hunter, 1980; Kloos and McCullough, 1982; Wagner, 1936; Table 5) and increasing resistance in these and other insects to synthetic insecticides reemphasize the need for a transdisease perspective in the design and operation of control programs involving plants.

Information is needed on the social organization of rural communities and the feasibility of producing and applying plant molluscicides in the framework of existing cooperative or state farms, manufacturing firms, and self-help associations. Unlike the synthetic molluscicides, whose use may not require community acceptance, plant molluscicides normally involve community collaboration in their development and use. The extent of this collaboration will depend largely on program objectives, health legislation, the infrastructure, and the willingness of local people to cooperate. The effectiveness of community programs in China may not be readily achieved in other endemic areas, but the support given by farmers in Egypt (El-Sawy et al., 1981) and Ethiopia (Lemma et al., 1978, 1979) seems encouraging. Of interest is a schistosomiasis control pilot project in Tanzania that includes focal mollusciciding with plant extracts from *Swartzia madagascariensis* and chemotherapy using the primary health care approach. The impact of snail control activities is monitored by community members using simple, locally made palm leaf traps (Suter, 1986). Some of the conditions necessary for establishing cooperative and self-help associations in Africa and the limits of their usefulness were described by Hamer (1981). Shortage of arable land in endemic areas does not necessarily limit the production of plant materials in localities where plant wastes, by-products from cultivated and wild plants, or effective aquatics are abundant. Viability of programs will depend largely on sound management, high motivation of personnel, technical support from governments, and the extent to which programs can be incorporated into rural development strategies.

Since the input required in terms of manpower and material may be substantial for long-term programs and quite promising plant species are available in most countries, it is imperative that cost/benefit analyses be made for all candidate plant materials. Although the savings from using indigenous rather than imported products and manpower does not require that plant molluscicides be strictly price competitive with synthetic compounds, at least initially, it is quite clear that low-cost products are most likely to be used in developing countries.

The need is urgent to standardize phytochemical screening and bioassay methods to permit more comprehensive evaluation of plants and better comparability of the results obtained by various investigators. Also, botanical nomenclature of promising plants requires standardization. Because the methodology developed by WHO (1965) for laboratory screening and evaluation of synthetic molluscicides does not cover plant cultivation and extraction, a comprehensive manual will have to be produced and criteria established for further evaluation of plant molluscicides (see Annex to this book). Some variables, including duration of plant extraction (Kloos et al., 1982), are neglected by most investigators. New research findings should be disseminated periodically to governments and designated research centers.

Each program, when adequately organized and funded (preferably for several years), should include at least one agronomist, botanist, phytochemist, toxicologist, malacologist, and social scientist (anthropologist, geographer, economist, or sociologist), along with a technical support staff. Cooperative agreements should be arranged with outside research groups, such as the Food and Agriculture Organization (FAO) and WHO, to make available their experience and certain materials or facilities that may be unavailable in the country where the program is to be developed.

In addition to improvements in foreign trade balance to be gained by national governments from the use of indigenous rather than imported pesticides and other products, economic and public health benefits will accrue to the people in participating communities. Many countries have begun actively promoting the growth of small-scale industries as a key strategy for reducing unemployment (Mabawanko, 1979). The relative simplicity of plant production, preparation, and application will not only permit the employment of local people but also reduce the problem of a chronic shortage of highly trained medical and scientific personnel (McCullough et al., 1980).

Another primary criterion in selecting plant molluscicide technologies should be the extent to which each technology can be incorporated into the total rural development program, thus avoiding a common problem caused by the introduction of technologies unsuitable for local cultures and economies (Lateef, 1980). Appropriate technology on any scale, whether of the grass roots or Western type, is technology that can be understood, implemented, used, and maintained by local people using local resources. Jequier (1981) pointed out that the appropriate technology concept has already won the support of leaders in industry and finance and that political push is needed to fully exploit its potentialities. The very nature of the concept – natural products, processed and applied in the country of origin – makes it imperative that governments, research institutes, and rural communities undertake collaborative programs designed to screen, cultivate, apply, and monitor the effects of carefully selected plant products.

Acknowledgments

The authors want to thank the World Health Organization and Dr. N. R. Farnsworth for assistance with computerized literature search.

REFERENCES

Abdel-Alim, M., and M. Kamel. Phytochemical and molluscicidal screening of three *Euphorbia* species. *Egypt. J. Bot.* 24:231–233 (1984).

Abou El-Hassan, A. A., H. A. Shoeb, A. S. Rafwan, M. A. Eman, and S. W. El-Amin. The molluscicidal properties of *Euphorbia lactea*. In *Proceedings of the Tenth International Congress of Tropical Medicine and Malaria*, Abstract No. 586. Manila, November 1980.

Abu-Shady, H., and T. O. Soine. The chemistry of *Ambrosia maritima* L. *J. Am. Pharm. Assoc.* 42:387–395 (1953).

Adams, R. P., and M. F. Balandrin. Application of biotechnical methods for the *in vitro* biosynthesis of bioactive saponins and propagation of endod. In *Phytocacca Dodecandra (Endod)*, ed. A. Lemma, D. Heyneman, and S. M. Silangwa, 133–137. Dublin: Tycooly International (1984).

Adesina, S. K., and C. O. Adewunmi. The isolation of molluscicidal agents from the root of *Clausena anisata* (Willd.) Oliv. In *Abstracts, Fourth International Symposium on Medicinal Plants*, 44–45. Ile-Ife, Nigeria, July 1981.

Adesina, S. K., C. O. Adewunmi, and V. O. Marquis. Phytochemical investigations of the molluscicidal properties of *Tetrapleura tetraptera* (Taub.) *J. Afr. Med. Plants* 3:7–15 (1980).

Adewunmi, C. O., and E. K. Adesogan. Anthraquinones and oruwacin from *Morinda lucida* as possible agents in fascioliasis and schistosomiasis control. *Fitoterapia* 55:259–263 (1984).

Adewunmi, C. O., and V. O. Marquis. Molluscicidal evaluation of some *Jatropha* species grown in Nigeria. *Q. J. Crude Drug Res.* 18:141–145 (1980).

——. Laboratory and field trials of the molluscicidal property of *Calliandra portoricensis* (Jacq) Benth. *Niger. J. Pharmacol.* 12:305–306 (1981a).

——. Laboratory evaluation of the molluscicidal properties of aridan, an extract from *Tetrapleura tetraptera* (Mimosaceae) of *Bulinus globosus. J. Parasitol.* 67:713–716 (1981b).

Adewunmi, C. O., and E. A. Sofowora. Preliminary screening of some plant extracts for molluscicidal activity. *Planta Med.* 39:57–64 (1980).

Adewunmi, C. O., S. K. Adesina, and V. O. Marquis. On the laboratory and field trials of *Tetrapleura tetraptera. Bull. Anim. Health Prod. Afr.* 30:89–94 (1982).

Adewunmi, C. O., A. O. Segun, and S. O. Asaolu. The effect of prolonged administration of low doses of the methanolic extract of *Bridelia atroviridis* on the development of *Bulinus globosus. Q. J. Crude Drug Res.* (in press).

Ahmed, E. H. M., A. K. Bashir, and Y. M. El-Kheir. Investigations of molluscicidal activity of certain Sudanese plants used in folk medicine: 4. *Planta Med.* 50:74–77 (1984).

Al-Azzawil, H. T., and H. B. Banna. Some enzyme histochemical study on certain mouse tissues after Bayluscide administration. In *Proceedings of the Tenth International Congress of Tropical Medicine and Malaria*, Abstract No. 579. Manila, November 1980.

Alzérreca, A., B. Arboleta, and G. Hart. Molluscicidal activity of natural products: The effect of *Solanum* glycosidic alkaloids on *Lymnaea cubensis* snails. *J. Agric. Univ. Puerto Rico* 57:69–72 (1981).

Amorin, J. P., and S. B. Pessoa. Experiencia de alguns vegetais como moluscicida. *Rev. Bras. Malariol. Doencas Trop.* 14:254–261 (1962).

Anantaraman, M. Biological control of aquatic snails. *Indian J. Vet. Sci.* 25:65–67 (1955).

Archibald, R. G. The use of the fruit of the tree *Balanites aegyptiaca* in the control of schistosomiasis in the Sudan. *Trans. R. Soc. Trop. Med. Hyg.* 27:207–211 (1933).

Azevedo, J. F., and L. Medeiros. L'action molluscicide d'une plante de l'Angola, la *Securidaca longipedunculata* Fresen, 1837. *Bull. Soc. Pathol. Exot.* 56:68–76 (1963).

Baalawy, S. S. Laboratory evaluation of the molluscicidal potency of a butanol extract of *Phytolacca dodecandra*. *Bull. WHO* 47:422–425 (1972).

Barbosa, F. S., and D. A. Mello. Molluscicidal activity of vegetal products. *Rev. Bras. Pesquisas Med. Biol* 2:264–266 (1969).

Barbosa, F. S., O. D. Calado, J. C. D. Moraes, and E. D. Almeida. Acao moluscicida sinergia da saponina do *Sapindus saponaria* e pentochlorophenato de sodio. *Pub. Avulsas Inst. Aggeu Magalhaes* 1(10):129–139 (1952).

Barros, D. A. D., J. L. C. Lopes, W. Vichnewski, J. N. Lopes, P. Kulanthaivel, and W. Herz. Sesquiterpene lactones in molluscicidal extract of *Eremanthus glomerulatus*. *Planta Med.* 1:38–39 (1985).

Beadle, L. C. *The Inland Waters of Tropical Africa: An Introduction to Tropical Limnology.* New York: Longman (1974).

Bollinger, W. H. Sustaining renewable plant resources: Techniques from applied botany. In *Renewable Resources: A Systematic Approach,* ed. C.-L. Enrique, 379–390. New York: Academic Press (1980).

Bousfield, J. D. Plant extracts and chemically triggered positive rheotaxis in *Biomphalaria glabrata* (Say), snail intermediate host of *Schistosoma mansoni* (Sambon). *J. Appl. Ecol.* 16:681–690 (1979).

Breuil, J., J. Moyroud, and P. Coulanges. Elements of the ecological battle against schistosomiasis in Madagascar (in French). *Arch. Inst. Pasteur Madagascar* 50:131–144 (1983).

Broberg, G. Observations on the mode of action of extracts of *Glinus lotoides* on the miracidia of *Fasciola gigantica* and *Schistosoma mansoni*. *Suomen Elainlaakarilehti* 86:146–147 (1980).

———. Molluscicidal effects of eucalyptus. *Vet. Rec.* May 29:526 (1982).

Brooker, R. M. Chalepensin, chalepin and chalepin acetate, three novel furoumarins from *Ruta chalepensis*. *Lloydia* 30:73 (1967).

Brossat, J. Y., P. Cerf, and P. Coulanges. Etudes préliminaires des propriétés molluscicides d'extraits de *Polygonum*. *Arch. Inst. Pasteur Madagascar* 48:281–291 (1979).

Cheng, T. H. Schistosomiasis in mainland China: A review of research and control programs since 1949. *Am. J. Trop. Med. Hyg.* 20:26–53 (1971).

Coates, D., and T. A. Redding-Coates. Ecological problems associated with irrigation canals in the Sudan with particular reference to the spread of bilharziasis, malaria and aquatic weeds and ameliorative role of fishes. *Int. J. Environ. Stud.* 16:207–212 (1981).

Cowper, S. G. The effect of certain inorganic and vegetable substances on the English pond snail *Planorbis corneus* (Linné, 1758). *Ann. Trop. Med. Parasitol.* 42:119–130 (1946).

———. *A Synopsis of African Bilharziasis,* 252–255. London: H. K. Lewis (1971).

Daffalla, A. A., and M. A. Amin. Laboratory and field evaluation of the molluscicidal properties of habat-el-mollok (*Croton* spp.). *East Afr. J. Med. Res.* 3:185–195 (1976).

Dahlgren, R. M. A revised classification of the angiosperms with comments on correlation between chemical and other characters. In *Phytochemistry and Angiosperm Phylogeny,* ed. D. A. Young and D. S. Seigler, 149–204. New York: Praeger (1981).

Deverall, B. J. Phytoalexins. In *Phytochemical Ecology*, ed. J. B. Harbourne, 217–233. New York: Academic Press (1972).

Domon, B., and K. Hostettmann. New saponins from *Phytolacca dodecandra*. *Helv. Chim. Acta* 67:1310–1315 (1984).

Domon, B., A. C. Dorsaz, and K. Hostettmann. High performance liquid chromatography of oleanane saponins. *J. Chromatogr.* 315:441–446 (1984).

Dossaji, S., and I. Kubo. Quercetin 3-(2'-galloylglucoside), a molluscicidal flavonoid from *Polygonum senegalense*. *Phytochemistry* 19:482 (1980).

Dossaji, S. F., M. G. Kairu, A. T. Gwonde, and J. T. Ouma. On the evaluation of the molluscicidal properties of *Polygonum senegalense* forma *senegalense*. *Lloydia* 40:290–293 (1977)

Dos Santos Filho, D., S. J. Sarti, W. Vichnewski, M. S. Bulhões, and H. de Freitas Leitão Filho, Molluscicidal activity on *Biomphalaria glabrata* of a diterpene lactone and a flavone isolated from *Baccharis trimera* (Less.). *Rev. Fac. Farmacol. Odontol. Ribeirão Preto* 17:(1):43–47 (1980).

Duncan, J. The toxicology of plant molluscicides. *Pharmacol. Ther.* 27:243–264 (1985).

El-Emam, M. A., A. M. Mohamed, H. A. Shoeb, and M. F. El-Shafiee. The effect of damsin and Bayluscide homologue on the oxygen consumption of *Biomphalaria alexandrina* and *Bulinus truncatus*. *Helminthologia* 18: 125–130 (1981).

El-Emam, M. A., S. M. El-Emam, and H. A. Shoeb. The molluscicidal activity of Euphorbiaceae: I. *Euphorbia lactea* (Haw). *Helminthologia* 19:227–236 (1982).

El-Kheir, Y. M., and M. S. El-Tohami. Investigation of molluscicidal activity of certain Sudanese plants used in folk-medicine. *J. Trop. Med. Hyg.* 82:237–247 (1979).

El-Sawy, M. F., K. Bassouny, and A. I. Magdoub. Biological combat of schistosomiasis: *Ambrosia maritima* (damsissa) for snail control. *J. Egypt. Soc. Parasitol.* 11:99–117 (1981).

Fang, M. Y. A plant highly effective for killing snails. *Shengwuxue Tongbao* 3:110 (1959).

Farnsworth, N. R., and G. A. Cordell. A review of some biologically active compounds isolated from plants as reported in the 1974–1975 literature. *Lloydia* 39:420–453 (1976).

Farnsworth, N. R., W. D. Loub, M. L. Quinn, G. A. Cordell, and D. D. Soejarto. A computerized natural products information system. *J. Nat. Prod.* 42:689 (1979).

Felker, P., and R. S. Bandursky. Uses and potential uses of leguminous trees for minimum energy input agriculture. *Econ. Bot.* 33:172–184 (1979).

Ferguson, F. K. *The Role of Biological Agents in the Control of Schistosome-Bearing Snails*. Atlanta, Georgia: Centers for Disease Control (1978).

Frischkorn, C. G. Cercarial activity of some essential oils of plants from Brazil. *Naturwissenschaften* 65:480–483 (1978).

Gafner, F., J. D. Msonthi, and K. Hostettmann. Phytochemistry of African plants: 3. Molluscicidal saponins from *Talinum tenuissimum* Dinter. *Helv. Chim. Acta* 68:555–558 (1985).

Galal, E. E., F. S. Salem, M. R. El-Sharkawi, and F. H. Youssef. Preliminary screening of the *Amaryllis vittata* (*H. vittatum*) as a molluscicide. *J. Drug Res. Egypt* 8:133–143 (1976).

Gevrey, K., S. Michael, and J. Euzeby. Mise en évidence de la toxicité d'un complexe algal sur la faune aquatique. *Vet. Med. Comp. Lyon* 74:191–194 (1976).

Glass, E. H., and H. D. Thurston. Traditional and modern crop protection in perspective. *Bioscience* 28(3):109–115 (1978).

Graham, K., E. A. Graham, and G. H. Towers. Cercaricidal activity of phenylheptatriyne and a-tertienyl, naturally occurring compounds in species of Asteraceae and Compositae. *Can. J. Zool.* 58:1955–1958 (1980).

Hamburger, M., M. Gupta, and K. Hostettmann. Coumarins from *Polygala panic-ulata*. *Planta Med.* 3:215–217 (1985).

Hamer, J. H. Preconditions and limits in the formation of associations: The self-help and cooperative movement in sub-Saharan Africa. *Afr. Stud. Rev.* 24:113–32 (1981).

Hostettmann, K. Saponins with molluscicidal activity from *Hedera helix* L. *Helv. Chim. Acta* 63:606–609 (1980).

——. The use of plant-derived compounds for the control of schistosomiasis. *Natur-wissenschaften* 71:347–351 (1984).

Hostettmann, K., M. Hostettmann-Kaldas, and K. Nakanishi. Molluscicidal saponins from *Cornus florida* L. *Helv. Chim. Acta* 61:1990–1995 (1978).

Hostettmann, K., H. Kizu, and T. Tomimori. Molluscicidal properties of various saponins. *Planta Med.* 44:34–35 (1982).

Hunter, J. M. Strategies in the control of river blindness. In *Conceptual and Metho-dological Issues in Medical Geography*, ed. M. S. Meade, 38–76. Chapel Hill, N.C.: Department of Geography, University of North Carolina (1980).

Hussein Ayoub, B. S. M. Molluscicidal properties of *Acacia nilotica*. *Planta Med.* 46:181–183 (1982).

——. Polyphenolic molluscicide from *Acacia nilotica*. *Planta Med.* 50:532 (1984a).

——. Effect of the galloyl group on the molluscicidal activity of tannins. *Fitoterapia* 55:343–345 (1984b).

——. Acacia extract as an algacide and molluscicide. Patentschrift CH 649683, Switzerland (no date).

Iarotski, L. S., and A. Davis. The schistosomiasis problem in the world: Results of a WHO questionnaire survey. *Bull. WHO* 59:115–127 (1981).

Jacobson, M. Neem research in the U.S. Department of Agriculture: Chemical, biological and cultural aspects. In *Proceedings of the First International Neem Conference*, 33–42. Rottach-Egern: The German Research Foundation (1980).

Jardin, C. *List of Foods Used in Africa*. 2nd ed. Nutrition Information Documents Series No. 2. Rome: Food and Agriculture Organization (1970).

Jequier, N. Appropriate technology needs political 'push.' *World Health Forum* 2:541–543 (1981).

Jewers, K., and T. A. King. Improvements relating to molluscicides. Patent No. 1277414. London: The Patent Office (1972).

Johns, T., K. Graham, T. Swain, and G. H. N. Towers. Molluscicidal activity of affinin and other isobutylamides from the Asteraceae. *Phytochemistry* 21:2737–2738 (1982).

Johnson, A. L. Structure elucidation and molluscicide evaluation of the major saponin from the berries of *Phytolacca americana*. *Diss. Abs.* 35B:731–732 (1974).

Johnson, A. L., and Y. Shimizu. Phytolaccinic acid, a new triterpene from *Phytolacca americana*. *Tetrahedron* 30:2033–2036 (1974).

Khand, M. A. J., and M. A. H. Qadri. Determination of lethal doses of *Artemisia* and *Taramira* oils in comparison with DDT and lindane against full grown larvae of *Anopheles stephensi* Liston (Culicidae). *Agric. Pakistan* 25(1):21–33 (1974).

Kloos, H., and F. S. McCullough. Plant molluscicides: A review. *Planta Med.* 46:195–208 (1982).

——. Ethiopian plants with proven and suspected molluscicidal activity: A new approach in plant evaluation. *J. Trop. Med. Hyg.* 88:189–196 (1985).

Kloos, H., K. C. Lim, and D. Heyneman. Preliminary screening of some Egyptian and Ethiopian plants for molluscicidal activity and observations on a possible method for natural snail control. Unpublished Report (1982).

Kloos, H., F. Waithaka Thiongo, J. H. Ouma, and A. E. Butterworth. Preliminary evaluation of some wild and cultivated plants for snail control in Machakos District, Kenya. *J. Trop. Med. Hyg.* 90: in press (1987).

in local snail control programs in Kenya. *Econ. Bot.* (in press).

Kubo, I., and A. Matsumoto. Molluscicides from olive *Olea europaea* and their efficient isolation by counter-current chromatographies. *J. Agric. Food Chem.* 32:687–688 (1984).

Kubo, I., J. A. Klocke, T. Matsumoto, and T. Kamikawa. Insecticidal and molluscicidal activities of isobutylamides isolated from *Fagara macrophylla* and their synthetic analogs. In *Pesticide Synthetics: Rational Approaches*. American Chemical Society Series 255:163–172 (1984a).

Kubo, I., T. Matsumoto, M. Kozuka, A. Chapya, and H. Naoki. Quinoziline alkaloids from the African medicinal plant *Calpurnia aurea*: Molluscicidal activity and structural study by 2D-NMR. *Agric. Biol. Chem.* 48:2839–2841 (1984b).

Lateef, N. V. *Crisis in the Sahel: A Case Study in Development Cooperation*, 125. Boulder, Colorado: Westview Press (1980).

Lemma, A. Laboratory and field evaluation of the molluscicidal property of *Phytolacca dodecandra*. *Bull. WHO* 42:597–617 (1970).

Lemma, A., and B. N. Ames. Screening for mutagenic activity of some molluscicides. *Trans. R. Soc. Trop. Med. Hyg.* 69:167–168 (1975).

Lemma, A., and P. Yau. Studies on the molluscicidal properties of endod (*Phytolacca dodecandra*), III. *Ethiop. Med. J.* 12:109–114 (1974).

Lemma, A., G. Brody, C. W. Newell, R. M. Parkhurst, and W. A. Skinner. Endod (*Phytolacca dodecandra*), a natural product molluscicide, increased potency with butanol extraction. *J. Parasitol.* 58:104–107 (1972).

Lemma, A., P. Goll, J. Duncan, and B. Mazengia. Control of schistosomiasis with use of endod in Adwa, Ethiopia: Results of a five-year study. In *Proceedings of the International Conference on Schistosomiasis*, Vol. 1, 415–436. Cairo: Ministry of Health (1978).

Lemma, A., D. Heyneman, and H. Kloos, eds. Studies on the molluscicidal and other properties of the endod plant *Phytolacca dodecandra*. Unpublished Report, Department of Epidemiology and International Health, University of California, San Francisco (1979).

Lemma, A., D. Heyneman, and S. M. Silangwa. *Phytolacca Dodecandra (Endod)*. Dublin: Tycooly International (1984).

Liu, H.-W., and K. Nakanishi. The structure of balanitins, potent molluscicides isolated from *Balanites aegyptiaca*. *Tetrahedron* 38:513–519 (1982).

Lugt, C. B. *Phytolacca Dodecandra Berries as a Means of Controlling Bilharzia-Transmitting Snails*. Addis Ababa: Litho Printers (1981).

Mabawanko, A. G. *An Economic Evaluation of Apprenticeship Training in Western Nigerian Small-Scale Industries*. African Rural Economy Paper No. 17. East Lansing, Michigan: Department of Agricultural Economics, Michigan State University (1979).

Maegraith, B. Schistosomiasis in China. *Lancet* 1(7013):208–214 (1958).

Mahran, G. H., M. Saleh, G. M. El-Hossary, H. M. Motawe, and A. M. Mohamed. A contribution to the molluscicidal activity of *Canna indica* L. Family Cannaceae as a method of control of *Schistosoma*. *Egypt. J. Bilharz.* 1:279–286 (1974).

Mahran, G. H., E. A. El-Hossary, M. Saleh, and H. M. Motawe. Isolation and identification of certain molluscicidal substances in *Canna indica* L. *J. Afr. Med. Plants* 1:107–119 (1977).

Mamo, E., W. A. Alamshed, L. Tesfaye, G. Chaltu, A. Lemma, and W. Y. Legesse. Studies on the toxocity of *Phytolacca dodecandra* to Ethiopian highland sheep. *Bull. Anim. Health Prod. Afr.* 27:79–86 (1979).

Manson-Bahr, P. H. *Manson's Tropical Diseases*. 14th ed., 716. Baltimore: Williams & Wilkins (1954).

Maradufu, A., and J. H. Ouma. A new chalcone as a natural product from *Polygonum senegalense. Phytochemistry* 17:823–824 (1977).

Marchant, Y. Y., F. Balaa, B. F. Abeysekera, and G. H. N. Towers. Molluscicidal activity of sesquiterpene lactones. *Biochem. Systematic Ecol.* 12:285–286 (1984).

Marston, A., and K. Hostettmann. Plant molluscicides. *Phytochemistry* 24:639–652 (1985).

Marston, A., J. D. Msonthi, and K. Hostettmann. Phytochemistry of African plants: 1. Naphthoquinones of *Diospyrus usambarensis*: Their molluscicidal and fungicidal activities. *Planta Med.* 50:279–280 (1984).

McCullough, F. S. Biological control of the snail intermediate hosts of human *Schistosoma* spp.: A review of its present status and future prospects. *Bull. WHO* 38:5–13 (1981).

McCullough, F. S., P. Gayral, J. Duncan, and J. D. Christie. Molluscicides in schistosomiasis control. *Bull. WHO* 58:681–689 (1980).

Medina, F. Studies on the molluscicidal activity of *Solanum mammosum* and the biology of *Fossaria cubensis* (Gastropoda). Ph.D. dissertation, University of Cincinnati (1984).

Medina, F. R., and L. S. Ritchie. Molluscicidal activity of the Puerto Rican weed, *Solanum nodiflorum* Jacquin, against snail hosts of *Fasciola hepatica. Econ. Bot.* 34:368–375 (1980).

Medina, F. R., and R. Woodbury. Terrestrial plants molluscicidal to lymnaeid hosts of *Fasciola hepatica* in Puerto Rico. *J. Agric. Univ. Puerto Rico* 63:366–376 (1979).

Morais, J. G., A. M. Almeida, F. S. Barbosa, and O. B. Calado, Acão moluscicida sinergica da saponina de *Sapindus saponaria* e pentachlorophenato de sodio. *Ann. 10th Congr. Bras. Hig.*, Belo Horizonte, 370 (1953).

Moss, M. O. Aflatoxin and related mycotoxins. In *Phytochemical Ecology*, ed. J. B. Harbourne, 124–144. New York: Academic Press (1972).

Mozley, A. Freshwater mollusca of the Tanganyika Territory and the Zanzibar Protectorate, and their relation to human schistosomiasis. *Trans. R. Soc. Edinburgh* 59:687–730 (1939).

———. *The Control of Bilharzia in Southern Rhodesia.* Salisbury: Rhodesian Printing and Publishing Co. (1944).

Msangi, A. S., and C. Zeller. Investigations in the molluscicidal effects of *Sapindus saponaria* on the bilharzia vector snail, *Bulinus (Physopsis) africanus. Proc. East Afr. Acad.* 3:52–57 (1965).

Muley, E. B. Biological and chemical control of the snail vector *Melania scabra* (Gastropoda: Prosobranchia). *Bull. Zool. Surv. India* 1:1–5 (1978).

Nakanishi, K. Recent studies on bioactive compounds from plants. *J. Nat. Prod.* 45:15–26 (1982).

Nakanishi, K., and I. Kubo. Studies on warburganal, muzigadial and related compounds. *Isr. J. Chem.* 16:28–31 (1977).

Oliveira Matos, M. E., and T. C. B. Tomassini. Wedelin, a saponin from *Wedelia scaberrima. J. Nat. Prod.* 45:836–840 (1983).

Otieno, L. H. Observations on the action of sisal waste on freshwater pulmonate snails. *East Afr. Agric. Forest. J.* 32:68–71 (1966).

Parkhurst, R. M., D. W. Thomas, W. A. Skinner, and L. W. Cary. Molluscicidal saponins of *Phytolacca dodecandra*: Lemmatoxin-C. *Indian J. Chem.* 11:1192–1195 (1973a).

———. Molluscicidal saponins of *Phytolacca dodecandra*: Oleanoglycotoxin-A. *Phytochemistry* 12:1437–1442 (1973b).

———. Molluscicidal saponins of *Phytolacca dodecandra*: Lemmatoxin. *Can. J. Chem.* 52:702–705 (1974).

Pereira, J. P., and C. Pereira de Souza. Ensaios preliminaires com *Anacardium occidentale* como moluscicida. *Cienc. Cult.* 26:1054–1057 (1974).

Pereira, J. P., C. Pereira de Souza, and M. M. Mendes. Molluscicidal properties of the *Euphorbia cotonifolia* L. *Rev. Bras. Pesquisas Med. Biol.* 18:135–143 (1980).

Pereira de Souza, C., E. V. Martins, E. P. Dias, and N. Katz. Vertebrate hormones influence on the reproduction of *Biomphalaria glabrata* and on *Schistosoma mansoni* infection. *Rev. Bras. Pesquisas Med. Biol.* 11:135–140 (1978).

Pezzuto, J. M., S. M. Swanson, and N. R. Farnsworth. Evaluation of the mutagenic potential of endod (*Phytolacca dodecandra*), a molluscicide of potential value for the control of schistosomiasis. *Toxicol. Lett.* 22:15–20 (1984).

Plank, H. K. All parts of the desert palm toxic to snails. *Annual Report of the U.S. Department of Agriculture,* 24. Mayaguez: AES Puerto Rico (1945).

Purchio, A., and R. Campos. Molluscicidal activity of aflatoxin. *Rev. Inst. Med. Trop. Sao Paulo* 13:236–238 (1970).

Ransford, O. N. Schistosomiasis in the Kota Kota District of Nyasaland. *Trans. R. Soc. Trop. Med. Hyg.* 41:617–625 (1948).

Rao, V. P., V. D. Chowdary, and R. Narayana. Effect of aqueous tobacco extract on the bioenergetic parameters of the host snail *Lymnaea luteola.* In *Proceedings of the Tenth International Congress of Tropical Medicine and Malaria,* Abstract No. 578. Manila, November 1980.

Renno, R. Contribuicão ao estudo das Characeae para o combate à esquistossome. *An. R. Acad. Farm.* (Madrid) 38:688–699 (1972).

Ritchie, L. S., G. V. Hillyer, and E. C. Cushing. Molluscicidal and cercaricidal activities of substances contained in tissues of certain plants. *Milit. Med.* 8:795–798 (1963).

Rothschild, M. Some observations on the relationship between plants, toxic insects and birds. In *Phytochemical Ecology,* ed. J. B. Harbourne, 1–12. New York: Academic Press (1972).

Rouquayrol, M. Z., M. P. Sousa, and F. J. A. Matos. Actividade moluscicida de *Pithecellobium multiflorum. Rev. Soc. Bras. Med. Trop.* 7:11–19 (1973).

Rouquayrol, M. Z., M. C. Fonteles, J. E. Alencar, F. J. A. Matos, and A. A. Craveiro. Molluscicidal activity of essential oils from the Brazilian northeast plants. *Rev. Bras. Pesquisas Med. Biol.* 13:135–143 (1980).

Roy, A. K. Control of root-knot (nematode) *Meloidogyne incognita* of jute with decaffeinated tea waste and water hyacinth compost. *Phytopathology* 32:365–368 (1979).

Ruffino, L. *Estudio sobre las propriedades molusquisidas de algunas plantas Venezolanas.* Maracay, Venezuela: Ministerio de Sanidad y Asistencia Social, Departamento Investigaciones Aplicadas, Servicio Investigaciones en Molusquisidas (1975).

Saleh, M. M., M. Shabana, and M. A. Torki. Phytochemical and molluscicidal studies on *Hedychium gardnerianum* and its possible use in control of bilharzia. *Planta Med.* 45:138–139 (1982).

Saliternik, Z., and G. Witenberg. Investigations on the control of bilharziasis vectors in Israel. *Bull. WHO* 21:161–177 (1959).

Seigler, D. S. Terpenes and plant phylogeny. In *Phytochemistry and Angiosperm Phylogeny,* ed. D. A. Young and D. S. Seigler, 117–148. New York: Praeger (1981).

Sherif, A. P., and M. F. El-Sawy. Molluscicidal action of Egyptian herbs: I. Laboratory experimentation. *Alexandria Med. J.* 8:139–148 (1962).

Shoeb, H. A., and M. A. El-Emam. The molluscicidal properties of substances gained from *Ambrosia maritima*. In *Proceedings of the International Conference on Schistosomiasis*, Vol. 1, 487–494. Cairo: Ministry of Health (1978).

Shoeb, H. A., and M. El-Sayed. A short communication on the molluscicidal properties of some plants from Euphorbiaceae. *Helminthologia* 21:33–54 (1984).

Silva, M. J. M., M. P. Sousa, and M. Z. Rouquayrol. Actividade moluscicida de plantas do nordeste Brasileiro. *Rev. Bras. Farm.* 52:117–123 (1971).

Singh, D. K., and R. A. Agarwal. Correlation of the anticholinesterase and molluscicidal activity of the latex of *Euphorbia royleana* on the snail *Lymnaea acuminata*. *J. Nat. Prod.* 47:702–705 (1984).

Soine, T. O. Naturally occurring coumarins and related physiological activities. *J. Pharm. Sci.* 53:231–260 (1964).

Sousa, M. P., and M. Z. Rouquayrol. Molluscicidal activity of plants from northeast Brazil. *Rev. Bras. Pesquisas Med. Biol.* 7:389–393 (1974).

Sousa, M. P., M. Z. Rouquayrol, and M. J. M. Silva. Actividade moluscicida de plantas do nordeste Brasileiro. *Rev. Bras. Farm.* 51:1–9 (1970).

Staba, E. J. Milestones in plant tissue culture systems for the production of secondary products. *J. Nat. Prod.* 48:203–209 (1985).

Sullivan, J. T., C. S. Richards, H. A. Lloyd, and G. Krishna. Anacardic acid: Molluscicide in cashew nut shell liquid. *Planta Med.* 44:175–177 (1982).

Suter, R. Studies on the transmission of *S. haematobium* in Kikawila village (Kilomboro District, Morogoro Region, Tanzania), before and after transmission control by selective mass treatment of the population and focal mollusciciding with a plant molluscicide (*Swartzia madagascariensis*). Ph.D. thesis, University of Basel, 1986.

Teesdale, C. Freshwater molluscs in the Coast Province of Kenya with notes on an indigenous plant and possible use in the control of bilharzia. *East Afr. Med. J.* 31:351–365 (1954).

Teh-Lung, S. 'Tea cake' as a cercaricidal and protective agent against cerariae of *Schistosoma japonicum*. *Chin. Med. J.* 77:580–581 (1958).

Tekle, A. Molluscicidal property of *Sesbania sesban*. *Ethiop. Med. J.* 15:131–132 (1977).

Tesfaigzi, N. Comparative screening of some molluscicidal suspected plants. B.Sc. thesis, Addis Ababa University (1978).

Tomassini, T. C. B., and M. E. Oliveira Matos. On the natural occurrence of 15a-tiglinoy-loxy-kaur-16-en-19-oic acid. *Phytochemistry* 18:663–664 (1979).

Torrealba, J. F., J. Y. Scorza, M. S. Sanabria, A. D. Vasquez, B. I. Ramos, B. Riccardi, and L. S. Jordan. Nota preliminar sobre la accion malaquisita del fruto de paraparo (*Sapindus saponaria*). *Gac. Med. Caracas* 61:299–307 (1953).

Vassiliades, G., and O. T. Diaw. Action molluscicide d'une souche senegalaise d'*Ambrosia maritima*: Essais en laboratoire. *Rev. Elev. Med. Vet. Pays Trop.* 33:401–406 (1980).

Vichnewski, W., S. J. Sarti, B. Gilbert, and W. Herz. Goyazensolide, a schistosomicidal heliangolide from *Eremanthus goyaensis*. *Phytochemistry* 15:191–193 (1976).

Vietmeyer, N. D. A front line against deforestation. *Ceres* Sept./Oct.:38–41 (1979).

Wagner, H., H. Nickl, and Y. Aynehch. Molluscicidal saponins from *Gundelia tournefortii*. *Phytochemistry* 23:2505–2508 (1984).

Wagner, V. A. The possibility of eradicating bilharzia by extensive planting of the tree *Balanites*. *South Afr. Med. J.* 10:10–11 (1936).

Warren, K. S., and P. A. Peters. Cercariae of *Schistosoma mansoni* and plants: Attempts to penetrate *Phaseolus vulgaris* produce a cercaricide. *Nature* 217:647–648 (1969).

Whittaker, R. H., and P. P. Feeny. Allelochemics: Chemical interaction between species. *Science* 171:757–770 (1971).

WHO. Molluscicide screening and evaluation. *Bull. WHO* 33:567–581 (1965).

Yasuraoka, J., Y. Irie, H. Takamura, J. Hashiguchi, M. J. Santos, and A. T. Santos. Laboratory and field assessment of the molluscicidal activity of 'gogo' (*Entada phaseoloides*) against the amphibious intermediate host of *Schistosoma japonicum. Japn. J. Exp. Med.* 47:483–487 (1977).

Yasuraoka, J., J. Hashiguchi, and E. A. Banez. Laboratory assessment of the molluscicidal activity of the plant *Croton tiglium* against *Oncomelania* snails. In *Proceedings of the Philippine-Japan Joint Conference on Schistosomiasis Research and Control,* 106. Manila: Japan International Cooperation Agency (1979a).

Yasuraoka, J., J. Hashiguchi, and B. S. Blas. Laboratory assessment of the molluscicidal activity of the plant *Jatropha curcas* against *Oncomelania* snails. In *Proceedings of the Philippine-Japan Joint Conference on Schistosomiasis Research and Control,* 110. Manila: Japan International Cooperation Agency (1979b).

Yohannes, L. W. Studies on the phytotoxicity of endod (*Phytolacca dodecandra*). In *Proceedings of the O.A.U. Symposium on Schistosomiasis,* 82–85. Addis Ababa: Organization of African Unity (1970).

Yohannes, L. W., T. Lemma, and A. Lemma. New approaches to endod (*Phytolacca dodecandra*) extraction. *Sinet (Ethiop. J. Sci.)* 2:121–127 (1979).

Young, D. A., and D. S. Seigler, eds. *Phytochemistry and Angiosperm Phylogeny.* New York: Praeger (1981).

4

BIOCHEMISTRY OF RECOGNIZED MOLLUSCICIDAL COMPOUNDS OF PLANT ORIGIN

T. O. Henderson

Department of Biological Chemistry, College of Medicine,

N. R. Farnsworth

Program for Collaborative Research in the Pharmaceutical Sciences, College of Pharmacy,

T. C. Myers

Department of Biological Chemistry, College of Medicine,

Health Sciences Center
University of Illinois at Chicago

A half-century ago, Archibald (1933) recognized that the establishment of an effective schistosomiasis control program in the southern and western districts of the Sudan would require an approach that combined treatment of infected persons with measures to eliminate the snail vectors of the disease. Archibald also understood that snail control measures that were effective in the Gezira district, an area of river and canal irrigation, might not succeed in the southern and western areas, where the snail habitats were rainwater courses and ponds that disappeared during the dry season. Furthermore, he knew that people in the arid western region opposed chemical treatment of the streams and ponds that were their only available

109

water sources. Thus it occurred to him that 'some control of the molluscan population in these areas might be effected if they were forested during the dry season with a tree shedding a fruit that possessed lethal properties for molluscs – preferably a fruit that was nontoxic to man and animals.' Archibald's studies, in which he determined that the fruit of the desert tree *Balanites aegyptiaca* was molluscicidal and tentatively identified the active compound as a saponin, pioneered present-day efforts to find and study plant molluscicides.

In this chapter, we describe the biochemistry of known plant molluscicides. Chapter 5 uses the information presented here as a guide for evaluating potential plant molluscicides.

ALKENYL PHENOLS

Pereira and Pereira de Souza (1975) reported that the hexane extract of the shells of the cashew nut *Anacardium occidentale* was toxic to the snail *Biomphalaria glabrata* and the human parasite *Schistosoma mansoni*. The lethal dose, median (LD_{50}) was 1.40 ppm for adult and 0.60 ppm for newly hatched *B. glabrata*. For the snail egg masses, the LD_{50} was 18 ppm, and *S. mansoni* was also in this range. Concentrations of the extract of 10 ppm or greater produced mortality rates of 100% in the guppy, *Lebistes reticulatus*.

Sullivan et al. (1982) isolated the molluscicidal compounds from the nut-shells and found that they were components of anacardic acid, which is a mixture of 6 to n-C_{15} alkylsalicylic acids having side chains with various degrees of unsaturation (Structure I). The triene compound was the most toxic against *B. glabrata* (LD_{50}, 0.35 ppm, 24 hours), while the diene and monoene compounds were less so (LD_{50}, 0.9 ppm and 1.4 ppm, respectively) and the saturated compound significantly less so (LD_{50}, >5 ppm). Neither the decarboxylated anacardic acid nor salicylic acid killed snails, and the authors concluded that both the carboxyl group and the unsaturated side chain were necessary for molluscicidal activity.

It should be noted that while these two studies demonstrated strong molluscicidal effects, these compounds are also potent allergenics and vesicants in humans; they are the 'poison' in poison oak, ivy, and sumac (Baer, 1979).

Structure I. Anacardic acid.

FURANOCOUMARINS AND COUMARINS

Schonberg and Latif (1954) demonstrated that the furanocoumarins bergapten, isopimpinellin, and xanthotoxin (Structure II-1, 2, and 3) were toxic to the snail *Biomphalaria boissi*; bergapten was the most effective, xanthotoxin the least. At 5 ppm (24 hours), bergapten killed 32 of 32 snails and isopimpinellin 22 of 32. Xanthotoxin required concentrations greater than 10 ppm to be effective. These results suggest that in the furanocoumarin series, a major requirement for optimal activity is the presence of at least one acidic proton.

Brooker et al. (1967) reported that the furanocoumarin chalepensin (Structure III) was an effective molluscicide when tested against *Australorbis glabratus (Biomphalaria glabrata)*. No concentrations or test conditions were given except for the duration of exposure (24 hours).

Use of the coumarins and furanocoumarins as molluscicides is problematic because of their pharmacologic and toxicologic effects – especially their hepatotoxic and photosensitizing effects – which have been extensively documented (see Chapter 5).

Structure II. 1. Bergapten: R_1 = OCH_3, R_2 = H. 2. Isopimpinellin: R_1 = R_2 = OCH_3. 3. Xanthotoxin: R_1 = H, R_2 = OCH_3.

Structure III. Chalepensin.

FLAVONOIDS AND ROTENOIDS

The flavonoid compounds, which occur both in the free form and as glycosides, are the largest group of naturally occurring phenols. They are distributed throughout nature but are especially abundant in the higher plant families Polygonaceae, Rutaceae, Leguminosae, Umbelliferae, and Compositae (Asteraceae). Many plants containing flavonoids are diuretics or antispasmodics, while certain isoflavones have estrogenic activity (Trease and Evans, 1978).

Dossaji et al. (1977) reported that crude aqueous methanol extracts of *Polygonum senegalense* were molluscicidal. Subsequently, two molluscicidal flavonoid compounds were isolated from *P. senegalense*. Maradufu and Ouma (1978) isolated the compound 2',4'-dihydroxy-3',6'-dime-thoxychalcone from the seeds and leaves (Structure IV). They reported that the compound was 100% lethal to *Biomphalaria pfeifferi* and *Biomphalaria sudanica* in less than six hours, at levels of 40 ppm. Dossaji and Kubo (1980) isolated quercetin 3-(2'-galloylglucoside) (Structure V) and found it to be an effective molluscicide. At 10 ppm, the compound killed 100% of *Lymnaea natalensis*, *B. pfeifferi*, and *B. glabrata* within 12 hours. These authors noted, however, that quercetin and other 3, 5,7-trihydroxyflavones are highly mutagenic.

Another flavone having molluscicidal activity, 3',5-dihydroxy-4',6, 7-trime-thoxyflavone (Structure VI), was isolated from the stems, flowers, and fruits of *Baccharis trimera* (Less) by Dos Santos Filho et al. (1980). These

Structure IV. 2',4'-Dihydroxy-3',6'-dimethoxychalcone.

Structure V. Quercetin-3-(2'-galloylglucoside).

Structure VI. 3′,5-Dihydroxy-4′,6,7-trimethoxyflavone.

workers reported that the compound caused 100% mortality in *B. glabrata* in six to 24 hours (concentration not specified).

Of the rotenone-containing plants, Mozley (1952) reported that *Tephrosia vogelii* was toxic to snails. However, Adewunmi and Sofowora (1980), in studies on extracts of the related species *Tephrosia uniflora,* determined that the latter was not molluscicidal. The reason for these differing results is not known.

Rotenone (Structure VII) is used in pure or crude form as a selective nonsystemic insecticide. The usual application is a 1% concentration in dust form, which has generally been thought to be nontoxic when administred orally to animals or ingested by humans. The acute LD_{50} of rotenone is 1.5 g/kg, which suggests low toxicity by the oral route. There is now evidence, however, that rotenone administered orally in single doses of 5 or 10 mg/kg to pregnant rats produced teratologic effects; a dose of 2.5 mg/kg failed to produce toxicity or teratogenicity (Khera et al., 1982). Haley (1978) has reviewed the literature on rotenone's biological effects.

SESQUITERPENES

Sesquiterpenes are ubiquitous in plants, especially in essential oil-bearing species in Compositae, Labiatae, Umbelliferae, and other families. The two

Structure VII. Rotenone.

molluscicides ambrosin and damsin from *Ambrosia maritima* are simple sesquiterpenes having an α, β-unsaturated lactone moiety. All sesquiterpenes with this moiety are either cytotoxic or tumor inhibiting; more importantly, they have allergenic effects in humans. Muzigadial and warburganal are nonlactonic simple sesquiterpene aldehydes; little can be said of their potential toxicity, but the plants from which they derive are used as spices in Africa, with no reported adverse effects.

Kubo, Nakanishi and colleagues (Kubo et al., 1976; Nakanishi and Kubo, 1977; Nakanishi, 1982) isolated the closely related drimane sesquiterpenes warburganal (Structure VIII-1) and muzigadial (Structure VIII-2) from the bark of the East African medicinal plant *Warburgia ugandensis* (Canellacea; called muziga in Swahili). These compounds were found to be toxic to the snails *B. glabrata*, *B. pfeifferi*, and *L. natalensis* at 5 to 10 ppm (two hours). They are also potent antifeedants against the African armyworm and effective antimicrobials at the μg/mL level. Warburganal is reported to be cytotoxic to KB cells and is hemolytic when injected into animals.

Shoeb and El-Emam (1978; see also Kloos and McCullough, 1981) reported that damsin and ambrosin (Structures IX-1 and 2, respectively) had molluscicidal activity against the snails *Biomphalaria alexandrina* and *Bulinus*

Structure VIII. 1. Warburganal. 2. Muzigadial.

Structure IX. 1. Damsin. 2. Ambrosin.

truncatus in the range of 8.5 to 13.5 ppm. A derivative, tribromodamsin, was effective in the same range. However, these and related compounds have been shown to be cytotoxic to KB cells and to cause contact dermatitis (Doskotch and Hufford, 1969; Lee et al., 1971; Mitchell et al., 1970; Mitchell and Dupuis, 1971; Kupchan et al., 1971; Torrance et al., 1975).

DITERPENES

Tomassini and Matos (1979) found that kaur-16-en-19-oic acid (Structure X-1) was molluscicidal against *B. glabrata,* and 15-α-tiglinoyloxy-kaur-16-en-19-oic acid (Structure X-2) was miracidicidal against *S. mansoni* miracidia. These compounds were isolated from *Wedelia scaberrima* Benth. (Brazil). Effective concentrations and experimental conditions were not given. It should be noted that Adesogan and Durodola (1976) found that kaurenoic acid had a very mild antibacterial activity.

Dos Santos Filho et al. (1980) reported the isolation and characterization of a diterpene lactone (Structure XI) from the stems, flowers, and fruits of *Baccharis trimera*; it was toxic to *B. glabrata*.

Structure X. 1. Kaur-16-en-19-oic acid. 2. 15-α-Tiglinoyloxy-kaurenoic acid.

Structure XI. Diterpene lactone from *Baccharis trimera*.

SAPONINS

Triterpenoid

Hostettmann (1980) found that a methanol extract of the berries of the common ivy *Hedera helix* L. killed *B. glabrata* at concentrations as low as 40 ppm. He subjected this material to fractionation and isolated four triterpenoid saponins, which were subsequently identified by mass spectrometry (MS) and ^{13}C nuclear magnetic resonance (NMR). All four contained the aglycone hederagenin and differed only in the sugars present at position 3 (Structures XII-1 to 4). The LD_{100} (24 hours) for these compounds against *B. glabrata* was 3, 15, 8, and 12 ppm, respectively. Compounds 1 and 3, containing arabinose, were more potent molluscicides than were glucosides 2 and 4.

The berries of the Ethiopian endod plant, *Phytolacca dodecandra,* have molluscicidal activity, as demonstrated in both laboratory and field trials (see Lemma et al., 1979; Kloos and McCullough, 1981; and references therein). Jewers and King (1972), in a patent filed in 1968, determined that the structure of the major molluscicide derived from the hydrolysis of endod berry extracts was a mixture of triterpenoid saponins having oleanolic acid as the aglycone and a glycoside chain of two to four sugars at position 3 (Structure XIII). About the same time, Powell and Whalley (1969) reported that hydrolysis of the crude saponins of endod berries yielded a product that, on subsequent analysis, was composed of the aglycones oleanolic acid (Structure XIV) and bayogenin (Structure XV).

Parkhurst and colleagues isolated and determined the structure of four glycosides of oleanolic acid from endod berries. The major compound, in terms of amount present in the crude saponin mixture, was found to be a

Structure XII. Hederagenin derivatives. 1. Rα = -L-Arabinopyranoside. 2. R = β-D-Glucopyranoside. 3. R = α-L-Rhamnopyranosyl-(1,2)-β-arabinopyranoside, 4. R = β-D-Glucopyranosyl-(1,2)-β-D-glucopyranoside (R = H = Hederagenin).

Structures XIII and XIV. Oleanolic glycoside (R = glycoside chain of 2–4 sugars) and oleanolic acid (R = H).

Structure XV. Bayogenin.

branched triglycoside, oleanoglycotoxin-A, which is 3-[2,4-di-O-(β-D-glu-copyranosyl)-β-D-glucopyranosyl]-olean-12-ene-28-oic acid (Structure XVI) (Parkhurst et al., 1973a). The most active molluscicidal compound in the crude endod saponin mixture, however, was a branched triglycoside containing two moles of glucose and one mole of galactose per mole of oleanolic acid. This compound, called lemmatoxin, had the suggested structure 3–0-[4'-0-(β-D-glucopyranosyl-3'-O-(β-D-galactopyranosyl)-β-D-glucopyranosyl]-olean-12-ene-28-oic acid (Structure XVII) (Parkhurst et al., 1974). Lemmatoxin was found to be two times more active than oleanoglycotoxin-A against *B. glabrata* (LD$_{90}$, 1.5 ppm, 24 hours). The two other compounds, together called lemmatoxin-C (Parkhurst et al., 1973b), had about half the molluscicidal activity of lemmatoxin. The lemmatoxin-C compounds are closely related, in that they are both linear trisaccharide derivatives of oleanolic acid that contain rhamnose as the terminal glycoside. NMR and mass spectral data suggest that one component (at least 70% of the total) may be rhamnose-glucose-glucose-oleanolic acid (Structure XVIII) having

Structure XVI. Oleanoglycotoxin-A.

Structure XVII. Lemmatoxin.

Structure XVIII. Lemmatoxin-C.

the configuration 3-O-α-L-rhamnopyranosyl-2'-O-β-D-glucopyranosyl-2'-O-β-D-glucopyranosyl)-olean-12-ene-28-oic acid. The investigators suggested that galactose may be present in a compound having the structure rhamnose-galactose-glucose-oleanolic acid. Oleanoglycotoxin-A, lemmatoxin, and lemmatoxin-C are reported to occur in the crude saponin mixture from *P. dodecandra* in a ratio of 18%, 16%, and 17%, respectively (Stolzenberg and Parkhurst, 1974). As an additional comment, Parkhurst et al. (1973b) noted that the synthetic β-glucoside of oleanolic acid has only one tenth the molluscicidal activity of lemmatoxin-C (that is, about 1/20 of the activity of lemmatoxin).

Hostettmann et al. (1982) studied the structure-activity relationships of a series of monodesmosidic and bidesmosidic triterpenes of the oleanane type, primarily representing oleanolic acid and hederagenin. From their data, it appears that the number, sequence, point of attachment, and configuration of sugars attached to the oleanane aglycone are less important than the fact that sugars attached at two positions on the aglycone render the resulting saponins inactive as molluscicides.

In addition to their molluscicidal effects, the crude saponins of *P. dode-candra* and lemmatoxin were found to be effective spermicides when tested against human sperm, and lemmatoxin-C was an extremely potent rat spermicide (Stolzenberg and Parkhurst, 1974). Furthermore, the crude mixed saponins and purified oleanoglycotoxin-A, lemmatoxin, and lemmatoxin-C were found to be effective blastocidal agents and abortifacients when introduced via intrauterine injection during early stages of pregnancy in rats (Stolzenberg et al., 1975, 1976). The synthetic cellobioside of oleanolic acid was not nearly as effective a blastocide as the natural triglycoside derivatives.

A disadvantage to the use of endod saponins as molluscicides is that the mixed saponins are lethal to fish at low concentrations. Lemma et al. (1972), in reporting a lethal concentration, 90% (24 hours) of 3 ppm against *B. glabrata* for a butanol extract of endod berries, stated that the extract was piscicidal at 1.5 to 3.6 ppm (LD_{90}, 24 hours) for three species of fish. They also reported that the extract was a potent emetic when administered intragastrically to monkeys, dogs, and cats at dosages as low as 5 mg/kg.

Spirostanol

The bark of the common North American dogwood *Cornus florida* L., upon extraction with methanol, was found to yield a product with strong molluscicidal activity (Kubo and Nakanishi, 1977). Hostettmann et al. (1978) subsequently isolated and characterized the active components from *C. florida* bark and found them to be glycosidic derivatives of the steroidal spirostane aglycone sarsapogenin having the structures sarsapogenin-O-β-D-xylopyranosyl-(1->2)-β-D-galactopyranoside and sarsapogenin-O-β-D-glu-copyranosyl-(1->2)-β-D-galactopyranoside (Structure XIX-1 and 2).

Structure XIX. Sarsapogenin glycosides. 1. R = Xylose. 2. R = Glucose.

Although *C. florida* may be of little use as a molluscicide against schisto-somiasis vectors because it does not grow in endemic areas (Kloos and McCullough, 1981), it may prove useful in the control of such other snail-transmitted diseases as liver fluke (*Fasciola* sp.) and swimmer's itch (schis-tosomal dermatitis).

Archibald (1933) described his studies in the Sudan on the effectiveness of parts of the desert palm *B. aegyptiaca* in killing schistosomal cercaria and miracidia and the snail vectors of schistosomiasis. He found that the berries, kernel, bark, and branches all possessed molluscicidal properties and specu-lated that this activity was probably due to the presence of saponins. He outlined a simple procedure for preparing an aqueous emulsion from the berries that was effective in the field against the snail vectors and the free-living stages of the parasite. He found that the emulsion also was lethal to fish and tadpoles. More recently, Nakanishi and colleagues isolated and characterized three molluscicidal saponins, called balanitin-1, 2, and 3, from aqueous methanol extracts of *B. aegyptiaca* (Liu and Nakanishi, 1981; Nak-anishi, 1982). These compounds all contain yamogenin as the aglycone (Structure XX); they differ from each other in the number and kinds of sugars present in the glycosidic moiety (Structure XXI).

Structure XX. Yamogenin.

Medina and Woodbury (1979), Medina and Ritchie (1980), and others have documented the molluscicidal effectiveness of extracts of Solanaceae plants, especially *Solanum nodiflorum* and *Solanum mammosum*. Alzérreca et al. (1981) reported isolating the aglycone solasodine (Structure XXII-1) from partially purified *S. mammosum* extracts that they had shown to be an effective molluscicide (LD_{95}, 25 ppm, 24 hours) against *Fossaria (Lymnaea) cubensis*, the snail vector of the liver fluke *Fasciola hepatica*. They noted that solasodine was the aglycone of both solasonine (Structure XXII-2) and solamargine (Structure XXII-3), both of which had been isolated from *S. mammosum*. Solasodine was not itself a molluscicide, possibly as a result of its low water solubility due to the absence of the sugar moiety. More recently,

Structure XXI. Balanitins. Balanitin-1 is shown; balanitin-2 has H instead of rhamnose-1, whereas balanitin-3 has H instead of rhamnose-2 and xylose at position 6' of glucose-2.

Alzérreca and Hart (1982) reported that tomatine (Structure XXIII) as well as a mixture of solasonine and solamargine was an effective molluscicide against both *F. (L.) cubensis* and *B. glabrata* (LD_{100}, 10 to 25 ppm, 24 hours for the former). Digitonin (Structure XXIV) was, too, but the spirosolane aglycones, solasodine and tomatidine (Structure XXV) were not. These authors speculated that the molluscicidal differences between saponins containing either solasodine or tomatidine had little to do with the types of sugars present in the glycosidic linkages. They concluded that it made little

Structure XXII. 1. Solasodine: R = H. 2. Solasonine: R = L-rhamnose + D-glucose + D-galactose. 3. Solamargine: R = 2(L-rhamnose) + D-glucose.

Structures XXIII and XXV. Tomatine and tomatidine. Tomatine: R = D-xylose + 2(D-glucose) + D-galactose. Tomatidine: R = H.

Structure XXIV. Digitonin.

difference whether nitrogen or oxygen was present in positions X or Y (Structure XXVI), since digitonin and the spirostane-type saponins from *C. florida* were also effective molluscicides (compare Hostettmann et al., 1982, and Structure XIX). Hostettmann et al. (1982), in testing 24 saponins for their effectiveness as molluscicides against *B. glabrata,* found tomatine to be effective at levels as low as 4 ppm (LD_{100}, 24 hours), whereas the steroid alkaloid saponin α-solanine was ineffective. Tomatine contains tomatadine, which has a spirosolane skeleton similar to that found in the active compounds solasonine and solamargine and is structurally related to the active spirostanol glycosides from *C. florida.* Thus it would appear that the molluscicidal activity of the steroid alkaloid saponin series depends as much on the aglycone as it does on the presence or absence of sugars.

Structure XXVI. X-Y structure (Alzérreca and Hart, 1982).

OTHERS

Johns et al. (1982) reported that affinin (spilanthol; N-isobutyl-2,6,8-decatrienamide, Structure XXVII), at concentrations above 50 ppm, immobilized *Physa occidentalis* Tryon after 60 minutes, with death occurring within 18 hours. At 150 ppm, cercarial emergence ceased, the snails were immobilized after 30 minutes, and carcariae were immobilized within 50 seconds. The crude pulverized roots of *Heliopsis longipes* (A. Gray) Blake, used as

Structure XXVII. Affinin.

the source of affinin by these workers, was also found to be a piscicide when tested against the guppy *L. reticulatus*. The amount of pulverized root required for these activities was not stated. Affinin has been isolated from the leaves of *Spilanthes oleracea* Jacq and the flowers of *Wedelia parviceps* Blake as well as from *H. longipes*.

Hussein Ayoub (1982) reported that acetone, ethanol, and aqueous extracts of the fruit of *Acacia nilotica* (L.) Willd. ex Del., as well as a spray-dried product of aqueous extracts designated TAN, killed both *B. truncatus* and *B. pfeifferi*. Of the extracts, the acetone was the most effective molluscicide, killing 100% of exposed snails at a concentration of 75 ppm (24 hours). The ethanol extract killed 100% at 150 ppm (24 hours), the aqueous extract at 200 ppm (24 hours). The TAN product was much more effective than the water extract, killing 100% of both snail species within six hours at a concentration of 60 ppm. The author found that TAN contained more than 56% condensed and hydrolyzable tannins and concluded that the molluscicidal activity was probably due to these compounds. Schaufelberger and Hostettmann (1983) confirmed these results and also found that crude extracts of typical tannin plants were highly potent molluscicides. According to these workers, the molluscicidal activity was due to hydrolyzable and condensed tannins.

Mahran et al. (1974) reported that extracts of *Canna indica* L. were lethal to *B. alexandrina*, the intermediate host in Egypt for *S. mansoni*. In subsequent work, they isolated two compounds that were potent molluscicides (Mahran et al., 1977). One was identified as triacontanol (Structure XXVIII); no details on the levels of the pure compound required to kill *B. alexandrina* were given. The second was tentatively identified as a long chain carboxylic acid containing both a secondary alcohol function and a ketone function.

$$CH_3(CH_2)_{11} \underset{\underset{CH_3}{|}}{CH} (CH_2)_{15} CH_2OH$$

Structure XXVIII. Triacontanol.

CONCLUSIONS

Of the 40 or so molluscicides of plant origin whose biochemical data we have summarized, several compounds are very potent, killing snails at levels of 1 ppm or lower. Unfortunately, a number of these compounds are also toxic to humans or other life forms (see also Chapter 5).

In general, taking into account both molluscicidal activity and toxicologic effects, the saponins appear to offer the greatest potential as molluscicides. As Chapter 5 points out, such structural variations as the presence or absence of polar substituents in aglycone moieties of the saponins produce marked differences in biological effects. Specific examples of this phenomenon, taken from Tschesche and Wulff (1973), are given in Table 1, which includes a summary of the hemolytic activity of a number of saponins. These data indicate that saponins having polar groups in rings D and E of the aglycones are significantly less hemolytic than those without polar groups in those rings. (See Structure XII for the lettering system used to denote the rings of the aglycones.) Such differential biological activity might be exploited in looking for saponins with maximal molluscicidal activity but minimal toxicity to other life forms around snail habitats.

On the other hand, in comparing the antitumor activity of selected saponins with their toxicity in rodents, Tschesche and Wulff (1973) found that the effective dose, 50% (ED_{50}) of these saponins as antitumor compounds in rats was only about half the LD_{50} (see Table 2). While that again demonstrates the differential biological activity of saponins, it also indicates the narrow margin of safety with which one often must deal. This point must be kept in mind when considering the effectiveness of plant (and other) molluscicides and their real or potential toxicity to humans and the environment in general.

Acknowledgments

Certain aspects of the data collection required to compile this report were supported in part by a contract from the National Cancer Institute (USA,

Table 1. Comparison of Saponin Hemolytic Activity[a]

Saponins With Polar Groups in Rings D and E		Saponins Without Polar Groups in Rings D and E	
Saponin	Hemolysis (μg/mL)	Saponin	Hemolysis (μg/mL)
Sarsaparilloside	>250	Parillin	4.0
Convallamaroside	>250	Convallamaroginin triglycoside	20.0
Lanatigoside	>250	Lanatigonin	0.6
Methylprotogracillin	>200	Gracillin	2.0
Hederacoside C	>250	α-Hederin	3.0
Hederacoside B	>250	β-Hederin	0.9
Aescinol	>250	Aescin	2.5
Dideacyltheasaponin	>250	Theasaponin	20.0

[a]From Tschesche and Wulff (1973).

Table 2. Comparison of Antitumor Activity and Toxicity of Selected Saponins[a]

Saponin	$ED_{50}(mg/kg)$[b]	$LD_{50}(mg/kg)$[c]
Aescin	20	36
Cyclamin	10	20
Senegin (mixture)	1.5	3
Primulasaponin	40	70
α-Hederin	4	10
Parillin	50	80

[a]From Tschesche and Wulff (1973).
[b]In rats bearing the Walker 256 carcinosarcoma. Dose required to decrease the tumor size by 50% following several intraperitoneal injections of the saponin.
[c]Intraperitoneal dosing in rats.

CM-97259), as well as by funds from the World Health Organization Special Programme for Research Development and Research Training in Human Reproduction (78135) and the UNDP/WORLD BANK/WHO Special Programme for Research and Training in Tropical Diseases. TOH and TCM gratefully acknowledge the research support of the Edna McConnell Clark Foundation (EMCF Grant No. 282–0044), and TOH also thanks Professor Emil Malek, Tulane University, New Orleans, and Professors Anne Frame and Pedro Bendezu, Inter American University of Puerto Rico, Metropolitan Campus, Hato Rey, for their patient introductions to, and instruction in, malacology and parasitology during his recent sabbatical leave. Special thanks are extended to Professor C. O. Adewunmi for providing valuable references and thoughts on the subject matter, and to Dr. K. E. Mott for his interest and encouragement. We also thank Ms. D. M. Lattyak for typing the manuscript on short notice.

REFERENCES

Adesogan, E., and J. I. Durodola. Antitumor and antibiotic principles of *Annona senegalensis*. *Phytochemistry* 15:1311–1312 (1976).
Adewunmi, C. O., and E. A. Sofowora. Preliminary screening of some plant extracts for molluscicidal activity. *Planta Med.* 39:57–65 (1980).
Alzérreca, A., and G. Hart. Molluscicidal steroid glycoalkaloids possessing stereoisomeric spirosolane structures. *Toxicol. Lett.* 12:151–155 (1982).
Alzérreca, A., B. Arboleta, and G. Hart. Molluscicidal activity of natural products: The effect of *Solanum* glycosidic alkaloids on *Lymnaea cubensis* snails. *J. Agric. Univ. Puerto Rico* 57:69–72 (1981).

Archibald, R. G. The use of the fruit of the tree *Balanites aegyptiaca* in the control of schistosomiasis in the Sudan. *Trans. R. Soc. Trop. Med. Hyg.* 27:207–211 (1933).

Baer, H. The poisonous Anacardiaceae. In *Toxic Plants,* ed. A. D. Kinghorn, 161–170. New York: Columbia University Press (1979).

Brooker, R. M., J. N. Eble, and N. A. Starkovsky. Chalepensin, chalepin and chalepin acetate, three novel furanocoumarins from *Ruta chalepensis. Lloydia* 30:73–77 (1967).

Doskotch, R. W., and C. D. Hufford. Damsin, the cytotoxic principle of *Ambrosia ambrosioides* (Cav.) Payne. *J. Pharm. Sci.* 58:186–188 (1969).

Dossaji, S., and I. Kubo. Quercetin 3-(2'-galloylglucoside), a molluscicidal flavonoid from *Polygonum senegalense. Phytochemistry* 19:482 (1980).

Dossaji, S. F., M. G. Kairu, A. T. Gondwe, and J. H. Ouma. On the evaluation of the molluscicidal properties of *Polygonum senegalense* forma *senegalense. Lloydia* 40:290–293 (1977).

Dos Santos Filho, D., S. J. Sarti, W. Vichnewski, M. S. Bulhoes, and H. de Freitas Leitao Filho. Molluscicidal activity on *Biomphalaria glabrata* of a diterpene lactone and a flavone isolated for *Baccharis trimera* (Less.). *Rev. Fac. Farmacol. Odontol. Ribeirao Preto* 17:43–47 (1980) (*Chem. Abs.* 96:2071e, 1980).

Haley, T. J. A review of the literature of rotenone. *J. Environ. Pathol. Toxicol.* 1:315–337 (1978).

Hostettmann, K. Saponins with molluscicidal activity from *Hedera helix* L. *Helv. Chim. Acta* 63:606–609 (1980).

Hostettmann, K., M. Hostettmann-Kaldas, and K. Nakanishi. Molluscicidal saponins from *Cornus florida* L. *Helv. Chim. Acta* 61:1990–1995 (1978).

Hostettmann, K., H. Kizu, and T. Tomimori. Molluscicidal properties of various saponins. *Planta Med.* 44:34–35 (1982).

Hussein Ayoub, S. M. Molluscicidal properties of *Acacia nilotica. Planta Med.* 14:181–183 (1982).

Jewers, K., and T. A. King. Improvements relating to molluscicides. Patent No. 1277417. London: The Patent Office (1972).

Johns, T., K. Graham, and G. H. Neil Towers. Molluscicidal activity of affinin and other isobutylamides from the Asteraceae. *Phytochemistry* 21:2737–2738 (1982).

Khera, K. S., C. Whalen, and G. Angers. Teratogenicity study on pyrethrum and rotenone (natural origin) and Ronnel in pregnant rats. *J. Toxicol. Environ. Health* 10:111–119 (1982).

Kloos, H., and F. McCullough. Plant molluscicides: A review. WHO Unpublished Report WHO/VBC/81.834; WHO/SCHISTO/81.59 (1981).

Kubo, I., and K. Nakanishi. Host plant resistance to pests. In *Symposium on the Chemical Basis for Host Plant Resistance to Pests,* 165. Symposium Series No. 62. Washington, D.C.: American Chemical Society (1977).

Kubo, I., Y.-W. Lee, M. Pettei, F. Pilkiewicz, and K. Nakanishi. Potent armyworm antifeedants from the East African *Warburgia* plants. *Chem. Communications,* 1013–1014 (1976).

Kupchan, S. M., M. A. Eakin, and A. M. Thomas. Tumor inhibitors: 69. Structure-cytotoxicity relationships among the sesquiterpene lactones. *J. Med. Chem.* 14:2105–2126 (1971).

Lee, K.-H., E. S. Huang, C. Piantadosi, J. S. Pagano, and T. A. Geissman. Cytotoxicity of sesquiterpene lactones. *Cancer Res.* 31:1649–1654 (1971).

Lemma, A., G. Brody, G. W. Newell, R. M. Parkhurst, and W. A. Skinner. Studies on the molluscicidal properties of endod (*Phytolacca dodecandra*): I. Increased potency with butanol extraction. *J. Parasitol.* 58:104–107 (1972).

Lemma, A., D. Heyneman, and H. Kloos, eds. Studies on the molluscicidal and other properties of the endod plant *Phytolacca dodecandra*. Unpublished Report, Department of Epidemiology and International Health, University of California, San Francisco (1979).

Liu, H.-W., and K. Nakanishi. A micromethod for determining the branching points of oligosaccharides based on circular dichroism. *J.A.C.S.* 103:7005–7006 (1981).

Mahran, G. H., M. Saleh, G. A. El-Hossary, H. M. Motawe, and A. M. Mohamed. A contribution to the molluscicidal activity of *Canna indica* L. family Cannaceae as a method for control of *Schistosoma. Egypt. J. Bilharz.* 1:279–286 (1974).

Mahran, G. H., G. A. El-Hossary, M. Saleh, and H. M. Motawe. Isolation of certain molluscicidal substances in *Canna indica* L. *J. Afr. Med. Plants* 1:107–119 (1977).

Maradufu, A., and J. H. Ouma. A new chalcone as a natural molluscicide from *Polygonum senegalense. Phytochemistry* 17:823–824 (1978).

Medina, F. R., and L. S. Ritchie. Molluscicidal activity of the Puerto Rican weed, *Solanum nodiflorum* Jacquin, against the snail hosts of *Fasciola hepatica. Econ. Bot.* 34:368–375 (1980).

Medina, F. R., and R. Woodbury. Terrestrial plants molluscicidal to lymnaeid hosts of *Fasciola hepatica* in Puerto Rico. *J. Agric. Univ. Puerto Rico* 63:366–376 (1979).

Mitchell, J. C., and G. Dupuis. Allergic contact dermatitis from sesquiterpenoids of the Compositae family of plants. *Br. J. Dermatol.* 84:139–150 (1971).

Mitchell, J. C., B. Fritig, B. Singh, and G. H. N. Towers. Allergic contact dermatitis from Frullania and Compositae. *J. Invest. Dermatol.* 54:233–239 (1970).

Mozley, A. *Molluscicides.* London: H. K. Lewis (1952).

Nakanishi, K. Recent studies on bioactive compounds from plants. *J. Nat. Prod.* 45:15–26 (1982).

Nakanishi, K., and I. Kubo. Studies on warburganal, muzigadial and related compounds. *Isr. J. Chem.* 16:28–31 (1977).

Parkhurst, R. M., D. W. Thomas, W. A. Skinner, and L. W. Cary. Molluscicidal saponins of *Phytolacca dodecandra*: Oleanoglycotoxin-A. *Phytochemistry* 12:1437–1442 (1973a).

———. Molluscicidal saponins of *Phytolacca dodecandra*: Lemmatoxin-C. *Indian J. Chem.* 11:1192–1195 (1973b).

———. Molluscicidal saponins of *Phytolacca dodecandra*: Lemmatoxin. *Can. J. Chem.* 52:702–705 (1974).

Pereira, J. P., and C. Pereira de Souza. Preliminary studies of *Anacardium occidentale* as a molluscicide. *Cienc. Cult.* 26:1054–1057 (1974). (*Chem. Abs.* 82:134008k, 1975).

Powell, J. W., and W. B. Whalley. Triterpenoid saponins from *Phytolacca dodecandra. Phytochemistry* 8:2105–2107 (1969).

Schaufelberger, D., and K. Hostettmann. On the molluscicidal activity of tannin containing plants. *Planta Med.* 48:105–107 (1983).

Schonberg, A., and N. Latif. Furochromones and coumarins: XI. The molluscicidal activity of bergapten, isopimpinellin and xanthotoxin. *J.A.C.S.* 76:6208 (1954).

Shoeb, H. A., and M. A. El-Emam. The molluscicidal properties of substances gained from *Ambrosia maritima*. In *Proceedings of the International Conference on Schistosomiasis*, Vol. 1, 487–494. Cairo: Ministry of Health (1978).

Stolzenberg, S. J., and R. M. Parkhurst. Spermicidal actions of extracts and compounds from *Phytolacca dodecandra. Contraception* 10:135–143 (1974).

Stolzenberg, S. J., R. M. Parkhurst, and E. J. Reist. Blastocidal and contraceptive actions of saponins from *Phytolacca dodecandra. Fed. Proc.* 34:339 (1975).

———. Blastocidal and contraceptive actions by an extract and compounds from endod (*Phytolacca dodecandra*). *Contraception* 14:39–51 (1976).

Sullivan, J. T., C. S. Richards, H. A. Lloyd, and G. Krishna. Anacardic acid: Molluscicide in cashew nut shell liquid. *Planta Med.* 44:175–177 (1982).

Tomassini, T. C. B., and M. E. O. Matos. On the natural occurrence of 15α-tiglinoyloxy-kaur-16-en-19-oic acid. *Phytochemistry* 18:663–664 (1979).

Torrance, S. J., R. M. Wiedhopf, and J. R. Cole. Ambrosin, tumor inhibitory agent from *Hymenoclea salsola* (Asteraceae). *J. Pharm. Sci.* 64:887–888 (1975).

Trease, G. E., and W. C. Evans. *Pharmacognosy.* London: Bailliere Tindall (1978).

Tschesche, R., and G. Wulff. Chemie und biologie der Saponine. *Fortschr. Chem. Org. Naturst.* 30:461–606 (1973).

5

PLANTS WITH POTENTIAL
MOLLUSCICIDAL ACTIVITY

N. R. Farnsworth

Program for Collaborative Research in the Pharmaceutical Sciences,
College of Pharmacy,

T. O. Henderson

Department of Biological Chemistry, College of Medicine,

D. D. Soejarto

Department of Medicinal Chemistry and Pharmacognosy, College of Pharmacy,

Health Sciences Center
University of Illinois at Chicago

Most plant-derived drugs have been discovered through scientific investigation of ethnomedical (folkloric) claims that a specific plant produces a useful biological effect. Random selection and mass screening have been the types of programs most widely used in the search for new plant-based drugs in recent years, but such programs are expensive and have been unproductive in terms of identifying useful drugs for humans. They have led, however, to the discovery of a large number of novel structures that are academically interesting and biologically active. Serendipity has also played a role in the discovery of useful drugs from plants. But, perhaps because none of these approaches are considered strictly scientific, they have failed to attract the attention of institutions, scientists, and government agencies interested in drug development, at least in terms of establishing serious research and development programs in connection with medicinal plants.

131

This chapter describes scientific methods short of bioassay that can be employed in developing countries to identify plants with potential molluscicidal activity. Such plants must either grow abundantly in areas where schistosomiasis is endemic or be amenable to large-scale cultivation in these areas. The active principles in the plants should be soluble in water so as to avoid expensive extraction procedures, and they should kill snails at levels that are not detrimental to the flora and other fauna, including people. Placing a molluscicide into the water supply of a region has serious short- and long-term implications. Systemic effects could result if people drink the treated water, or cutaneous reactions if they wash their clothes or bathe in it. Also to be considered is the degradation or metabolic products produced when an active plant molluscicide is ingested by fish that people may subsequently consume.

Successful drug development programs start with a large number of active entities. Thus several plants must be available as candidates for extensive toxicologic, agronomic, and ecologic studies. How can these candidates be identified in the absence of bioassay data? Obviously, it would be impossible to acquire and test all the many thousands of plant species for molluscicidal activity. Nor would one wish to, since many are not found in areas where the snail vectors of schistosomiasis thrive or cannot be grown in sufficient quantities in such areas.

The extracts of a number of plants appear to elicit molluscicidal activity, based on the results of *in vitro* tests or field studies, or both (Chapter 3). Indeed, the molluscicidal principles in a few of these active plants have been structurally characterized (Table 1). But drug development is a complex process. By the time the most promising of the known active plants have been studied for field effectiveness and for their effects on the flora and fauna, all could be found unsuitable for general use.

When a search program is set up, one must anticipate all the problems associated with biological variation. But plants present no greater numbers of variability problems than have been encountered and overcome in the area of drug bioassay with synthetic chemical compounds. Any method employed to identify active molluscicidal plants will be imperfect; some plants will be overlooked, while others will be pursued but will eventually be determined to be of no practical interest.

That the product to be used as a molluscicide is not a pure chemical compound responsible for the effect is unimportant. Chemotaxonomic data (that is, data on the relationship of secondary plant constituents to phylogeny) are often useful in predicting that certain classes of metabolites will occur in plant taxa, even though their presence has not been demonstrated experimentally. If the chemotaxonomic marker being employed has been shown to have molluscicidal activity, parallel biotaxonomic analogies may be postulated.

Table 1. Plant-Derived Molluscicides of Known Structure.

Chemical Class	Compound Name[a]
Aliphatic amides	Affinin
Alkenyl phenols	Anacardic acids
Alkanols	Triacontan-1-ol
Furanocoumarins	Bergapten, chalepensin, isopimpinellin, xanthotoxin
Flavonoids and rotenoids	2',4'-Dihydroxy-3',6'-dimethoxychalcone; 3',5-dihydroxy-4',6',7-trimethoxyflavone; quercetin-3-(2''-galloylglucoside); rotenone
Sesquiterpenes	Ambrosin, damsin, muzigadial, warburganal
Diterpenes	Diterpene dilactone from *Baccharis trimera*; ent-kaur-16-en-19-oic acid
Monodesmosidic triterpene saponins	Hederagenin-3-O-ara
	Hederagenin-3-O-ara-(1→2)-glu
	Hederagenin-3-O-ara-(1→2)-rha
	Hederagenin-3-O-ara-(1→2)-rha-(1→3)-rib
	Hederagenin-3-O-ara-(1→2)-rha-(1→3)-rib-(1→4)-glu
	Hederagenin-3-O-ara-(1→2)-rha-(1→3)-xyl
	Hederagenin-3-O-ara-(1→2)-rha-(1→3)-xyl-(1→4)-glu
	Hederagenin-3-O-glu
	Hederagenin-3-O-glu-(1→2)-glu
	Hederagenin-3-O-rha-(1→2)-ara
	Lemmatoxin
	Lemmatoxin-C
	Oleanoglycotoxin
	Oleanolic acid-3-O-ara-(1→2)-glu
	Oleanolic acid-3-O-ara-(1→2)-rha-(1→3)-rib
	Oleanolic acid-3-O-ara-(1→2)-rha-(1→3)-xyl-(1→4)-glu
	Oleanolic acid-3-O-rha
	Phytolaccagenin-3-O-glu-(1→2)-xyl
	Primulic acid
Monodesmosidic spirostanol saponins	Balanitin 1,2 and 3; digitonin
	Sarsapogenin-3-O-gal
	Sarsapogenin-3-O-glu-(1→2)-gal
	Sarsapogenin-3-O-xyl-(1→2)-gal
Spirosolane (steroid alkaloid) saponins	Solamargine; solasonine; tomatine

[a]Ara = α-L-arabinopyranosyl; gal = β-D-galactopyranosyl; glu = β-D-glucopyranosyl; rha = α-L-rhamnopyranosyl; rib = β-D-ribopyranosyl; xyl = β-D-xylopyranosyl.

From an inspection of molluscicidal test data on plant extracts, it is evident that plant parts vary considerably in their degree of activity (Chapter 3). Secondary metabolites do not always accumulate in plant parts to the same degree. Indeed, in some cases molluscicidal activity is absent from all but one plant part. Thus one cannot make predictions as to the best plant to study intensively if only a single plant part is tested. Further, considerable variation in the level of secondary metabolites found in plants must be expected. Many factors are responsible for this variation, including genetics, time of collection, geographic area, temperature, amount of rainfall, cultivated versus wild collections, and season. To some degree, an organized program can overcome most of these problems and preclude the missing of active species.

INFORMATION REQUIRED

To begin, one must survey the literature for several types of information that can eventually be analyzed and used for predictive purposes. Part of this information has already been gathered.

Test Results

A systematic search of the scientific literature for information on plants that have been tested for molluscicidal activity has been completed. (Information on plants that affect cercarial penetration and those with antischistosomal activity has also been collected but will not be used in the present discussion.) All relevant data, including both positive and negative test results, have been computerized in the NAPRALERT system (Farnsworth et al., 1981). Data were judged positive if a plant extract killed snails at concentrations of 100 ppm or less. The equivocal activity point was arbitrarily set at concentrations of greater than 100 ppm.

Known Compounds

A literature search was also carried out to identify all plant-derived substances of known structure that have been reported either to have molluscicidal activity or to lack such activity (Chapter 4). These substances have been entered in the NAPRALERT data base and are listed in Table 1.

Plants Containing Known Compounds

Ideally, a literature search should be carried out to identify all plants that are reported to contain not only any of the specific natural molluscicidal compounds listed in Table 1 but also any of their analogues or derivatives, since in most cases the analogues elicit biological effects qualitatively similar to those elicited by the parent compound.

To some extent, the NAPRALERT data base was useful for this purpose. Its information comes from the world scientific literature and has been collected in a systematic manner since 1975. Some pre-1975 data of special interest have also been computerized: complete literature surveys on about 500 genera of plants, representing thousands of species. Some areas of biological activity are extensively covered. For example, all data concerning plants and plant-derived compounds that have been reported to affect any aspect of mammalian reproduction, and similar reports since about 1900 on antitumor and cytotoxic activity, have been collected and are included in the NAPRALERT data base.

These data make it possible to anticipate which plants will be active molluscicides even though they have not been tested for molluscicidal activity; the concentration of the active constituent will be the major determinant. Thus, if a developing country needs a plant molluscicide but no plant indigenous to that country has shown molluscicidal activity, the data will tell which plants contain an active molluscicidal chemical and would be likely candidates for bioassay and perhaps for subsequent practical use.

Ethnomedical Data

Another type of data is undoubtedly important in identifying potential molluscicidal plants, but we have not used it here. The literature is full of ethnomedical information, which often is thought by scientific investigators to be invalid. However, few concrete data have been assembled to support either positive or negative opinions on this subject. Experimental and ethnomedical data on plants reported to have piscicidal, insecticidal, and anthelmintic (antiparasitic) properties should be collected, since these plants might also have molluscicidal properties.

Plants Containing Saponins

Since it is known that certain types of triterpene and steroid saponins elicit pronounced molluscicidal activity (Hostettmann et al., 1982), it would seem reasonable to identify all plants that have been tested for the presence or absence of saponins in order to provide additional predictive information that can be analyzed for the identification of potential molluscicidal plants. The term *saponin*, however, deserves clarification.

The textbook definition of *saponin* usually includes any material that forms a persistent froth when shaken with water, hemolyzes red blood cells, forms complexes with cholesterol, and is toxic to cold-blooded animals. It is generally believed that saponins are not absorbed following oral administration to mammalian species, including humans. However, many examples can be cited in which a plant extract being tested for the presence of saponins conforms to most but not all of these properties. A modification of the definition is usually employed when substances of known structure are considered; for example, saponins are glycosides of triterpene or steroid (spirostanol) sapogenins or both. Their froth-forming ability makes it apparent that they affect surface tension. Their hemolytic activity is generally considered to result from their formation of complexes with cholesterol in red blood cell membranes, which causes a collapse of the cell and the release of hemoglobin. Their toxic effect on fish is a result of paralysis of the gills and consequent interference with respiratory functions (Farnsworth, 1966).

Quantitatively, saponins differ markedly in hemolytic activity and fish toxicity. For example, the saponin lanatigonin hemolyzes red blood cells at a concentration of 0.6 µg/mL, whereas several other saponins are not hemolytic even at 250 µg/mL (Tschesche and Wulff, 1973). In general, saponins with polar substituents in rings D and E are less hemolytic than those without (Tschesche and Wulff, 1973). Similar differences can be expected with respect to molluscicidal and piscicidal activity.

When comparing the saponins' antitumor activity with their toxicity in rodents, on the other hand, the margin of safety may be rather narrow. From limited data available, the effective dose, 50% (ED_{50}) of saponins for tumor inhibition in rats seems to be only about half their lethal dose, median (LD_{50}) (Tschesche and Wulff, 1973). This point must be kept in mind when considering the potential human toxicity of plant-derived molluscicides.

More than 10000 species of higher plants had been reported tested for the presence of saponins by 1966 (Farnsworth, 1966). Simple tests were employed, including determination of a persistent froth after adding a small amount of plant to water and shaking the mixture. Fish-toxicity, hemolytic-effect, and chemical spot tests were also made. Although some of the positive test results undoubtedly represent false positives, the data serve as a good starting point for identifying taxa in which saponins, and hence potential molluscicides, occur in nature. Since 1966, an estimated 10000 more species have been reported tested for the presence of saponins. The problems encountered in such testing have been reviewed by Farnsworth (1966).

PRELIMINARY FINDINGS

We are now able to demonstrate the utility of data from plants tested for molluscicidal activity and, at least partially, for known compounds and plants containing known compounds, with a view to predicting taxa with potential molluscicidal activity.

Data on the Monocotyledoneae (monocots) have been assembled in Table 2 and on the Dicotyledoneae (dicots) in Table 4. We have generally used the classification system of Engler (1964), and taxa are arranged alphabetically. According to Engler, there are about 53 monocot and 292 dicot families. We have included in Tables 2 and 4 the estimated total number of genera and species in each family for which data pertinent to plant molluscicides have been reported and the estimated number of species in each genus.

These figures place the tabulated data in proper perspective, making it apparent that the number of taxa tested for molluscicidal activity to date is very small and that no valid conclusions can be established relative to phylogenetic relationships of molluscicidal plants. Tables 2 and 4 also include a brief notation on the general geographic distribution of each family and genus. Data on both number of species and geographic distribution have been taken primarily from Willis and Airy Shaw (1973), with secondary reference to Engler (1964) if necessary.

Tables 2 and 4 represent a compilation of positive and negative data reported in the scientific literature from testing of plant extracts for molluscicidal activity. Most of the data were derived from references cited in Chapter 3. Data on plant part tested, type of extract employed, and species of snail tested are not included but can be found in Chapter 3. Test results were considered positive if a plant extract exerted any percentage kill at a concentration of 100 ppm or less, weak if more than 100 ppm of the extract were required for snail kill, and negative if there was no kill at any test level. If several parts of a plant were tested, the results recorded were for the most active test. Plants listed as not tested were included if the literature indicated that a chemical substance or substances of known structure (Table 1) had been detected in or isolated from the species, or if a chemical substance with a predictable molluscicidal effect – for example, monodesmosidic triterpene, monodesmosidic spirostanol saponins, or spirosolane glycosides (Alzérreca and Hart, 1982; Hostettmann, 1980; Hostettmann et al., 1978, 1982) – had been isolated from the species.

We have not been able to document every species known to contain one or more of the active or potentially active molluscicides listed in Table 1. However, we have consulted not only the NAPRALERT data base but also the major reviews on the occurrence of triterpene and steroid saponins (Agarwal and Rastogi, 1974; Basu and Rastogi, 1967; Boiteau et al., 1964; Chandel and Rastogi, 1980; Mahato et al., 1982; Tschesche and Wulff, 1973) and spirosolane glycoalkaloids (Roddick, 1974; Schreiber, 1968). Together, these sources should identify this type of information comprehensively to the present time.

Footnotes to Tables 2 and 4 point out the significance of terms used to designate the type of molluscicide present in designated species, when it is known. All compounds whose names are italicized have actual or predictable molluscicidal properties.

Monocot Taxa

It can be seen in Table 2 that of 54 monocot species reported tested for molluscicidal activity, 19 were active – five at 100 ppm or less and 14 at

Table 2. Monocots Tested for Molluscicidal Activity or Reported to Contain Active Molluscicides

Family/Genus/Species	Molluscicidal Activity[a]	Type of Molluscicide Present[b]	Geographic Distribution[c]
AGAVACEAE (20/670)[d]			Trop. & subtrop.
Agave (300)[e]			S. U.S. to trop. S. Am.
A. *americana* L.	±	*Steroid*	
A. *angustifolia* Haw.	−	None reported	
A. *cantala* Roxb.	nt	Steroid	
A. *fourcroyodes* Lem.	−	None reported	
A. *lecheguilla* Torrey	nt	Steroid	
A. *schottii* Engelm.	nt	Steroid	
A. *sisalana* Perrine	±	Steroid	
Beschorneria (10)			Mex.
B. *yuccoides* C. Koch	nt	*Steroid*	
Hesperaloe (2)			S. U.S.; Mex.
H. *parviflora* (Torrey) Coult.	nt	Steroid	
Polianthes (13)			Mex.; Trinidad
P. *tuberosa* L.	nt	Steroid	
Sansevieria (60)			Trop. & S. Afr.; Madag.; Arabia
S. *guineensis* Willd.	−	None reported	
S. *liberica* Gerome & Labroy	−	None reported	
S. *trifasciata* Prain	±	None reported	
Yucca (40)			U.S.; Mex.; W.I.
Y. *aloifolia* L.	−	None reported	
Y. *filamentosa* L.	nt	*Steroid*	
Y. *filifera* Engelm.	nt	*Steroid*	
Y. *gloriosa* L.	nt	*Steroid*	
Y. *pallida* McKelvey	±	None reported	
Y. *schidgera* Roezl ex Ortgies	±	None reported	
Y. *schottii* Engelm.	nt	*Steroid*	
ALISMATACEAE (13/90)			Cosmop.
Sagittaria (20)			Cosmop., esp. Am.
S. *lancifolia* L.	−	None reported	
AMARYLLIDACEAE (85/1,000)			U.S.; trop. or subtrop.
Amaryllis (1)			S. Afr.
A. *vittata* L'Herit.	±	None reported	
ARACEAE (115/2,000)			Trop. & temp.
Colocasia (8)			Indomal.; Polynes.
C. *esculenta* (L.) Schott	−	None reported	
Dieffenbachia (30)			Trop. Am.; W.I.

Table 2 *cont'd.*

Family/Genus/Species	Molluscicidal Activity[a]	Type of Molluscicide Present[b]	Geographic Distribution[c]
D. seguine (Jacquin) Schott	−	None reported	
Philodendron (275)			Warm Am.; W.I.
P. dubium Chod. & Visch.	−	None reported	
P. scandens ssp. *oxycardium*[f]	nt	*Alkenyl phenols*	
Pistia (1)			Trop. & subtrop.
P. stratiotes L.	±	None reported	
Syngonium (20)			C. & trop. S. Am.; W.I.
S. podophyllum Schott	−	None reported	
BROMELIACEAE (60/1,400)			Trop. Am.; W.I.
Bromelia (40)			Trop. Am.; W.I.
B. pyramidalis Sims	−	None reported	
B. sp.	−	None reported	
CANNACEAE (1/55)			Am.
Canna (55)			Trop. & subtrop. Am.
C. indica L.	±	*Triancontan-1-ol*	
C. sp.	−	None reported	
COMMELINACEAE (38/550)			Mostly trop. & subtrop.
Commelina (23)			Trop. & subtrop.
C. virginica L.	−	None reported	
Rhoeo (1)			Mex.; C. Am.; W.I.
R. discolor Hance	−	None reported	
CYPERACEAE (90/4,000)			Worldwide
Cyperus (55)			Trop. & warm temp.
C. alternifolius L.	−	None reported	
C. giganteus Vahl	−	None reported	
C. rotundus L.	±	*Triterpene*	
Eleocharis (200)			Cosmop.
E. interstincta (Vahl) R. & S.	−	None reported	
DIOSCOREACEAE (5/750)			Trop. & warm temp.
Dioscorea (600)			Trop. & subtrop.
D. aculeata L. (= *D. esculenta* (Lour.) Burk.)	−	None reported	

Cont'd.

Table 2 *cont'd.*

Family/Genus/Species	Molluscicidal Activity[a]	Type of Molluscicide Present[b]	Geographic Distribution[c]
D. caucasica Lipsky	nt	*Steroid*	
D. composita Hemsl.	nt	Steroid	
D. deltoidea Wallich	nt	*Steroid*	
D. floribunda Mart. & Gal.	nt	*Steroid*	
D. gracillima Miq.	nt	*Steroid*	
D. nipponica Makino	nt	*Steroid*	
D. polystachya Turcz.	nt	*Steroid*	
D. prazeri Prain & Burkill	nt	*Steroid*	
D. rotundata Poir.	±	None reported	
D. sativa L.	nt	*Steroid*	
D. septemloba Thunb.	nt	*Steroid*	
D. tenuipes Franch. & Savat.	nt	*Steroid*	
D. tokoro Makino ex Miyabe	−	*Steroid*	
D. zingiberensis C.H. Wright	nt	Steroid	
Rajania (25)			W.I.
R. cordata L.	−	None reported	
GRAMINEAE (620/ 10,000)			Worldwide
Cymbopogon (60)			Trop. & subtrop.;
C. nardus (L.) Rendle (= *Andropogon nardus*　　L.)	−	None reported	Afr. & As.
Cynodon (10)			Trop. & subtrop.
C. dactylon (L.) Pers.	−	None reported	
Eriochloa (20)			Trop. & subtrop.
E. polystachya HBK.	−	None reported	
Gynerium (1)			Mex. to subtrop. S. Am.
G. sagittatum Beauvois	−	None reported	
Melinis (18)			Trop. S. Am. & W.I. (1); trop. & S. Afr. & Madag. (17)
M. minutiflora Beauvois	±(V)	None reported	
Panicum (500)			Trop. & warm temp.
P. maximum Jacquin	−	None reported	
P. muticum Forsk. (= *Brachiaria mutica* (Forsk.) Stapf)	−	None reported	
Pennisetum (130)			Warm regions

Table 2 *cont'd.*

Family/Genus/Species	Molluscicidal Activity[a]	Type of Molluscicide Present[b]	Geographic Distribution[c]
Pennisetum (130)			Warm regions
P. *americanum* (L.) Leeke	–	None reported	
IRIDACEAE (60/800)			Trop. & temp.; chiefly S. Afr. & trop. Am.
Tigridia (12)			Mex. to Chile
T. *pavonia* Ker-Gawl.	–	None reported	
LILIACEAE (250/3,700)			Cosmop.
Allium (450)			N. hemisph.
A. *erubescens* K. Koch	nt	*Steroid*	
A. *giganteum* Regel	nt	Steroid	
A. *karataviense* Regel	nt	*Steroid*	
A. *narcissiflorum* Villars	nt	*Steroid*	
A. *turcomanicum* Regel	nt	*Steroid*	
Aloe (275)			Trop. & S. Afr.; Madag.; Arabia
A. *ferox* P. Miller	–	None reported	
A. *vera* (L.) Burm.f. (= A. *barbadensis* P.Miller)	–	None reported	
Anemarrhena (1)			N. China
A. *asphodeloides* Bge.	nt	*Steroid*	
Asparagus (300)			Old World, mostly in dry places
A. *adscendens* Roxb.	nt	*Steroid*	
A. *cochinchinensis* (Lour.) Merr.	nt	Steroid	
A. *curillus* Buch.-Ham. ex Roxb.	nt	*Steroid*	
A. *gonocladus* Baker	nt	Triterpene	
A. *officinalis* L.	nt	*Steroid*	
A. *racemosus* Willd.	nt	*Steroid*	
Aspidistra (8)			E. As.
A. *elatior* Bl.	nt	*Steroid*	
Chlorogalum (7)			Calif.
C. *pomeridianum* Kunth	nt	Steroid type	
Convallaria (2)			N. temp.
C. *keiskei* Miq.	nt	*Steroid*	
C. *majalis* L.	nt	*Steroid*	

Cont'd.

Table 2 *cont'd.*

Family/Genus/Species	Molluscicidal Activity[a]	Type of Molluscicide Present[b]	Geographic Distribution[c]
Dipcadi (55)			Medit.; Afr.; Madag.; Penins. India
D. fesoghlense Baker	+	None reported	
Eriospermum (80)			Afr.
E. abyssinicum Baker	±	None reported	
Fritillaria (85)			N. temp.
F. stenanthera (= Rhinopetalum stenantherum* Regel)[f]	nt	*Solanidine glycoside*	
Heloniopsis (4)			Japan; Formosa
H. orientialis (Thunb.) Koidz.	nt	*Steroid*	
Hosta (10)			China; Japan
H. ovata (= *Funkia ovata* Spreng.)[f]	nt	*Steroid*	
Metanarthecium (5)			Japan; Formosa
M. luteo-viride Maxim.	nt	*Steroid*	
Ophiopogon (20)			Himal. to Japan & Philipp. Is.
O. japonicus (L.f.) Ker-Gawl.	nt	*Steroid*	
Paris (20)			N. paleotemp.
P. formosana Hayata	nt	Steroid	
P. polyphylla J.E.Smith	nt	*Steroid*	
P. quadrifolia L.	nt	*Steroid*	
P. tetraphylla A.Gray	nt	*Steroid*	
P. verticillata M.-Bieb.	nt	Steroid	
Polygonatum (50)			N. temp.
P. latifolium Desf.	nt	*Steroid*	
P. multiflorum All.	nt	Steroid type	
P. stenophyllum Maxim.	nt	Steroid type	
Ruscus (3)			Madeira; W. & C. Eur.; Medit. to Iran
R. aculeatus L.	nt	*Steroid*	
R. hypoglossum L.	nt	Steroid	
R. hyrcanus Woronov	nt	*Steroid*	
Smilax (350)			Trop. & subtrop.
S. aristolochiaefolia P.Miller	nt	*Steroid*	
S. aspera L.	nt	*Steroid*	
S. excelsa L.	nt	*Steroid*	

Table 2 cont'd.

Family/Genus/Species	Molluscicidal Activity[a]	Type of Molluscicide Present[b]	Geographic Distribution[c]
Trillium (30)			W. Himal. to Japan & Kamchatka; N. Am.
T. erectum L.	nt	Steroid	
T. kamtschaticum Ledeb.	nt	Steroid	
T. smallii Maxim.	nt	Steroid	
T. tschonoskii Maxim.	nt	Steroid	
Veratrum (25)			N. temp.
V. nigrum L.	nt	Steroid	
MUSACEAE (2/42)			Trop. Afr.; As.; Australia
Heliconia (80)			Trop. Am.
H. psittacorum L.	−	None reported	
ORCHIDACEAE (735/ 17,000)			Cosmop.
Cymbidium (40)			Trop. As.; Australia
C. giganteum Wallich	nt	Triterpene	
Eulophia (200)			Pantrop.
E. guineensis Lindley	±	None reported	
PONTEDERIACEAE (7/ 30)			Trop.
Eichhornia (7)			S.E. U.S. to Argent.; W.I.
E. natans Solms	+	None reported	
TACCACEAE (2/31)			Trop.
Tacca (30)			Trop., esp. S.E. As.
T. cheancer[f]	nt	Steroid	
TYPHACEAE (1/10)			Temp. & trop.
Typha (10)			Temp. & trop.
T. domingensis Pers.	−	None reported	
XYRIDACEAE (4/270)			Trop. & subtrop., mostly Am.
Xyris (250)			Trop. & subtrop.
X. anceps Lamk.	+	None reported	
ZINGIBERACEAE (49/ 1,500)			Trop., chiefly Indomal.
Alpinia (250)			Warm As.; Polynes.
A. purpurata K.Schum.	−	None reported	
A. speciosa K.Schum.	−	None reported	
Costus (150)			Trop.
C. speciosus (Koenig) Smith	nt	Steroid	

Cont'd.

Table 2 *cont'd.*

Family/Genus/Species	Molluscicidal Activity[a]	Type of Molluscicide Present[b]	Geographic Distribution[c]
Hedychium (50)			Madag.; Indomal.; S.W. China
H. coronarium Koenig	+	None reported	
H. gardnerianum Roscoe	+	*Triterpene*	
Zingiber (90)			E. As.; Indomal.; N. Australia
Z. officinale Roscoe	±	None reported	

[a]+ = Test extract reported active at 100 ppm or less; ± = active at more than 100 ppm; − = inactive; nt = extract not tested but potentially active due to the type of chemical constituent known to be present.
[b]*Triterpene* (monodesmosidic), triterpene (bidesmosidic), triterpene type (structure not elucidated to the point where sugar chains and their attachment are known), *steroid* (monodesmosidic spirostanol type), steroid (bidesmosidic spirostanol type), steroid type (structure not elucidated to the point where sugar chains and their attachment are known), *steroid alkaloid* (monodesmosidic), steroid alkaloid (aglycone), other terms used are self-explanatory (if italicized, the substance indicated has molluscicidal properties); (V) = water saturated with essential oil from the plant was tested.
[c]General geographic distribution for the family or genus according to Willis and Airy Shaw (1973) or Engler (1964); Medit. includes the entire Mediterranean basin, S. Eur. from Portugal to the Balkan Peninsula, Asia Minor, Palestine, and N. Afr. from Morocco to Egypt; E. As. covers China, Korea, and Japan; S.E. As. covers Burma, Thailand, and Indochina (Vietnam, Laos, and Cambodia) and often the adjacent parts of S. China from Yunnan to Hong Kong and Hainan; Indomal. covers the area from tropical India east to New Guinea; C. Am. covers Guatemala to Panama. Other geographic regions should be self-explanatory.
[d]Number of genera/species in the family according to Willis and Airy Shaw or Engler.
[e]Number of species in the genus according to Willis and Airy Shaw or Engler.
[f]The validity of this binomial has not been confirmed.

more than 100 ppm. An additional 50 monocot species not tested for molluscicidal activity are known to contain active molluscicidal constituents. All but three of the 50 contain monodesmosidic triterpene or spirostanol saponins or both. These data, though limited, are impressive. But they are not surprising, since it is generally believed that steroid saponins occur in plants always as glycosides, never as the free aglycone (Farnsworth, 1966), and the monodesmosidic types are encountered more frequently than the bidesmosidic types.

Of particular interest is the frequent occurrence of monodesmosidic spirostanol saponins in the genera *Allium*, *Dioscorea*, *Paris*, *Ruscus*, *Smilax*, and *Trillium*; it is highly probable that species of these genera have molluscicidal activity.

A resume of the most promising monocot taxa is presented in Table 3.

Table 3. Monocot Families Promising for Molluscicidal Activity

Family	Known Species (Estimated)	Active Species vs. Species Tested	Not Tested but Predicted Active Species
Agavaceae	670	5/10	5
Cannaceae	55	1/2	1
Cyperaceae	4,000	1/3	1
Dioscoreaceae	750	1/4	11
Liliaceae	3,700	2/4	31
Orchidaceae	17,000	1/1	1
Xyridaceae	270	1/1	0
Zingiberaceae	1,500	3/5	2

Dicot Taxa

Data presented in Table 4 are more complex than those presented in Table 2. Some 512 species of dicots have been tested for molluscicidal activity, of which 108 were active, 161 weakly active, and 241 inactive; one tested both active and inactive, and one tested both weakly active and inactive. The 512 species represent only 91 of the estimated 292 dicot families. Some 384 dicot species not tested but with known active molluscicidal constituents were identified.

Thirty-seven dicot families seem most interesting as sources of molluscicides, based on the incomplete data at hand (Table 5). Caution must be used if one extrapolates data such as these, since they represent only a small fraction of the known species from some of the most accessible taxa. Also, toxicologic problems could preclude interest in several families that on the surface may appear interesting.

Table 4. Dicots Tested for Molluscicidal Activity or Reported to Contain Active Molluscicides.

Family/Genus/Species	Molluscicidal Activity[a]	Type of Molluscicide Present[b]	Geographic Distribution[c]
ACANTHACEAE (250/ 2,500)[d]			Trop. As., Afr., Am.; also Medit., U.S., Australia
Acanthus (50)[e]			S. Eur.; trop. & subtrop. As.; Afr.
A. *ilicifolius* L.	−	*Triterpene*	
Brillantaisia (40)			Trop. Afr.; Madag.
B. *vogeliana* Benth.	+	None reported	
Crossandra (50)			Trop. Afr.; Madag.; Saudi Arabia
C. *flava* Hook.	+	None reported	
C. *infundibuliformis* Nees	−	None reported	
Justicia (300)			Trop. & subtrop. Afr.; Madag.
J. *simplex* D.Don	nt	*Triterpene*	
J. *verticillata*[f]	−	None reported	
Lankesteria (7)			Trop. Afr.; Madag.
L. *elegans* T.Anders.	±	None reported	
Ruellia (5)			Trop. & subtrop. Am.
R. *equisetiformis*[f]	−	None reported	
Stethoma (5)			Trop. S. Am.
S. *pectoralis* Rafin.	−	None reported	
ACERACEAE (3/200)			N. temp. & trop. mts.
Acer (200)			N. temp. (esp. in hill districts) & trop. mts.; many in China & Japan; few S.E. As.; 1 in W. Malaysia
A. *negundo* L.	nt	Triterpene type	
AIZOACEAE (130/1,200)			Esp. S. Afr.; also trop. Afr. & As.; Australia; W. U.S.; S. Am.
Glinus (12)			Trop. & subtrop.
G. *lotoides* L.	nt	*Triterpene*	
AMARANTHACEAE (65/ 850)			Trop. & temp.
Achyranthes (100)			Trop. & subtrop.; mostly Afr. & As.
A. *aspera* L.	−	*Triterpene*	
A. *bidentata* Bl.	±	Triterpene type	
Amaranthus (60)			Trop. & temp.
A. *paniculatus* L.	−	None reported	
A. *spinosus* L.	nt	*Triterpene*	

Table 4 *cont'd.*

Family/Genus/Species	Molluscicidal Activity[a]	Type of Molluscicide Present[b]	Geographic Distribution[c]
ANACARDIACEAE (60/600)			Chiefly trop.; also Medit.; E. As., Am.
Anacardium (15)			Trop. Am.
A. *occidentale* L.	+	*Anacardic acids (alkenyl phenols)*	
Astronium (15)			C. & trop. S. Am.; W.I.
A. *fraxinifolium* Schott	±	None reported	
Gluta (13)			Madag. (1); Indomal. (12)
G. *renghas* L.	nt	*Alkenyl phenols*	
Holigarna (8)			Indomal.
H. *arnottiana* Hook.f.	nt	*Alkenyl phenols*	
Mangifera (40)			S.E. As.; Indomal.
M. *indica* L.	−	None reported	
Melanorrhoea (20)			S.E. As.; Malay Penins.; Borneo
M. *usitata* Wallich	nt	*Alkenyl phenols*	
Metopium (3)			Florida; Mex., W.I.
M. *toxiferum* (L.) Krug & Urban	nt	*Alkenyl phenols*	
Pentaspadon (5)			Malaysia; Solomon Is.
P. *motleyi* Hook.f.	nt	*Alkenyl phenols*	
P. *officinalis* Holmes ex King	nt	*Alkenyl phenols*	
Schinus (30)			Mex. to Argent.
S. *terebinthifolius* Raddi	−(V)	None reported	
Semecarpus (50)			Indomal.; Micrones.; Solomon Is.
S. *anacardium* L.f.	nt	*Alkenyl phenols*	
S. *heterophylla* Bl.	nt	*Alkenyl phenols*	
S. *travancorica* Bedd.	nt	*Alkenyl phenols*	
Smodingium (1)			S. Afr.
S. *argutum* E. Meyer	nt	*Alkenyl phenols*	
Spondias (12)			Indomal.; S.E. As.; trop. Am.
S. *lutea* L.	±	None reported	
Tapirira (15)			Mex. to S. Am.
T. *guianensis* Aubl.	+	None reported	
Toxicodendron (15)			E. As.; N. & S. Am.

Cont'd.

Table 4 *cont'd.*

Family/Genus/Species	Molluscicidal Activity[a]	Type of Molluscicide Present[b]	Geographic Distribution[c]
T. diversilobum (T. & G.) Greene	nt	*Alkenyl phenols*	
T. radicans (L.) Kuntze *T. striatum*[f]	±	*Alkenyl phenols*	
	nt	*Alkenyl phenols*	
T. succedaneum (L.) Kuntze	nt	*Alkenyl phenols*	
T. vernicifluum (Stokes) Barkley	nt	*Alkenyl phenols*	
T. vernix (L.) Kuntze	nt	*Alkenyl phenols*	
ANNONACEAE (120/ 2,100)			Chiefly trop; esp. Old World
Annona (120)			Warm, esp. Am.
A. *glabra* L.	−	None reported	
A. *muricata* L.	−	None reported	
A. *senegalensis* Pers.	+	*ent−Kaur−16−e n−19−oic acid*	
A. sp.	−	None reported	
A. sp.	±	None reported	
A. *squamosa* L.	+	None reported	
Cleistopholis (5)			Trop. W. Afr.
C. *patens* Engl. & Diels	±	None reported	
Xylopia (150)			Trop., esp. Afr.
X. *quintasii* Pierre ex Engl. & Diels	−	None reported	
APOCYNACEAE (180/ 1,500)			Mostly trop., a few temp.
Allamanda (15)			Trop. S. Am.; W.I.
A. *cathartica* L.	+	None reported	
Anodendron (20)			Sri Lanka; Japan; Formosa; Hainan; Malaysia; Solomon Is.
A. *affine* Druce	nt	*Sterol glycoside*	
Hunteria (16)			Trop. Afr.; S. India; Sri Lanka; Andam.; S. China; S.E. As.; Malay Penins.; Anambas Is.

Table 4 *cont'd.*

Family/Genus/Species	Molluscicidal Activity[a]	Type of Molluscicide Present[b]	Geographic Distribution[c]
H. umbellata Hallier f.	±	None reported	
Ichnocarpus (18)			S. China; Indomal.
I. frutescens R.Br.	nt	*Triterpene*	
Nerium (3)			Medit. to Japan
N. oleander L.	−	None reported	
Peschiera (25)			Trop. Am.; W.I.
P. affinis[f]	±	None reported	
Plumeria (7)			Warm Am.
P. bracteata A.DC.	±	None reported	
Rauvolfia (100)			Trop.
R. caffra Sonder	+	None reported	
R. ternifolia[f]	±	None reported	
Strophanthus (60)			Trop. Afr., Madag.; Indomal.
S. sarmentosus DC.	−	None reported	
Voacanga (25)			Trop. Afr.; Madag.; Malaysia
V. africana Stapf ex S.Elliot	+	None reported	
AQUIFOLIA-CEAE (3/500)			Trop. & subtrop.
Ilex (400)			Cosmop. (exc. N. Am.)
I. cornuta Lindl. & Paxt.	−	*Triterpene*	
I. latifolia Thunb.	−	*Triterpene*	
ARALIACEAE (55/700)			Chiefly trop.; esp. Indomal.; trop. Am.
Aralia (35)			Indomal.; E. As.; N. Am.
A. balfouriana Hort. ex André	−	None reported	
A. mandschurica Rupr.	nt	*Triterpene*	

Cont'd.

Table 4 *cont'd.*

Family/Genus/Species	Molluscicidal Activity[a]	Type of Molluscicide Present[b]	Geographic Distribution[c]
Eleutherococcus (15)			Himal. to Japan
E. senticosus (Rupr. & Maxim.) Maxim.	nt	*Triterpene*	
Fatsia (2)			Japan; Formosa
F. japonica Decne. & Planch.	nt	*Triterpene*	
Hedera (15)			Canary Is.; W. & C. Eur.; Medit. to Himal.; Japan; Queensl.
H. caucasigena[f]	nt	*Triterpene*	
H. colchica K. Koch	nt	*Triterpene*	
H. helix L.	+	*Hederagenin−3 −O−β−D− glucopyranoside Hederagenin−3 −O−β−D− glucopyranosyl− (1→2)−β−D− glucopyranoside Hederagenin−3 −O−α−L− rhamnopyrano- syl−(1→2)−α− L−arabinopy- ranoside*	
H. pastuchovii Woronov ex Grossheim	nt	*Triterpene*	
H. rhombea Sieb. & Zucc.	nt	*Triterpene*	
Kalopanax (1)			E. As.
K. septemlobus (Thunb.) Koidz.	nt	*Triterpene*	
Panax (8)			Trop. & E. As.; N. Am.
P. ginseng C.A. Meyer	nt	*Triterpene*	
P. japonicus (T. Nees) C.A. Meyer	nt	*Triterpene*	
P. pseudoginseng Wallich	nt	*Triterpene*	

Table 4 *cont'd.*

Family/Genus/Species	Molluscicidal Activity[a]	Type of Molluscicide Present[b]	Geographic Distribution[c]
P. pseudoginseng Wallich var. *notoginseng* (Burk.) Hoo & Tseng	nt	*Triterpene*	
P. quinquefolius L.	nt	*Triterpene*	
P. repens Maxim.	nt	*Triterpene*	
P. trifolius L.	nt	*Triterpene*	
Polyscias (80)			Paleotrop.
P. guilfoylei (Cogn. & Marchal) L.H. Bailey	+	None reported	
Tetrapanax (1)			S. China; Formosa
T. papyrifer K. Koch	nt	*Triterpene*	
Schefflera (200)			Trop. & subtrop.
S. capitata (C.B. Clarke) Harms	nt	*Triterpene*	
ASCLEPIADA- CEAE (130/2,000)			Trop. & subtrop.; rare elsewhere
Asclepias (120)			Am., esp. U.S.
A. curassavica L.	+	None reported	
Calotropis (6)			Trop. Afr.; As.
C. gigantea (L.)R.Br.	±	None reported	
Cryptostegia (2)			Madag.
C. grandiflora R.Br.	+	None reported	
BALSAMINA- CEAE (5/600)			Euras.; Afr.; N. Am.
Impatiens (500)			Trop. & temp. Euras. & Afr.; Madag.; India; N. & C. Am.
I. balsamina L.	–	None reported	
I. sultani Hook.f.	–	None reported	
BEGONIACEAE (5/920)			Trop.
Begonia (900)			Trop. & subtrop., esp. Am.
B. sp.	–	None reported	

Cont'd.

Table 4 *cont'd.*

Family/Genus/Species	Molluscicidal Activity[a]	Type of Molluscicide Present[b]	Geographic Distribution[c]
BERBERIDA-CEAE (16/650)			N. temp.; trop. mts.; S. Am.
Caulophyllum (2)			N.E. As.; N. Am.
C. *robustum* Maxim.	nt	*Triterpene*	
C. *thalictroides* (L.) Michx.	nt	*Triterpene*	
Leontice (7)			S.E. Eur. & E. As.
L. *eversmannii* Bge.	nt	*Triterpene*	
BETULACEAE (2/95)			N. temp.; trop. mts.; Andes; Argent.
Alnus (35)			N. temp.; S. to Assam & Indoch. & Andes
A. *serrulatoides*[f]	nt	*Triterpene*	
BIGNONIACEAE (120/650)			Trop.; a few temp.
Crescentia (5)			Trop. Am.
C. *cujete* L.	−	None reported	
Kigelia (10)			Trop. Afr. & Madag.
K. *africana* (Lamk.) Benth.	±	None reported	
Tabebuia (100)			Mex. to Argent.; W.I.
T. *caraiba* (Mart.) Bureau	±	None reported	
BIXACEAE (1/1)			Trop. Am.
Bixa (1)			Trop. Am.
B. *orellana* L.	−	None reported	
BOMBACACEAE (20/180)			Esp. Am.
Bombax (18)			Trop. Afr.; As.
B. *costatum* Pellegrin & Vuillet	+	None reported	
Montezuma (1)			W.I. (Puerto Rico)
M. *speciosissima* Sessé & Moc.	−	None reported	
Pachira (2)			Trop. Am.
P. *aquatica* Aubl.	−	None reported	
BORAGINA-CEAE (100/2,000)			Trop. & temp.; esp. Medit.
Anchusa (50)			Eur.; N. Afr.; W. As.
A. *officinalis* L.	nt	*Triterpene*	

Table 4 *cont'd.*

Family/Genus/Species	Molluscicidal Activity[a]	Type of Molluscicide Present[b]	Geographic Distribution[c]
Bourreria (50)			Warm Am.; W.I.
B. *succulenta* Jacquin	–	None reported	
Cordia (250)			Warm regions
C. *corymbosa*[f]	–	None reported	
C. *obliqua*[f]	nt	*Triterpene*	
Heliotropium (250)			Trop. & temp.
H. *indicum* L.	–	None reported	
Tournefortia (150)			Trop. & subtrop.
T. *hirsutissima* L.	–	None reported	
BURSERACEAE (16/500)			Trop.
Bursera (80)			Trop. Am.
B. *simaruba* (L.) Sarg.	±	None reported	
Dacryodes (30)			Trop.
D. *edulis* (G.Don) H.J. Lam	–	None reported	
Protium (90)			Madag. to Malaysia & trop. Am.
P. *heptaphyllum* March.	±(V)	None reported	
P. *sp.*	–(V)	None reported	
CACTACEAE (150/2,000)			Esp. trop. Am., reaching Brit. Columbia & Patagonia; Andes (to 3,000 m)
Cactus (40)			C. & trop. S. Am.; W.I.
C. *sp.*	–	None reported	
Hylocereus (23)			Mex. to Peru
H. *trigonus* Safford	–	None reported	
Pereskia (20)			Mex. to trop. S. Am.; W.I.
P. *grandifolia* Haworth	nt	*Triterpene*	
CAMPANULA-CEAE (70/2,000)			Temp. & subtrop.; trop. mts.
Lobelia (250)			Mostly trop. & subtrop., esp. Am.
L. *longiflora* A.DC.	–	None reported	

Cont'd.

Table 4 *cont'd.*

Family/Genus/Species	Molluscicidal Activity[a]	Type of Molluscicide Present[b]	Geographic Distribution[c]
Platycodon (1)			N.E. As.
P. *grandiflorum* DC.	nt	*Triterpene*	
CANELLACEAE (5/16)			S. Am.; W.I.; E. Afr.; Madag.; with marked discontinuity
Warburgia (3)			Trop. E. Afr.
W. *stuhlmannii* Engl.	+	*Muzigadial; warburganal*	
W. *ugandensis* Sprague	+	*Muzigadial; warburganal*	
CANNABACEAE (2/3)			N. temp.
Cannabis (1)			C. As.
C. *sativa* L.	±	None reported	
CAPPARIDA-CEAE (46/800)			Trop. & warm temp.
Cleome (150)			Trop. & subtrop.
C. *aculeata* L.	−	None reported	
C. sp.	−	None reported	
C. *spinosa* Jacquin	−	None reported	
Crataeva (9)			Trop., exc. Australia & N. Caled.
C. *tapia* L.	−	None reported	
CARICACEAE (4/55)			Trop. Am.; Afr.
Carica (45)			Warm Am.
C. *papaya* L.	−	None reported	
CARYOPHYLLA-CEAE (70/1,750)			Cosmop.
Acanthophyllum (150)			S.W. & C. As.; Siber.
A. *gypsophiloides* Regel	nt	*Triterpene*	
A. *paniculatum* Regel & Herder	nt	*Triterpene*	
Agrostemma (2)			Euras.
A. *githago* L.	nt	*Triterpene*	
Dianthus (300)			Euras.; Afr., esp Medit.
D. *barbatus* L.	nt	*Triterpene*	
D. *deltoides* L.	nt	*Triterpene*	
D. *superbus* L. var. *longicalicinus*	nt	*Triterpene*	
Gypsophila (125)			Temp. Euras. (esp. E. Medit.); S. Egypt; 1 Australia; N.Z.

Table 4 *cont'd.*

Family/Genus/Species	Molluscicidal Activity[a]	Type of Molluscicide Present[b]	Geographic Distribution[c]
G. acutifolia Steven ex Spreng.	nt	Triterpene type	
G. bicolor Grossheim	nt	Triterpene type	
G. pacifica Komarov	nt	*Triterpene*	
G. paniculata L.	nt	*Triterpene*	
G. patrinii Ser.	nt	*Triterpene*	
G. struthium L.	nt	*Triterpene*	
G. trichotoma Wenderoth	nt	*Triterpene*	
Herniaria (35)			Eur., Medit. to Afghan.; S. Afr.
H. glabra L.	nt	*Triterpene*	
Lychnis (12)			Temp. Euras.
L. viscaria L.	nt	Triterpene type	
Saponaria (30)			Temp. Euras.; chiefly Medit.
S. officinalis L.	nt	*Triterpene*	
Silene (500)			N. temp., esp. Medit.
S. nutans L.	nt	*Triterpene*	
Spergularia (40).			Cosmop.; mostly halophytes
S. arbuscula I.M.Johnston	nt	Triterpene type	
Vaccaria (4)			C. & E. Eur.; Medit.; temp. As.
V. segetalis (Neck.) Garcke	nt	Triterpene type	
Viscago (=Silene) (500)			N. temp., esp. Medit.
V. behen Hornem. (= *Silene behen* L.)	nt	*Triterpene*	
V. viscosa Moench (= *Silene viscosa* Pers.)	nt	Triterpene type	
CASUARINA-CEAE (2/65)			Mascarenes; S.E. As. to N.E. Australia & Polynes.

Cont'd.

Table 4 *cont'd.*

Family/Genus/Species	Molluscicidal Activity[a]	Type of Molluscicide Present[b]	Geographic Distribution[c]
Casuarina (45)			E. Afr.; Mascarenes; S.E. As.; Malaysia; Australia; Polynes.
C. equisetifolia J.R. & G.Forst.	+	None reported	
CELASTRA-CEAE (55/850)			Trop. & temp.
Hippocratea (1)			S.E. U.S. & C. Mex. to trop. S. Am.; W.I.
H. volubilis[f]	−	None reported	
CHENOPODIA-CEAE (102/1,400)			Halophytic regions, worldwide
Beta (6)			Eur.; Medit.
B. vulgaris L.	nt	*Triterpene*	
Chenopodium (150)			Temp.
C. ambrosioides L. (= *C. anthelminticum* L.)	±	*Triterpene*	
Spinacia (3)			E. Medit. to C. As. & Afghan.
S. oleracea L.	nt	*Triterpene*	
CHRYSOBA-LANACEAE (10/400)			Trop. & subtrop.
Acioa (33)			N.E. S. Am. (3); trop. Afr. (30)
A. barteri Engl.	+	None reported	
A. rudatisii Wildem.	+	None reported	
Hirtella (95)			C. & trop. S. Am.; W.I.; trop. E. Afr.; Madag.
H. americana L.	±	None reported	
Licania (136)			Primarily S.E. U.S. to trop. S. Am.
L. tomentosa Fritsch. (= *Moquilea tomentosa* Benth.)	+	None reported	
COCHLOSPERM-ACEAE (2/25)			Trop.
Cochlospermum (20)			Trop., mostly xeroph.

Table 4 *cont'd.*

Family/Genus/Species	Molluscicidal Activity[a]	Type of Molluscicide Present[b]	Geographic Distribution[c]
C. insigne St. Hil.	+(V)	None reported	
C. regium Pilger	−(V)	None reported	
COMBRETA- CEAE (20/600)			Trop. & subtrop.
Buchenavia (20)			Trop. S. Am.; W.I.
B. capitata Eichl.	−	None reported	
Bucida (4)			S. Florida; C. Am.; W.I.
B. buceras Vellozo	−	None reported	
Combretum (250)			Trop., exc. Australia
C. dolichopetalum Engl. & Diels	±	None reported	
C. fragrans F. Hoffm. (= *C. ghasalense* Engl. & Diels)	+	None reported	
C. glutinosum Guillem. & Perrot.	−	None reported	
C. leprosum Mart.	±	None reported	
C. micranthum G.Don	−	None reported	
C. molle R.Br. ex G.Don	nt	*Triterpene*	
Terminalia (250)			Trop.
T. arjuna Wight & Arnott	nt	*Triterpene*	
T. avicennioides Guillem. & Perrot.	±	None reported	
T. catappa L.	±	None reported	
T. chebula (Gaertn.) Retz.	±	None reported	
T. macroptera Guillem. & Perrot.	±	None reported	
T. mollis M. Laws.	+	None reported	
COMPOSITAE (900/ 13,000)			Cosmop.

Cont'd.

Table 4 *cont'd.*

Family/Genus/Species	Molluscicidal Activity[a]	Type of Molluscicide Present[b]	Geographic Distribution[c]
Ageratum (60)			Trop. Am.
A. conyzoides L.	−(V)	None reported	
Ambrosia (40)			Cosmop., mostly Am.
A. ambrosioides (Cav.) Payne (= *Franseria ambrosioides* Cav.)	nt	*Ambrosin, damsin*	
A. arborescens Lamk.	nt	*Damsin*	
A. artemisiifolia L.	nt	*Damsin*	
A. chenopodiifolia L.	nt	*Damsin*	
A. cumanensis HBK.	nt	*Ambrosin, damsin*	
A. deltoidea (Torrey) Payne	nt	*Damsin*	
A. dumosa (= *Franseria dumosa*)[f]	nt	*Ambrosin*	
A. hispida Pursh	nt	*Ambrosin, damsin*	
A. jamaicensis[f]	nt	*Ambrosin, damsin*	
A. maritima L.	+	*Ambrosin, damsin*	
A. psilostachya DC.	nt	*Ambrosin, damsin*	
A. senegalensis DC.	+	None reported	
Artemisia (400)			N. temp.; S. Afr.; S. Am.
A. kurramensis Qazilb.	±	None reported	
Baccharis (400)			Am., esp. campos
B. trimera DC.	±	*Flavonoid; diterpene lactone*	
Bidens (230)			Cosmop.
B. bipinnata L.	±(V)	None reported	
B. pilosa L.	−	None reported	
Calendula (30)			Medit. to Iran
C. erecta[f]	−	None reported	
C. officinalis L.	nt	*Triterpene*	
Carthamus (13)			Medit.; Afr.; As.
C. tinctorius L.	±	None reported	

Table 4 *cont'd.*

Family/Genus/Species	Molluscicidal Activity[a]	Type of Molluscicide Present[b]	Geographic Distribution[c]
Chrysanthellum (5)			Trop.
C. procumbens A.Richard	nt	*Triterpene*	
Elephantopus (32)			Trop.
E. scaber L.	±(V)	None reported	
Eupatorium (1,200)			Mostly Am., a few in Eur.; As.; Afr.
E. odoratum L.	+	None reported	
E. sp.	±(V)	None reported	
Helianthus (110)			Am.
H. annuus L.	nt	*Triterpene*	
Heliopsis (12)			Am.
H. longipes	+	*Affinin*	
Hymenoclea (4)			S.W. U.S.; Mex.
H. monogyra T. & G. ex A.Gray	nt	*Ambrosin*	
H. salsola T. & G. ex A.Gray	nt	*Ambrosin*	
Iva (15)			N. & C. Am.; W.I.
I. xanthifolia Nutt.	nt	*Ambrosin*	
Mikania (250)			Trop. Am.; W.I.; 2 spp. S. Afr.
M. cordifolia Willd.	–	None reported	
M. fragilis Urban	–	None reported	
Parthenium (15)			Am.; W.I.
P. bipinnatifidum (Ortega) Rollins	nt	*Ambrosin, damsin*	
P. hysterophorus L.	–	*Triterpene, ambrosin, damsin*	
P. incanum HBK.	nt	*Ambrosin, damsin*	
Pectis (70)			S. U.S. to Brazil; W.I.; Galapagos
P. apodocephala Baker	±(V)	None reported	
Pluchea (50)			Trop. & subtrop.
P. odorata Cass.	–	None reported	

Cont'd.

Table 4 cont'd.

Family/Genus/Species	Molluscicidal Activity[a]	Type of Molluscicide Present[b]	Geographic Distribution[c]
Pseudelephantopus (3)			C. & trop. S. Am.
P. spicatus Rohr	−	None reported	
Spilanthes (60)			Am.; Afr.; Malaysia; N. Australia
S. oleracea L.	+	Affinin	
Tagetes (50)			Warm Am.
T. patula L.	−	None reported	
Verbesina (150)			Warm Am.
V. diversifolia DC.	−(V)	None reported	
Vernonia (100)			Am.; Afr.; As; Australia
V. plumbaginifolia Fenzl	−	None reported	
Wedelia (70)			Trop. and warm temp
W. parviceps S.F. Blake	+	Affinin	
W. scaberrima Benth.	−	None reported	
W. trilobata Hitchc.	−	None reported	
CONVOLVULA- CEAE (55/1,560)			Trop. & temp.
Calonyction (16)			Neotrop.
C. album House	+	None reported	
Ipomoea (500)			Trop. & warm temp.
I. bona-nox L.	−	None reported	
I. fistulosa Mart. ex Choisy	−	None reported	
I. rosea[f]	−	None reported	
I. tiliacea (Willd.) Choisy	−	None reported	
I. triloba L.	−	None reported	
Operculina (25)			Trop.
O. macrocarpa Urban	±	None reported	
CORNACEAE (12/100)			N. & S. temp. & trop. mts.
Cornus (4)			C. & S. Eur. to Cauc.; E. As.; Calif.

Table 4 *cont'd.*

Family/Genus/Species	Molluscicidal Activity[a]	Type of Molluscicide Present[b]	Geographic Distribution[c]
C. florida L.	+	*Sarsapogenin−3 −O−β−D− xylopyranosyl− (1→2)−β−D− galactopyranoside Sarsapogenin−3 −O−β−D− glucopyranosyl −(1→2)−β−D− galactopyranoside Sarsapogenin−3 −O−β−D− galactopyranoside*	
CRASSULA-CEAE (35/1,500)			Cosmop.; chiefly S. Afr.
Crassula (300)			Cosmop.; esp. S. Afr.
C. sp.	−	None reported	
Kalanchoe (100)			Trop. & S. Afr. to China & Java; 1 trop. S. Am.
K. pinnata Pers. (= *Bryophyllum pinnatum* Kurz)	−	None reported	
CUCURBITA-CEAE (110/640)			Primarily trop.
Acanthosicyos (2)			S. trop. Afr.
A. horrida Welw.	nt	*Triterpene*	
Bryonia (4)			Eur.; As.; N. Afr.; Canary Is.
B. alba L.	nt	*Triterpene*	
B. dioica Jacquin	nt	*Triterpene*	
Cayaponia (47)			Warm Am.; 1 trop. W. Afr. & Madag.; 1 Java
C. americana Cogn.	−	None reported	
Citrullus (3)			Afr.; Medit.; trop. As.
C. colocynthis (L.) Schrad.	nt	*Triterpene*	

Cont'd.

Table 4 *cont'd.*

Family/Genus/Species	Molluscicidal Activity[a]	Type of Molluscicide Present[b]	Geographic Distribution[c]
C. ecirrhosus Cogn.	nt	*Triterpene*	
C. lanatus (Thunb.) Matsum. & Nakai	nt	*Triterpene*	
C. naudinianus Hook.f	nt	*Triterpene*	
Coccinia (30)			Trop. & S. Afr.; 1 trop. India & Malaysia
C. hirtella Cogn.	nt	*Triterpene*	
Cucumis (25)			Mostly Afr.; few As.; 1 introd. trop. Am.
C. quinqueloba[f]	nt	*Triterpene*	
C. sativus L.	nt	*Triterpene*	
Cucurbita (15)			Am.
C. foetidissima HBK.	±	*Triterpene*	
C. hemsleya[f]	nt	*Triterpene*	
C. pepo L.	−(sd)	*Triterpene*	
Ecballium (1)			Medit.
E. elaterium (L.) A.Richard	nt	*Triterpene*	
Echinocystis (15)			Am.
E. fabacea Naudin	nt	*Triterpene*	
E. lobata T. & G.	nt	*Triterpene*	
E. wrightii Cogn.	nt	*Triterpene*	
Gynostemma (2)			E. As.; Indomal.
G. pentaphylla Makino	nt	*Triterpene*	
Hemsleya (1)			E. Himal.; China
H. amabilis Diels	nt	*Triterpene*	
Luffa (16)			Trop.
L. aegyptiaca P.Miller	nt	*Triterpene*	
L. operculata (L.) Cogn.	+	None reported	
Momordica (45)			Paleotrop.
M. charantia L.	±	*Triterpene*	
M. cochinchinensis (Lour.) Spreng.	nt	*Triterpene*	

Table 4 *cont'd.*

Family/Genus/Species	Molluscicidal Activity[a]	Type of Molluscicide Present[b]	Geographic Distribution[c]
M. grosvenori Swingle	nt	*Triterpene*	
Peponium (20)			Afr.; Madag.
P. mackenii Engl.	nt	*Triterpene*	
Wilbrandia (2)			Trop. S. Am.
W. sp.	±	None reported	
DATISCACEAE (1/2)			Dry W. Euras. & dry N. Am.; Medit.
Datisca (2)			to Himal. & C. As. (1); S.W. U.S. & N.W. Mex. (1)
D. glomerata Baill.	nt	*Triterpene*	
DILLENIACEAE (10/400)			Trop. & subtrop.
Curatella (2)			Trop. Am., W.I.
C. americana L.	+	None reported	
DIPSACACEAE (8/150)			Chiefly temp. Euras. & trop. & S. Afr.
Dipsacus (15)			Euras.; Medit.; trop. Afr.
D. azureus Schrenk	nt	Triterpene type	
Scabiosa (100)			Temp. Euras.; Medit.; mts. E. Afr.
S. songarica Schrenk	nt	*Triterpene*	
EMPETRACEAE (3/10)			N. temp., Andes; Falkl. Is.; Tristan
Empetrum (2)			N. temp. & Arctic; S. Andes; Falkl. Is.; Tristan; on moors
E. sibiricum Vassiliev	nt	*Triterpene*	
ERICACEAE (50/1,350)			Cosmop.; exc. in deserts; usu. confined to high altitude regions in tropics; almost absent from Australasia

Cont'd.

Table 4 *cont'd.*

Family/Genus/Species	Molluscicidal Activity[a]	Type of Molluscicide Present[b]	Geographic Distribution[c]
Gaultheria (200)			Circumpacif. (W. to W. Himal. & S. India; E. N. Am.; 8 in E. Brazil
G. procumbens L.	+	None reported	
Lyonia (30)			Himal.; E. As.; N. Am.; W.I.
L. ovalifolia Hort. ex Gard. var. *elliptica*	nt	*Triterpene*	
EUPHORBIA-CEAE (300/5,000)			Cosmop.; except Arctic
Acalypha (450)			Trop. & subtrop.
A. hispida Burm.f.	−	None reported	
A. ornata Hochst. ex A. Richard	±	None reported	
Alchornea (70)			Trop.
A. floribunda Muell.−Arg.	−	None reported	
Antidesma (170)			Old World trop. & subtrop.
A. laciniatum Muell.−Arg.	−	None reported	
Bridelia (60)			Afr.; As.
B. atroviridis Muell.−Arg.	+	None reported	
Codiaeum (15)			Malaysia, Polynes.; N. Australia
C. variegatum (L.) Bl.	±	None reported	
Croton (750)			Trop. & subtrop.
C. compressus Lamk.	±(V)	None reported	
C. macrostachys Hochst. ex. A.Richard	+	None reported	
C. micans Swartz	±(V)	None reported	
C. micans Swartz aff.	±(V)	None reported	

Table 4 *cont'd.*

Family/Genus/Species	Molluscicidal Activity[a]	Type of Molluscicide Present[b]	Geographic Distribution[c]
C. mucronatus Willd.	±(V)	None reported	
C. mucronifolius Muell.–Arg. aff.	±(V)	None reported	
C. rhamnifolius HBK. aff.	±(V)	None reported	
C. sonderianus Muell.–Arg.	±(V)	None reported	
C. sp.	−(V)	None reported	
C. sp.	±(V)	None reported	
C. tiglium L.	+	None reported	
C. zehntneri Pax & Hoffm.	±(V)	None reported	
C. zehntneri Pax & Hoffm. aff.	±(V)	None reported	
Cyrtogonone (1)			Trop. Afr.
C. argentea (Pax) Prain	+	None reported	
Euphorbia (2,000)			Cosmop.; chiefly subtrop. & warm temp.
E. candelabrum Trémaut ex Kotschy	+	None reported	
E. cotinifolia L.	+	None reported	
E. geniculata Ortega	nt	*Triterpene*	
E. gymnoclada Boiss.	±	None reported	
E. lactea Haworth	+	None reported	
E. poissonii Pax	−	None reported	
E. pulcherrima Willd. ex Klotzsch	−	None reported	
E. sp.	−	None reported	
E. tenuifolia Lamk.	−	None reported	
Hevea (12)			Trop. Am.
H. brasiliensis (Willd. ex A.Juss) Muell.–Arg.	−	None reported	

Cont'd.

Table 4 *cont'd.*

Family/Genus/Species	Molluscicidal Activity[a]	Type of Molluscicide Present[b]	Geographic Distribution[c]
Hura (2)			Mex. to trop. S. Am.; W.I.
H. crepitans L.	−	None reported	
Jatropha (175)			Trop. & subtrop.; N. Am.; S. Afr.
J. aceroides Hutchins.	+	None reported	
J. aethiopica Muell.−Arg.	±	None reported	
J. curcas L.	+	None reported	
J. gossypiifolia L.	±	None reported	
J. grandifolia[f]	−	None reported	
J. podagrica Hook.	+	None reported	
Manihot (170)			S.W. U.S. to trop. S. Am.
M. esculenta Crantz	−	None reported	
M. glaziovii Muell.−Arg.	±	None reported	
Phyllanthus (600)			Trop. & subtrop., excl. Eur. & N. As.
P. acidus (L.) Skeels	−	None reported	
P. lathyroides HBK.	−	None reported	
P. niruri L.	−	None reported	
P. nobilis Muell.−Arg.	−	None reported	
Putranjiva (200)			S. Afr.; E. As.
P. roxburghii Wallich	nt	*Triterpene*	
Ricinus (1)			Trop. Afr. & As.
R. communis L.	±	None reported	
Sapium (120)			Trop. & subtrop. (in Am., S. to Patag.)
S. laurocerasum Desf.	−	None reported	
S. sebiferum (L.) Roxb.	−	None reported	
Sebastiania (95)			Trop. Am.; Atl. U.S.; trop. W. Afr. (1); India to S. China & Australia; W. Malaysia (3)

Table 4 *cont'd.*

Family/Genus/Species	Molluscicidal Activity[a]	Type of Molluscicide Present[b]	Geographic Distribution[c]
S. chamaelea Muell.–Arg.	–	None reported	
EUPTELEA-CEAE (1/2)			E. As.
Euptelea (2)			Assam (2); S.W. & C. China; Japan
E. polyandra Sieb. & Zucc.	nt	Triterpene type	
FLACOURTIA-CEAE (93/1,000)			Trop. & subtrop.
Casearia (160)			Trop.
C. guianensis G.Don	±	None reported	
Paropsia (10)			Trop. Afr.; Madag.; Sumatra; Mal. Penins.
P. guineensis Oliver	–	None reported	
GESNERIACEAE (120/2,000)			Mostly trop. & subtrop.
Gesneria (50)			Trop. Am.; W.I.
G. albiflora Kuntze	–	None reported	
GUTTIFERAE (49/900)			Mostly trop. & subtrop., a few temp.
Calophyllum (12)			Mostly trop. As. & Pacif.; Madag.; trop. Am.; W.I.
C. calaba L.	–	None reported	
HIPPOCASTANACEAE (2/15)			N. temp.; S. Am.
Aesculus (13)			Mostly N. Am.; S. & E. As.; Eur.
A. hippocastanum L.	nt	*Triterpene*	
A. indica Colebr. ex Wallich	nt	*Triterpene*	
A. turbinata Bl.	nt	*Triterpene*	
HYDROPHYLLACEAE (18/250)			Cosmop.; exc. Australia
Hydrolea (20)			Trop. Am.; Afr.; As.
H. spinosa L.	±	None reported	

Cont'd.

Table 4 *cont'd.*

Family/Genus/Species	Molluscicidal Activity[a]	Type of Molluscicide Present[b]	Geographic Distribution[c]
LABIATAE (180/3,500)			Cosmop., chief center Medit. region
Coleus (150)			Paleotrop.
C. *blumei* Benth.	−	None reported	
Hyptis (400)			Warm. Am; W.I.
H. *pectinata* Poit.	+	None reported	
H. sp.	−(V)	None reported	
H. *suaveolens* (L.) Poit.	−(V)	None reported	
Leonotis (41)			Trop. & S. Afr.; 1 pantrop.
L. *nepetaefolia*[f]	−	None reported	
Ocimum (150)			Trop. & warm temp., esp. Afr.
O. *basilicum* L.	−	None reported	
O. *canum* Sims	+	None reported	
O. *gratissimum* L.	±(V)	None reported	
O. *officinalis*[f]	−	None reported	
O. sp.	−(V)	None reported	
LARDIZABALACEAE (8/135)			Himalayas to Japan; Chile
Akebia (5)			E. As.
A. *quinata* Decne.	±	*Triterpene*	
Stauntonia (15)			Burma to Formosa & Japan
S. *hexaphylla* Decne.	nt	*Triterpene*	
LAURACEAE (32/2,500)			Trop. & subtrop.; chief centers S.E. As.; Brazil
Dicypellium (1)			Brazil
D. *caryophyllatum* Nees	±(V)	None reported	
Ocotea (400)			Trop. & subtrop. Am; few trop. & S. Afr.; Mascarenes
O. sp.	±(V)	None reported	
Persea (150)			Trop.
P. *americana* P.Miller (=P. *gratissima* Gaertn.f.)	−,+	None reported	
LEGUMINOSAE (600/12,000)			Cosmop.
Abrus (12)			Trop.
A. *precatorius* L.	nt	*Triterpene*	

Table 4 *cont'd.*

Family/Genus/Species	Molluscicidal Activity[a]	Type of Molluscicide Present[b]	Geographic Distribution[c]
Acacia (800)			Trop. & subtrop.
A. *albida* Delile	−	None reported	
A. *arabica* (Lamk.) Willd. (= A. *nilotica* (L.) Delile)	+	None reported	
A. *concinna* DC.	+	Triterpene type	
A. *dudgeoni* Craib	+	None reported	
A. sp	+	None reported	
Aeschynomene (150)			Trop. & subtrop.
A. *americana* L.	−	None reported	
Albizia (150)			Warm Old World
A. *abyssinica*[f]	nt	*Triterpene*	
A. *anthelmintica* Brongn.	nt	*Triterpene*	
A. *ferruginea* (Guillem. & Perrot). Benth.	nt	*Triterpene*	
A. *lebbek* (L.) Benth.	±	*Triterpene*	
A. *procera* (Roxb.) Benth.	nt	*Triterpene*	
Andira (35)			Trop. Am.; Afr.
A. *inermis* HBK.	±	None reported	
A. *retusa* HBK.	+	None reported	
Astragalus (2,000)			Cosmop., exc. Australia
A. *glyciphyllos* L.	nt	*Triterpene*	
Bauhinia (300)			Warm regions
B. *forficata* Link	±	None reported	
Brachystegia (30)			Trop. Afr.
B. *eurycoma* Harms	−	None reported	
Caesalpinia (100)			Trop. & subtrop.
C. *coriaria* (Jacquin) Willd.	+	None reported	
C. *ferrea* K. von Mart.	±	None reported	
C. *pyramidalis* Tul.	±	None reported	
C. *sepiaria* Roxb.	−	None reported	
Cajanus (1)			Trop. As.; Afr.
C. *cajan* (L.) Huth	±	None reported	
Calliandra (100)			Madag.; warm As., Am.
C. *portoricensis* Benth.	+	None reported	
Calopogonium (10)			C. & S. Am.; W.I.

Cont'd.

Table 4 *cont'd.*

Family/Genus/Species	Molluscicidal Activity[a]	Type of Molluscicide Present[b]	Geographic Distribution[c]
C. velutinum (Benth.) Amshoff (= *Stenolobium velutinum* Benth.)	+	None reported	
Cassia (600)			Trop. & warm temp. (exc. Eur.)
C. alata L.	−	None reported	
C. apoucouita Aubl.	±	None reported	
C. hoffmanseggii K.von Mart. ex Benth.	±	None reported	
C. italica Lamk. ex F.W. Andrews	−	None reported	
C. mimosoides L.	−	None reported	
C. occidentalis L.	−	None reported	
C. sericea Swartz	−	None reported	
C. siamea Lamk.	−	None reported	
C. singueana Delile	±	None reported	
C. splendida Vogel	−	None reported	
C. tora L.	−	None reported	
C. trachypus K.von Mart. ex Benth.	±	None reported	
Centrosema (45)			Am.
C. pubescens Benth.	−	None reported	
Clitoria (40)			Trop. & subtrop.
C. ternatea L.	−	None reported	
Copaifera (30)			Mostly trop. Am.; some trop. Afr.
C. lansdorfii Desf.	±	None reported	
C. mopane J.Kirk ex Benth.	+	None reported	
Crotalaria (600)			Trop. & sutrop.
C. pallida Ait. (= *C. striata* DC.)	−	None reported	
C. retusa L.	−	None reported	
Dalbergia (300)			Trop. & subtrop.; S. Afr.
D. monetaria L.f. (= *Ecastaphyllum monetaria* Pers.)	−	None reported	
D. retusa Baill.	+	None reported	
Delonix (3)			Trop. Afr.; Madag.; As.
D. regia (Boj. ex Hook.) Raf.	±	None reported	

Table 4 *cont'd.*

Family/Genus/Species	Molluscicidal Activity[a]	Type of Molluscicide Present[b]	Geographic Distribution[c]
Derris (80)			Trop.
D. amaripensis[f]	−	None reported	
D. amazonica Killip	nt	*Rotenone*	
D. brevipes Baker	nt	*Rotenone*	
D. chinensis Benth.	nt	*Rotenone*	
D. elliptica (Roxb.) Benth.	+	*Rotenone*	
D. ferruginea (Roxb.) Benth.	nt	*Rotenone*	
D. malaccensis Prain	nt	*Rotenone*	
D. negrensis Benth.	nt	*Rotenone*	
D. obtusa[f]	−	None reported	
D. rariflora (Benth.) Macbride	nt	*Rotenone*	
D. uliginosa (Willd.) Benth. (= *D. trifoliata* Lour.)	nt	*Rotenone*	
D. urucu (Killip & Smith) Macbride (= *Lonchocarpus urucu* Killip & Smith)	nt	*Rotenone, triterpene*	
Desmodium (450)			Trop. & subtrop.
D. canum (J.F. Gmelin) Schinz ex Thellung	−	None reported	
D. ramosissimum G.Don	−	None reported	
D. sp.	−	None reported	
Detarium (4)			Trop. Afr.
D. senegalense J.F.Gmelin	±	None reported	
Dialium (41)			Primarily trop. Afr.; Madag.; Malaysia; 1 trop. S. Am.
D. guineense Willd.	+	None reported	
Dichrostachys (20)			Trop. Afr. to Australia, esp. Madag.
D. cinerea Wight & Arnott (= *D. glomerata* (Forsk.) Chiov.)	+	None reported	

Cont'd.

Table 4 *cont'd.*

Family/Genus/Species	Molluscicidal Activity[a]	Type of Molluscicide Present[b]	Geographic Distribution[c]
Dioclea (51)			Mostly trop. Am., 1 trop. Afr. & As.
D. *reflexa* Hook.f.	+	None reported	
Dolichos (150)			Warm regions
D. *lablab* L.	±	None reported	
Entada (30)			Warm regions
E. *africana* Guill. & Perrot	nt	*Rotenone*	
E. *phaseoloides* (L.) Merr.	+	None reported	
E. *sudanica* Schweinf.	nt	*Rotenone*	
Enterolobium (10)			Trop. Am.; W.I.
E. *contortisiliquum* Morong	±	Triterpene type	
Erythrina (100)			Trop. & subtrop.
E. *coromandelianum*[f]	−	None reported	
E. *velutina* Willd.	−	None reported	
Geoffroea (6)			Trop. Am.; W.I.
G. *spinosa* Jacquin	−	None reported	
Gleditsia (11)			Trop. & subtrop.
G. *horrida* Willd.	nt	*Triterpene*	
G. *japonica* Miq.	nt	*Triterpene*	
G. *triacanthos* L.	nt	*Triterpene*	
Gliricidia (10)			Trop. Am.; W.I.
G. *sepium* HBK. (= *Lonchocarpus sepium* DC.)	−	None reported	
Glycine (10)			Trop. & warm temp.; Afr. & As.
G. *max* (L.) Merr.	−	*Triterpene*	
Glycyrrhiza (18)			Temp. & subtrop. Am.; temp. Euras.; N. Afr.; S.E. Australia
G. *echinata* L.	nt	*Triterpene*	
G. *glabra* L.	nt	*Triterpene*	
G. *uralensis* Fischer ex DC.	nt	*Triterpene*	
Hymenaea (25)			Mex.; Cuba; trop. S. Am.
H. *courbaril* L.	−	None reported	
Indigofera (700)			Warm regions
I. *kerstingii* Harms	±	None reported	
I. *secundiflora* Poir.	±	None reported	

Table 4 *cont'd.*

Family/Genus/Species	Molluscicidal Activity[a]	Type of Molluscicide Present[b]	Geographic Distribution[c]
I. spicata Forsk.	+	None reported	
I. suffruticosa P.Miller	+	None reported	
Inga (200)			Trop. & subtrop. Am.; W.I.
I. vera Willd.	±	None reported	
Lonchocarpus (150)			Trop. Am.; W.I.; Afr.; Australia
L. cyanescens (Schumach. & Thonn.) Benth.	nt	*Rotenone*	
L. floribundus Benth.	nt	*Rotenone*	
L. hedyosmus Miq.	nt	*Rotenone*	
L. longifolius Benth. ex K.von Mart. (= *Derris* longifolia Benth.)	nt	*Rotenone*	
L. nicou (Aubl.) DC.	nt	*Rotenone*	
L. spruceanus Benth.	nt	*Rotenone*	
L. utilis A.C. Smith	nt	*Rotenone*	
Medicago (100)			Temp. Euras.; Medit.; S. Afr.
M. sativa L.	nt	*Triterpene*	
Millettia (180)			Trop. & subtrop., a few Am.
M. dura[f]	nt	*Rotenone*	
M. ferruginea Hochst.	nt	*Rotenone*	
M. pachycarpa Benth.	nt	*Rotenone*	
M. taiwaniana Dunn	nt	*Rotenone*	
Mimosa (500)			Trop. & subtrop. Am.; few in Afr. & As.
M. malacocentra K. von Mart. ex Benth.	±	None reported	
Mundulea (31)			Old World trop., primarily Madag.
M. sericea (Willd.) A.Cheval. (= *M. suberosa* Benth.)	nt	*Rotenone*	
Neorautanenia (3)			Trop. Afr.
N. ficifolia (Benth.) C.A. Smith	nt	*Rotenone*	

Cont'd.

Table 4 *cont'd.*

Family/Genus/Species	Molluscicidal Activity[a]	Type of Molluscicide Present[b]	Geographic Distribution[c]
Ononis (75)			Canary Is.; Medit.; Eur. to C. As.
O. *repens* L.	nt	*Triterpene*	
O. *spinosa* L.	nt	*Triterpene*	
Ormosia (50)			Trop.
O. *krugii* Urban	−	None reported	
Peltophorum (12)			Trop.
P. *pterocarpum* (DC.) Backer ex Heyne	−	None reported	
Periandra (8)			C. Am.; W.I.; Brazil
P. *dulcis* K.von Mart. ex Benth.	nt	*Triterpene*	
P. *mediterranea* (Vell.) Taub.	nt	*Triterpene*	
Phaseolus (240)			Trop. & subtrop.; chiefly Am.
P. *vulgaris* L.	−	*Triterpene*	
Piliostigma (3)			Trop. Afr.; Indomal.; Queensland
P. *reticulatum* (DC.) Hochst.	−	None reported	
Piptadenia (11)			Mex. to trop. S. Am.
P. *biuncifera* Benth.	±	None reported	
P. *macrocarpa* Benth.	±	None reported	
P. *moniliformis* Benth.	±	None reported	
Piscidia (10)			Florida; Mex.; W.I.
P. *erythrina* L. (= P. *piscipula* (L.) Sarg.)	nt	*Rotenone*	
P. *mollis* J.N. Rose	nt	*Rotenone*	
Pisum (6)			Medit.; W. As.
P. *sativum* L.	nt	*Triterpene*	
Chloroleucum (= *Pithecellobium*) (200)	±	None reported	Trop.
C. *multiflorum* Benth.	+	None reported	
C. *polycephalum* Benth.	−	None reported	
Plathymenia (3)			Brazil; Argent.
P. *reticulata* Benth.	±	None reported	
Platymiscium (30)			Mex. to trop. S. Am.
P. *piliferum* Taub.	−	None reported	
Podalyria (25)			S. Afr.
P. *tinctoria* Lamk.	nt	*Triterpene*	
Poiretia (7)			Trop. Am.
P. sp.	±(V)	None reported	

Table 4 *cont'd.*

Family/Genus/Species	Molluscicidal Activity[a]	Type of Molluscicide Present[b]	Geographic Distribution[c]
Prosopis (43)			Primarily warm Am.; 1 trop. Afr.; 2 Cauc. to India
P. sp.	±	None reported	
Psoralea (130)			Trop. & subtrop.
P. corylifolia L.	−	*Xanthotoxin*	
Pueraria (43)			Himal. to Japan; S.E. As.; Malaysia; Pacif.
P. hirsuta Kurz	−	None reported	
Samanea (20)			Mex. to trop. S. Am.; trop. Afr.
S. saman (Jacquin) Merr. (= *Pithecellobium saman* = *Albizia saman* (Jacquin) Benth.)	±	*Triterpene*	
Schrankia (30)			Warm. Am.
S. leptocarpa DC.	−	None reported	
Sesbania (50)			Trop. & subtrop.
S. sesban (L.) Merr.	+	Triterpene type	
Stryphnodendron (15)			Trop. Am.
S. coriaceum Benth.	±	None reported	
Stylosanthes (50)			Trop. & subtrop. Am.; Afr.; trop. As.
S. viscosa Swartz	+	None reported	
Swartzia (100)			Trop. Am.; Afr.
S. madagascariensis Desv.	+	*Rotenone*	
Sweetia (12)			S. Am.
S. dasycarpa Benth.	−	None reported	
Tamarindus (1)			Trop. Afr.
T. indica L.	±	None reported	
Tephrosia (300)			Trop. & subtrop.
T. candida (Roxb.) DC.	nt	*Rotenone*	
T. falciformis Ramasw.	nt	*Rotenone*	
T. macropoda Harv.	nt	*Rotenone*	
T. obovata Merr.	nt	*Rotenone*	
T. piscatoria Pers. (= *P. purpurea* Pers.)	nt	*Rotenone*	
T. strigosa (Dalzell)	nt	*Rotenone*	

Cont'd.

Table 4 *cont'd.*

Family/Genus/Species	Molluscicidal Activity[a]	Type of Molluscicide Present[b]	Geographic Distribution[c]
Santapau & Maheshwari			
T. toxicara (Swartz)	±	*Rotenone*	
Pers. (= *T. sinapou* (Buchoz) A.Chev.)			
T. uniflora Pers.	−	*Rotenone*	
T. villosa Pers.	nt	*Rotenone*	
T. virginiana (L.) Pers.	nt	*Rotenone*	
T. vogelii Hook.f.	+	None reported	
Tetrapleura (2)			Trop. Afr.
T. tetraptera Taub.	+	None reported	
Torresea (2)			Brazil
T. cearaensis Allemão	±	None reported	
Trifolium (300)			Temp. & subtrop. (not S.E. As. or Australia)
T. alpinum L.	nt	*Triterpene*	
Trigonella (135)			Medit.; Eur.; As.; S. Afr.; Australia
T. foenum−graecum L.	nt	*Steroid*	
Wisteria (10)			E. As.; E. N. Am.
W. chinensis DC.	nt	*Triterpene*	
Zornia (75)			Trop., esp. Am.
Z. brasiliensis Vog.	−(V)	None reported	
LOGANIACEAE (7/130)			Trop., a few warm temp.
Anthocleista (14)			Trop. Afr.; Madag.
A. vogelli Planch.	−	None reported	
Antonia (1)			Trop.
A. ovata Pohl	±(V)	None reported	
Strychnos (200)			Trop.
S. parvifolia A.DC.	±	None reported	
LYTHRACEAE (25/550)			All zones except frigid
Lagerstroemia (50)			Paleotrop.
L. indica L.	−	None reported	
MALPIGHIACEAE (60/800)			Trop., esp. S. Am.
Byrsonima (120)			C. & S. Am.; W.I.
B. sericea DC.	+	None reported	
Mascagnia (60)			Trop. Am.
M. rigida Griseb.	−	None reported	
MALVACEAE (75/1,000)			Trop. & temp.
Hibiscus (300)			Trop. & subtrop.
H. cannabinus L.	−	None reported	

Table 4 *cont'd.*

Family/Genus/Species	Molluscicidal Activity[a]	Type of Molluscicide Present[b]	Geographic Distribution[c]
Urena (6)			Trop. & subtrop.
U. lobata L.	–	None reported	
MELASTOMATACEAE			Trop. & subtrop.
(240/3,000)			
Miconia (700)			Trop. Am.; W.I.;
M. prasina DC.	–	None reported	1 W. Afr.
MELIACEAE (50/1,400)			Warm regions
Azadirachta (2)			
A. indica A.Juss.	+	None reported	Indomal.
Cedrela (7)			
C. odorata L.	±	None reported	Mex. to trop. S. Am.
Ekebergia (15)			
E. senegalensis Juss.	±	None reported	S. & trop. Afr.; Madag.
Guarea (170)			Mostly trop. Am.;
G. trichilioides L.	–	None reported	20 spp. Afr.
Swietenia (8)			
S. mahagoni (L.) Jacquin	nt	*Triterpene*	Trop. Am.; W.I.
Toona (15)			
T. sinensis M.Roem.	±	None reported	Trop. As., Australia
MELIANTHACEAE (2/15)			Trop. & S. Afr.
Bersama (2)			Trop. & S. Afr.
B. yangambiensis L.Toussaint	nt	*Triterpene*	Trop. & S. Afr.
MENISPERMACEAE (65/350)			Warm regions
Cissampelos (30)			
C. pareira L.	–	None reported	Trop.
MENYANTHACEAE (5/33)			N. & S. temp.; trop. S.E. As.
Nymphoides (20)			Trop. & temp.
N. humboldtiana (Kunth) Hoehne	–	None reported	
MORACEAE (53/1,400)			Trop. & subtrop.; a few temp.
Brosimum (50)			Trop. & temp. S. Am.
B. gaudichaudii Trécul	–	None reported	
Chlorophora (12)			Trop. Am.; Afr.; Madag.

Cont'd.

Table 4 *cont'd.*

Family/Genus/Species	Molluscicidal Activity[a]	Type of Molluscicide Present[b]	Geographic Distribution[c]
C. tinctoria (L.) Gaudich. ex B. & H.	−	None reported	
Dorstenia (170)			Trop.
D. cayapia Vellozo	±	None reported	
Ficus (800)			Warm regions, chiefly Indomal.; Polynes.
F. glumosa Delile	+	None reported	
F. indica L.	−	None reported	
MYRSINACEAE (35/ 1,000)			Trop. & subtrop.; S. Afr.; N.Z.
Myrsine (7)			Azores; Afr. to China
M. africana L.	nt	Triterpene	
Wallenia (25)			W.I.
W. yunquensis Mez	nt	Triterpene	
MYRTACEAE (100/ 3,000)			Chiefly Australia; trop. Am.
Eucalyptus (25)			Australia; N. Caled.
E. alba Reinw. ex Bl.	±	None reported	
E. bicolor A.Cunn. ex Hook.	±	None reported	
E. citriodora Hook.	±	None reported	
E. globulus Labill.	±	None reported	
E. longifolia Link	±	None reported	
E. pilularis DC.	±	None reported	
E. robusta Sm.	±	None reported	
E. saligna Sm.	±	None reported	
E. triantha Link	±	None reported	
Eugenia (1000)			Trop. & subtrop.
E. jambos L. (=Syzygium jambos (L.) Alst.)	−	None reported	
E. michelii Lamk. (=E. uniflora L.)	−(V)	None reported	
E. sp.	±(V)	None reported	
Myrcia (500)			Trop. S. Am.; W.I.
M. polyantha[f]	+(V)	None reported	
Pimenta (18)			Trop. Am.; W.I.
P. racemosa (P.Miller) J.W. Moore	−	None reported	
Psidium (140)			Trop. Am.; W.I.
P. guajava L.	−	None reported	

Table 4 *cont'd.*

Family/Genus/Species	Molluscicidal Activity[a]	Type of Molluscicide Present[b]	Geographic Distribution[c]
Syzygium (500)			Paleotrop.
S. *cumini* (L.) Skeels	–	None reported	
S. *jambos* (L.) Alst.	–(V)	None reported	
NYCTAGINACEAE			Mostly trop. & esp.
(30/ 290)			Am.
Boerhaavia (40)			Trop. & subtrop.
B. *coccinea* P.Miller	–	None reported	
B. *diffusa* L.	–	None reported	
Bougainvillea (18)			S. Am.
B. *glabra* Choisy	–	None reported	
Mirabilis (60)			Am.
M. *jalapa* L.	–	None reported	
OCHNACEAE (300)			Trop.
Lophira (2)			Trop. W. Afr.
L. *alata* Banks ex Gaertn.f.	+	None reported	
Ouratea (300)			Trop.
O. *fieldingiana* Engl.	–	None reported	
OLACACEAE (25/250)			Trop.
Olax (55)			Trop. Afr.; Madag.; Indomal.; Australia
O. *andronensis* Baker	nt	*Triterpene*	
O. *gambecola* Baill.	nt	Triterpene type	
O. *glabriflora* Danguy	nt	*Triterpene*	
O. *psittacorum* Vahl	nt	*Triterpene*	
O. *subscorpioides* Oliver	nt	Triterpene type	
Ximenia (15)			Trop. Am.; trop. & S. Afr.; trop. As.; Australia
X. *americana* L.	+	None reported	
OLEACEAE (29/600)			Esp. temp. & trop. As.
Jasminum (300)			Cosmop., esp. temp. & trop. As., trop. & subtrop. (exc. N. Am.)
J. *sambac* (L.) Ait.	–	None reported	
Ligustrum (50)			Eur. to N. Iran; E. As.; Indomal. to New Hebrides

Cont'd.

Table 4 *cont'd.*

Family/Genus/Species	Molluscicidal Activity[a]	Type of Molluscicide Present[b]	Geographic Distribution[c]
L. japonicum[f]	nt	*Triterpene*	
L. obtusifolium Sieb. & Zucc.	nt	*Triterpene*	
L. ovalifolium Hassk.	nt	*Triterpene*	
ONAGRACEAE (21/640)			Temp. & trop.
Ludwigia (75)			Cosmop., esp. trop. Am.
L. angustifolia[f]	±	None reported	
L. leptocarpa (Nutt.) Hara	−	None reported	
OPILIACEAE (8/60)			Trop.; esp. As.
Agonandra (10)			Mex. to trop. S. Am.
A. brasiliensis B. & H.	+	None reported	
PAPAVERACEAE (26/200)			Chiefly N. temp.
Argemone (92)			W. & E. U.S.; Mex.; W.I.
A. mexicana L.	±	None reported	
PASSIFLORACEAE (12/600)			Trop. & warm temp.
Adenia (92)			Trop. & subtrop. Afr. to Malaysia
A. lobata Engl.	−	None reported	
Passiflora (500)			Chiefly Am.; a few in As. & Australia; Madag. (1)
P. edulis Sims	nt	*Triterpene*	
P. sp.	−(V)	None reported	
PHYTOLACCACEAE (12/100)			Chiefly trop. Am. & S. Afr.
Petiveria (1)			Warm Am.; W.I.
P. alliacea L.	−	None reported	
P. sp.	−	None reported	
Phytolacca (35)			Trop. & subtrop.
P. abyssinica Hoffm.	nt	*Triterpene*	
P. acinosa Roxb.	−	*Triterpene*	
P. americana L.	+	*Phytolaccagenin xyloside*	
P. dioica L.	nt	*Triterpene*	
P. dodecandra L'Hérit.	+	*Lemmatoxin; lemmatoxin C; Oleanoglycotoxin A*	
P. esculenta Van Houtte	nt	*Triterpene*	

Table 4 *cont'd.*

Family/Genus/Species	Molluscicidal Activity[a]	Type of Molluscicide Present[b]	Geographic Distribution[c]
P. icosandra L.	±	None reported	
P. insularis Nakai	nt	*Triterpene*	
P. japonica Makino	nt	*Triterpene*	
P. octandra L.	+	*Triterpene*	
P. rivinoides Kunth & Bouché	±	None reported	
PIPERACEAE (4/2,000)			Trop.
Piper (1,900)			Trop.
P. aduncum L.	−	None reported	
P. cubeba L.f.	−	None reported	
P. kadsura[f]	−	None reported	
P. longum L.	−	None reported	
P. marginatum Jacquin	−; ±(V)	None reported	
P. peltatum L.	−	None reported	
P. sp.	±(V)	None reported	
P. tuberculatum Jacquin	±	None reported	
PITTOSPORACEAE (9/ 200)			Trop. & subtrop. Afr.; As.; Australia; N.Z.; Pacif.
Pittosporum (150)			Trop. & subtrop. Afr.; As.; Australia; N.Z.; Pacif.
P. neelgherrense Wight & Arnott	nt	*Triterpene*	
PLANTAGINACEAE (3/ 270)			Cosmop.
Plantago (265)			Cosmop.
P. major L.	−	None reported	
POLEMONIACEAE (15/ 300)			Chiefly N. Am.; a few in Chile; Peru; Eur.; N. As.
Polemonium (50)			Temp. Euras.; N. Am.; Mex.; Chile (2)
P. caeruleum L.	nt	*Triterpene*	
POLYGALACEAE (12/ 800)			Cosmop., exc. N.Z.; Polynes. & Arctic zone
Monnina (150)			Mex. to Chile
M. exaltata	−(V)	None reported	

Cont'd.

Table 4 *cont'd.*

Family/Genus/Species	Molluscicidal Activity[a]	Type of Molluscicide Present[b]	Geographic Distribution[c]
A.W. Benn.			
Polygala (600)			Cosmop., exc. N.Z.; Polynes.; Arctic zone
P. *arenaria* Willd.	nt	Triterpene type	
P. *paniculata* L.	−	None reported	
P. *senega* L.	nt	*Triterpene*	
P. *senega* L. var. latifolia	nt	*Triterpene*	
Torrey & Gray			
P. *tenuifolia*[f]	nt	*Triterpene*	
Securidaca (80)			Trop.; exc. Australia
S. *lanceolata* A.St.Hill	−	None reported	
S. *longepedunculata* Fresen.	+	None reported	
POLYGONACEAE (40/800)			Chiefly N. temp.; a few trop.; Arctic and s. hemisph.
Antigonon (8)			Trop. & subtrop. Am.
A. *leptopus* Hook. Arnott	−	None reported	
Coccoloba (150)			Trop. & subtrop. Am.
C. *cordifolia* Meisn.	±	None reported	
Polygonum (300)			Chiefly N. temp.; a few trop.; Arctic and s. hemisph.
P. *glabrum* Willd.	±	None reported	
P. *meisnerianum* Cham. & Schltdl.	±	None reported	
P. *mite* Schrank	±	None reported	
P. *multiflorum* Thunb.	±	None reported	
P. *orientale* L.	−	None reported	
P. *perfoliatum* L.	−	None reported	
P. *punctatum* Elliot	−	None reported	
P. *salicifolium* Brouss. ex Willd.	±	None reported	
P. *senegalense* Meisner	±	*2',4'-Dihydroxy-3',6'-dimethoxy-chalcone; Quercetin-3-(2'-galloyl-glucoside)*	
P. *senegalense* fma. albotomentosum	±	None reported	

Table 4 *cont'd.*

Family/Genus/Species	Molluscicidal Activity[a]	Type of Molluscicide Present[b]	Geographic Distribution[c]
P. senegalense fma. *senegalense*	±	*Cyanogenic glucoside*	
P. setulosum A.Richard	±	None reported	
P. tomentosum Willd.	±	None reported	
P. zambesicum[f]	±	None reported	
Triplaris (25)			Trop. S. Am.
T. gardneriana Wedd.	±	None reported	
PORTULACACEAE (19/ 580)			Cosmop.; but esp. Am.
Portulaca (200)			Trop. & subtrop.
P. oleracea L.	−	None reported	
PRIMULACEAE (20/1,000)			Cosmop., but esp. N. temp.
Anagallis (28)			W. Eur.; Afr.; Madag.; pantrop. (1); S. Am. (2)
A. arvensis L.	nt	*Triterpene*	
Cyclamen (15)			Eur.; Medit. to Iran
C. europaeum L.	nt	*Triterpene*	
C. neapolitanum Tenore	nt	Triterpene type	
Lysimachia (200)			Esp. E. As. & N. Am.
L. thyrsiflora L. (=*Naumburgia thyrsiflora* Reichb.)	nt	*Triterpene*	
L. hederifolium[f]	nt	Triterpene type	
L. neapolitanum[f]	nt	*Triterpene*	
Primula (500)			N. hemisph.
P. elatior Hill	nt	*Triterpene*	
P. officinalis Jacquin	nt	Triterpene type	
P. sieboldii E.Morren	nt	*Triterpene*	
PROTEACEAE (62/1,050)			Trop. As.; Malaysia; Australia; N. Caled.; N.Z.; trop. S. Am.; Chile; mts. trop. Afr.; S. Afr.; Madag.
Cardwellia (1)			Queensland
C. sublimis F.v.Muell.	nt	*Alkenyl phenols*	
Grevillea (190)			E. Malaysia; New Hebrid.; N. Caled.; Australia

Cont'd.

Table 4 *cont'd.*

Family/Genus/Species	Molluscicidal Activity[a]	Type of Molluscicide Present[b]	Geographic Distribution[c]
G. hilliana F.v.Muell.	nt	*Alkenyl phenols*	
G. banksii R.Br.	nt	*Alkenyl phenols*	
G. pteridifolia Knight	nt	*Alkenyl phenols*	
G. pyramidalis A.Cunn. ex Meisn.	nt	*Alkenyl phenols*	
G. robusta A.Cunn. ex R.Br.	nt	*Alkenyl phenols*	
Hakea (100)			Australia
H. persiehana F.v.Muell.	nt	*Alkenyl phenols*	
Opisthiolepis (1)			Queensland
O. heterophylla L.S. Smith	nt	*Alkenyl phenols*	
Persoonia (60)			Australia; N.Z.
P. elliptica R.Br.	nt	*Alkenyl phenols*	
P. linearis Andr.	nt	*Alkenyl phenols*	
Petrophila (35)			Australia
P. shirleyae F.M.Bailey	nt	*Alkenyl phenols*	
PUNICACEAE (1/2)			Socotra; Balkans to the Himal.
Punica (2)			Socotra (1); Balkans to the Himal. (1)
P. granatum L.	±	None reported	
RANUNCULACEAE (50/1,900)			Chiefly N. temp.
Anemone (150)			Cosmop.
A. narcissiflora[f]	nt	*Triterpene*	
A. rivularis Buch.–Ham. ex DC.	±	*Triterpene*	
Caltha (32)			Arctic & N. temp. (20); N.Z. & temp. S. Am. (12)
C. palustris L.	nt	*Triterpene*	
C. silvestris Vorosh.	nt	*Triterpene*	
Cimicifuga (15)			N. temp.
C. acerina Tanaka	nt	*Triterpene*	
C. dahurica Maxim.	nt	*Triterpene*	
C. japonica Spreng.	nt	*Triterpene*	
C. simplex Wormsk. ex DC.	nt	*Triterpene*	
Clematis (250)			Cosmop., chiefly temp.
C. chinensis Retz.	±	*Triterpene*	
C. mandschurica Rupr.	nt	*Triterpene*	

Table 4 *cont'd.*

Family/Genus/Species	Molluscicidal Activity[a]	Type of Molluscicide Present[b]	Geographic Distribution[c]
C. songorica Bge.	nt	*Triterpene*	
C. vitalba L.	nt	*Triterpene*	
Pulsatilla (30)			Temp. Euras.
P. cernua Bercht. & Presl	nt	*Triterpene*	
P. chinensis C.Muell.	±	*Triterpene*	
P. koreana Nakai	nt	*Triterpene*	
P. nigricans Stoerck ex DC.	nt	Triterpene type	
RHAMNACEAE (58/900)			Cosmop.
Hovenia (5)			Himal. to Japan
H. dulcis Thunb.	nt	*Triterpene*	
Maesopsis(1)			Trop. E. Afr.
M. eminii Engl.	+	None reported	
Ziziphus (100)			Trop. Am.; Afr.; Medit.; Indomal.; Australia
Z. joazeiro K.v.Mart.	±	None reported	
Z. jujuba P.Miller	−	*Triterpene*	
Z. rugosa Lamk.	nt	*Triterpene*	
Z. undulata Reissek	−	None reported	
Z. vulgaris Lamk.	nt	*Triterpene*	
ROSACEAE (100/2,000)			Cosmop.
Crataegus (200)			N. temp.
C. pentagyna Waldst. & Kit.	nt	Triterpene type	
Poterium (25)			Temp. Euras.
P. lasiocarpum Boiss. & Hausskn. ex Boiss.	nt	*Triterpene*	
P. polygamum Waldst. & Kit.	nt	*Triterpene*	
Quillaja (3)			Temp. S. Am.
Q. saponaria Molina	+	*Triterpene*	
Rosa (250)			N. temp. & trop. mts.
R. sp.	−	None reported	
Sanguisorba (2–3)			N. temp.
S. officinalis L.	nt	*Triterpene*	
RUBIACEAE (500/6,000)			Mostly trop.; a few temp. & Arctic
Anthocephalus (2)			C. As.
A. cadamba Miq.	nt	*Triterpene*	

Cont'd.

Table 4 *cont'd.*

Family/Genus/Species	Molluscicidal Activity[a]	Type of Molluscicide Present[b]	Geographic Distribution[c]
Borreria (150)			Warm regions
B. octodon Hepper	−	None reported	
B. ocymoides DC.	−	None reported	
B. verticillata G.F.W.Meyer	−	None reported	
Canthium (200)			Paleotrop.
C. subcordatum DC.	+	None reported	
Chiococca (20)			S. Florida; W.I.; trop. Am.
C. alba Hitchc.	−	None reported	
Cinchona (40)			Andes
C. calisaya Wedd.	nt	*Triterpene*	
C. pubescens Vahl	nt	*Triterpene*	
Coffea (40)			Paleotrop., esp. Afr.
C. spathicalyx K.Schum.	±	None reported	
Coutarea (7)			Trop. Am.; W.I.
C. hexandra K.Schum.	+	None reported	
Duggena (22)			Trop. afr.
D. hirsuta Britton	+	None reported	
Fadogia (60)			Trop. Afr.
F. erythrophloea (K.Schum. & K.Krausse) Hutch. & Dalziel	±	None reported	
Galium (400)			Cosmop.
G. hirtum Lamk. (= *Mollugo hirta* Thunb.)	nt	Triterpene	
G. spergula[f]	nt	*Triterpene*	
Gardenia (250)			Paleotrop.
G. ternifolia Schumach. & Thonning	±	None reported	
G. vogelii Hook.f.	+	None reported	
Genipa (6)			Warm Am.; W.I.
G. americana L.	±	None reported	
Gonzalagunia (35)			Trop. Am.; W.I.
G. hirsuta K.Schum.	−	None reported	
Guettarda (81)			Mostly trop. Am., also N. Caled.
G. americana[f]	−	None reported	
G. angelica Mart. ex Muell.−Arg.	−	None reported	

Table 4 *cont'd.*

Family/Genus/Species	Molluscicidal Activity[a]	Type of Molluscicide Present[b]	Geographic Distribution[c]
Hamelia (40)			Mex. to Paraguay; W.I.
H. erecta Jacquin	−	None reported	
Ixora (400)			Trop.
I. coccinea L.	−	None reported	
Massularia (1)			Trop. E. Afr.
M. acuminata (G.Don) Bullock ex Hoyle	−	None reported	
Mitragyna (12)			Trop. Afr.; As.
M. stipulosa Kuntze	±	None reported	
Morinda (80)			Trop.
M. lucida Benth.	+	None reported	
Nauclea (35)			Trop. Afr.; As.; Polynes.
N. diderrichii Merr.	±	None reported	
Oldenlandia (300)			Warm regions
O. affinis DC.	±	None reported	
Psychotria (700)			Warm regions
P. brachiata DC.	−	None reported	
P. grandis Swartz	−	None reported	
Randia (300)			Trop.
R. aculeata L.	−	None reported	
R. dumetorum Lamk.	nt	*Triterpene*	
R. nilotica Stapf	+	None reported	
R. vestita S.Moore	±	None reported	
Rothmannia (20)			Trop. & S. Afr.
R. urcelliformis (Schweinf. ex Hiern) Bullock ex Robyns	±	None reported	
R. whitefieldii (Lindl.) Dandy	+	None reported	
Rytigynia (70)			Trop. Afr.
R. canthioides Robyns	−	None reported	
Sherbournea (10)			Trop. Afr.
S. millenii (Wernh.) Hepper	−	None reported	
Tocoyena (20)			Mex.; trop. S. Am.; W.I.
T. formosa K.Schum.	−	None reported	

Cont'd.

Table 4 *cont'd.*

Family/Genus/Species	Molluscicidal Activity[a]	Type of Molluscicide Present[b]	Geographic Distribution[c]
RUTACEAE (150/900)			Trop. & temp., esp. S. Afr. & Australia
Boenninghausenia (1)			Assam to Japan
B. *albiflora* (Hook.) Reichb. ex Meisn.	nt	*Bergapten*	
Citrus (12)			S. China; E. As. (8); Indomal.
C. *aurantiifolia* (Christm.) Swingle	−	None reported	
C. *limon* (L.) Burm.f.	−(V)	None reported	
C. *medica* L.	±	None reported	
C. *tangerina* Hort. ex Tanaka	−	None reported	
C. *wilsonii* Tanaka	−	None reported	
Fagara (250)			Trop.
F. *lemairei* Wildem.	−	None reported	
F. *rhoifolia* Engl. (= *Zanthoxylum rhoifolium* Lamk.)	±	None reported	
F. sp.	±(V)	None reported	
Orixa (1)			Japan
O. *japonica* Thunb.	nt	*Bergapten, isopimpinellin, xanthotoxin*	
Pilocarpus (22)			Trop. Am.; W.I.
P. *microphyllus* Stapf	−(V)	None reported	
P. *pauciflorus* St.-Hil.	−(V)	None reported	
P. *pauciflorus* St.-Hil. aff.	−(V)	None reported	
P. sp.	±(V)	None reported	
Ptelea (3)			U.S.; Mex.
P. *trifoliata* L.	nt	*Bergapten*	
Ruta (60)			Medit.; temp.
R. *chalepensis* L.	−	*Bergapten, chalepensin, xanthotoxin*	
R. *graveolens* L.	nt	*Bergapten*	
R. *microcarpa* Svent. apúd agulló et al.	nt	*Bergapten*	
R. *oreojasme* Webb & Benth.	nt	*Bergapten, isopimpinellin, xanthotoxin*	

Table 4 *cont'd.*

Family/Genus/Species	Molluscicidal Activity[a]	Type of Molluscicide Present[b]	Geographic Distribution[c]
R. pinnata L.f.	nt	*Bergapten, isopimpinellin, xanthotoxin*	
SAPINDACEAE (150/ 2,000)			Trop. & subtrop.
Chytranthus (30)			Trop. Afr.
C. macrobotrys (Gilb.) Exell & Mendonca	nt	*Triterpene*	
Cupania (55)			Warm regions
C. americana L.	−	None reported	
Koelreuteria (8)			China, Formosa, Fiji
K. paniculata Laxm.	nt	Triterpene	
Lecaniodiscus (3)			Trop. Afr.
L. cupanioides Planch.	nt	*Triterpene*	
Magonia (2)			Brazil; Bolivia
M. pubescens St.-Hil.	+	None reported	
Paullinia (180)			Warm. Am.
P. pinnata L.	−	None reported	
Sapindus (13)			Trop. & subtrop. As.; Pacif.; Am.
S. emarginatus Vahl (= *S. trifoliatus* L.)	+	Triterpene type	
S. mukorossi Gaertn.	nt	*Triterpene*	
S. saponaria L.	±	*Triterpene*	
S. utilis Trab.	nt	*Triterpene*	
Serjania (215)			S. U.S. to trop. S. Am.
S. polyphylla Poir. ex Steud.	−	None reported	
S. sp.	+	None reported	
Talisia (50)			C. & trop. S. Am.; Trinidad
T. esculenta Radlk.	−	None reported	
Xanthoceras (2)			N. China
X. sorbifolia Bge.	nt	*Triterpene*	
X. sp.	nt	*Triterpene*	
SAPOTACEAE (75/800)			Trop.
Butyrospermum (1)			N. trop. Afr.
B. paradoxum (Gaertn.f.) Hepper	+	None reported	
Madhuca (85)			Indoch.; Indomal.; Australia

Cont'd.

Table 4 *cont'd.*

Family/Genus/Species	Molluscicidal Activity[a]	Type of Molluscicide Present[b]	Geographic Distribution[c]
M. latifolia (Roxb.) Macbr. (= *Bassia latifolia* Roxb.)	nt	*Triterpene*	
M. longifolia (L.) Macbr.(= *Bassia longifolia* L.)	nt	*Triterpene*	
Pradosia (12)			S. Am.
P. lactescens Radlk.	nt	*Triterpene*	
SCROPHULARIACEAE (220/3,000)			Cosmop.
Capraria (4)			Warm Am.; W.I.
C. biflora L.	±	None reported	
Digitalis (30)			Eur.; Medit.; Canary Is.
D. lanata Ehrh.	nt	*Steroid*	
D. purpurea L.	nt	*Steroid*	
Gratiola (20)			N. & S. temp.; trop. mts.
G. officinalis L.	nt	*Triterpene*	
Herpestis (100)			Warm regions
H. gardnerianum[f]	+	*Triterpene*	
H. monniera HBK. (= *Bacopa monnieri* (L.) Pennell)	nt	*Triterpene*	
Scoparia (20)			Trop. Am.
S. dulcis L.	±	None reported	
Verbascum (360)			N. temp. Euras.
V. nigrum L.	nt	*Triterpene*	
V. phlomoides L.	nt	*Triterpene*	
V. thapsus L.	nt	*Rotenone; triterpene*	
SIMAROUBACEAE (20/ 120)			Trop. & subtrop.
Brucea (10)			Paleotrop.
B. antidysenterica Lamk.	nt	*Triterpene*	
B. javanica (L.) Merr.	nt	*Triterpene*	
Simarouba (7)			Trop. Am.
S. amara Aubl.	nt	*Triterpene*	
S. glauca DC.	nt	*Triterpene*	
S. versicolor St.-Hil.	±	*Triterpene*	

Table 4 *cont'd.*

Family/Genus/Species	Molluscicidal Activity[a]	Type of Molluscicide Present[b]	Geographic Distribution[c]
SOLANACEAE (90/2,000)			Trop. & temp.; chief center is C. & S. Am.
Capsicum (50)			C. & S. Am.
C. *annuum* L.	nt	*Steroid*	
C. *frutescens* L.	+	None reported	
Cestrum (150)			Warm. Am.; W.I.
C. *diurnum* L.	±	*Steroid*	
C. *laurifolium* L'Hérit.	+	None reported	
C. *macrophyllum* Vent.	+	None reported	
C. *parqui* L'Hérit.	nt	*Solasonine*	
Cyphomandra (30)			C. & S. Am.; W.I.
C. *betacea* (Cav.) Sendt.	nt	*Solasodine glycoside*	
Datura (10)			Trop. & warm temp.
D. sp.	−	None reported	
D. *suaveolens* H.&B. ex Willd.	−	None reported	
Lycopersicon (7)			Pacif.; S. Am.; Galapagos
L. *cheesmanii* Riley	nt	*Tomatine*	
L. *chilense* Dunal	nt	*Tomatine*	
L. *esculentum* P.Miller	±	*Steroid; tomatine*	
L. *glandulosum* C.H. Muller	nt	*Tomatine*	
L. *hirsutum* Humb. & Bonpl.	nt	*Tomatine*	
L. *peruvianum* (L.) P.Miller	nt	*Tomatine*	
Nicotiana (46)			Extratrop. N. & S. Am.; Australia; Polynes.
N. *plumbaginifolia* Viv.	nt	*Solasodine glycoside*	
N. *tabacum* L.	±	None reported	
Physalis (100)			Cosmop.; esp. Am.
P. *angulata* L.	−	None reported	
Solanum (1,700)			Trop. & temp.
S. *ciliatum* Lamk.	−	None reported	
S. *hispidum* Pers.	nt	*Steroid*	
S. *intrusum* Soria	nt	*Steroid*	
S. *mammosum* L.	+	*Solasonine, solamargine*	

Table 4 *cont'd.*

Family/Genus/Species	Molluscicidal Activity[a]	Type of Molluscicide Present[b]	Geographic Distribution[c]
S. *melongena* L.	−	*Solasonine*	
S. *nigrum* L.	nt	*Solasonine, solamargine*	
S. *nodiflorum* Jacquin	+	*Solasonine*	
S. *paniculatum* L.	−	*Steroid*	
S. *polyadenium* Greenm.	nt	*Steroid; tomatine*	
S. *schimperianum* Hochst.	nt	*Solasodine glycoside*	
S. *seaforthianum* Andrews	−	None reported	
S. *torvum* Swartz	±	*Solasonine*	
S. spp.	nt	At least 100 additional *Solanum* species contain *spirosolane glycosides*	
STERCULIACEAE (60/700)			Chiefly trop.
Guazuma (4)			Trop. Am.
G. *ulmifolia* Lamk.	±	None reported	
STYRACACEAE (12/180)			Three centers: E. As. to W. Malaysia; S.E. U.S.; Mex. to trop. S. Am.
Styrax (130)			Warm Euras.; Malaysia; Am.
S. *japonica* Sieb. & Zucc.	nt	*Triterpene*	
S. *officinalis* L.	+	None reported	
THEACEAE (16/500)			Trop. & subtrop.
Camellia (82)			Indomal.; China; Japan
C. *oleifera* Abel	nt	Triterpene type	
C. *sasanqua* Thunb.	nt	Triterpene type	
C. *sinensis* (L.) Kuntze (=*Thea oleosa* Lour. = *Thea sinensis* L.)	nt	Triterpene type	
Schima (15)			E. Himal. to Formosa; Micrones.; W. Malaysia
S. *argentea* E.Pritz. ex Diels	±	Triterpene type	
S. *mertensiana* Koidz.	nt	*Triterpene*	

Table 4 *cont'd.*

Family/Genus/Species	Molluscicidal Activity[a]	Type of Molluscicide Present[b]	Geographic Distribution[c]
Ternstroemia (100)			Trop. (2 Afr.; 1 Queensl.)
T. japonica Thunb.	nt	Triterpene type	
THYMELAEACEAE (50/500)			Temp. & trop., esp. in Afr.
Gnidia (100)			Trop. & S. Afr.; Madag.; S.W. Saudi Arabia; W. coast of India; Sri Lanka
G. kraussiana Meisn. (= *Lasiosiphon kraussianus* (Meisn.) Hutch. & Dalziel)	+	None reported	
TILIACEAE (50/450)			Trop. & temp.
Corchorus (100)			Warm regions: chiefly S.E. As. and Brazil
C. sp.	−	None reported	
TROPAEOLACEAE (2/92)			Mex. to temp. S. Am.
Tropaeolum (90)			Mex. to temp. S. Am.
T. majus L.	−	None reported	
TURNERACEAE (7/120)			Chiefly trop. Am. & Afr.
Turnera (60)			Trop. & subtrop. Am.; S.W. Afr. (1)
T. sp.	±	None reported	
T. sp.	−	None reported	
ULMACEAE (15/200)			Trop. & temp.
Trema (30)			Trop. & subtrop.
T. guineensis Priemer	−	None reported	
Ulmus (45)			N. temp.; S. to Himal.; Indoch. & Mex.
U. macrocarpa Hance	−	None reported	
UMBELLIFERAE (275/2,850)			Cosmop.; chiefly N. temp.
Ammi (10)			Azores; Madeira; Medit.; temp. W. As.
A. majus L.	+	*Bergapten, isopimpinellin, xanthotoxin*	
Bupleurum (150)			Eur.; As.; Afr.; N. Am.

Cont'd.

Table 4 *cont'd.*

Family/Genus/Species	Molluscicidal Activity[a]	Type of Molluscicide Present[b]	Geographic Distribution[c]
B. falcatum L.	nt	*Triterpene*	
B. longeradiatum Turcz.	nt	*Triterpene*	
B. rotundifolium L.	nt	*Triterpene*	
Centella (40)			Afr.; Australia; N.Z.; Am.
C. asiatica (L.) Urban	−	*Triterpene*	
C. coriacea Nannf.	nt	*Triterpene*	
Eryngium (230)			Trop. to temp. (exc. trop. & S. Afr.)
E. bromeliaefolium F.Delaroche	nt	*Triterpene*	
E. foetidum L.	−	None reported	
E. sp.	−(V)	None reported	
Hydrocotyle (100)			Trop. & temp.
H. vulgaris L.	nt	*Triterpene*	
Ladyginia (1)			C. As.
L. bucharica Lipsky	nt	*Triterpene*	
URTICACEAE (45/550)			Trop. to temp.
Boehmeria (100)			Trop. & N. subtrop.
B. caudata Swartz	nt	*Triterpene*	
B. nivea (L.) Gaudich.	±	None reported	
Cecropia (100)			Trop. Am.
C. carbonaria K.von Mart. ex Miq.	−	None reported	
Urera (35)			Hawaii; warm Am.; trop. & S. Afr.; Madag.
U. baccifera Gaudich.	−	None reported	
U. caracasana Griseb.	−	None reported	
U. sp.	−	None reported	
VALERIANACEAE (13/400)			Eur.; As.; Afr.; Am.
Patrinia (20)			C. As. & Himal. to E. As.
P. intermedia Roem. & Schult.	nt	*Triterpene*	
P. scabiosaefolia Link	nt	*Triterpene*	
VERBENACEAE (75/3,000)			Almost all trop. & subtrop.
Citharexylum (115)			U.S. to Argent.
C. fruticosum L.	−	None reported	

Table 4 *cont'd.*

Family/Genus/Species	Molluscicidal Activity[a]	Type of Molluscicide Present[b]	Geographic Distribution[c]
Clerodendrum (400)			Trop. & subtrop.
C. *dusenii* Guerke	−	None reported	
C. *fragrans* (Vent.) R.Br.	±	None reported	
Lantana (150)			Trop. Am., W.I.; trop. & S. Afr.
L. *camara* L.	±	None reported	
Lippia (220)			Trop. Am.; Afr.
L. *aristata* Schauer	+(V)	None reported	
L. *dulcis* Trevir.	−	None reported	
L. *sidoides* Cham. aff.	+(V)	None reported	
L. sp.	±(V)	None reported	
L. sp.	−(V)	None reported	
L. *thymoides* Mart. & Schauer	±(V)	None reported	
Petitia (2)			W.I.
P. *domingensis* Jacquin	−	None reported	
Stachytarpheta (100)			Am.; some spp. trop. weeds
S. *cayenensis* (L.C. Richard) Vahl	−	None reported	
Vitex (250)			Trop. & temp.
V. *agnus-castus* L.	−(V)	None reported	
V. *oxycuspis* Baker	+	None reported	
VITACEAE (12/700)			Mostly trop. & subtrop.
Cissus (350)			Trop., rarely subtrop.
C. *oreophila* Gilg & Brandt	−	None reported	
C. *populnea* Guillem. & Perrot.	−	None reported	
C. *sicyoides* L.	−	None reported	
ZYGOPHYLLACEAE (25/240)			Trop. & subtrop.
Balanites (25)			Trop. Afr. to Burma
B. *aegyptiaca* (L.) Delile	+	*Balanitin 1; balanitin 2 balanitin 3*	
B. *maughamii* Sprague	+	None reported	
B. *roxburghii* Planch.	+	*Steroid*	

Cont'd.

Table 4. *cont'd.*

Family/Genus/Species	Molluscicidal Activity[a]	Type of Molluscicide Present[b]	Geographic Distribution[e]
Kallstroemia (23)			Chiefly S. U.S.; W.I. to Argent.; also N. to N.E. Australia
K. *pubescens* (G.Don) Dandy	nt	*Steroid*	
Larrea (2)			Temp. S. Am.
L. *divaricata* Cav.	nt	Triterpene type	
Peganum (6)			Medit. to Mongolia; S. U.S.; Mex.
P. *harmala* L.	±	None reported	
Tribulus (20)			Trop. to subtrop.
T. *terrestris* L.	±	*Steroid*	

[a]+ = test extract was reported active at 100 ppm or less; ± = active at more than 100 ppm; − = inactive; nt = extract not tested but potentially active due to the type of chemical constituent known to be present; (V) = water saturated with essential oil from the plant was tested; (sd) = seed extract was tested.

[b]*Triterpene* (monodesmosidic), triterpene (bidesmosidic), triterpene type (structure not elucidated to the point where sugar chains and their attachment are known), *steroid* (monodesmosidic spirostane type), steroid (bidesmosidic spirostane type), steroid type (structure not elucidated to the point where sugar chains and their attachment are known), *steroid alkaloid* (monodesmosidic), steroid alkaloid (aglycone), other terms used are self-explanatory (if underlined, the substance indicated has molluscicidal properties).

[c]General geographic distribution for the family or genus indicated according to Willis and Airy Shaw (1973) or Engler (1964); Medit. includes the entire Mediterranean basin, S. Eur. from Portugal to the Balkan Peninsula, Asia Minor, Palestine, and N. Afr. from Morocco to Egypt; C. As. covers the Kazakh, Turkmen, Uzbek, Kirgiz, and Tadzhik Republics and the Chinese province of Sinkiang; E. As. covers China, Korea, and Japan; S.E. As. covers Burma, Thailand, and Indochina (Vietnam, Laos, and Cambodia) and often the adjacent parts of S. China from Yunnan to Hong Kong and Hainan; W. Malaysia covers the Malay Peninsula, Sumatra, Java, Borneo, and the Philippine Is.; C. Am. covers Guatemala to Panama. Other geographic regions should be self-explanatory.

[d]Number of genera/species in the family according to Willis and Airy Shaw or Engler.

[e]Number of species in the genus according to Willis and Airy Shaw or Engler.

[f]The validity of this binomial has not been confirmed.

Predicted Active Species

It seems logical to predict that plants not tested for molluscicidal activity but known to contain constituents with demonstrated molluscicidal activity would be active if tested. Knowing the concentration of the active constituent or constituents in the plant or plant part would be important in making a valid prediction. From data available and presented in Tables 2 and 4, which list only a few of the plants known to contain active molluscicides, there is evidence that this exercise may be of value in identifying what we term predicted active species. For example, 566 species of angiosperms have been

Table 5. Dicot Families Promising for Molluscicidal Activity

Family	Known Species (Estimated)	Active Species vs. Species Tested	Not Tested but Predicted Active Species
Acanthaceae	2,500	3/8	1
Anacardiaceae	600	4/7	15
Annonaceae	2,100	4/8	–
Apocynaceae	1,500	7/9	2
Araliaceae	700	2/3	17
Asclepiadaceae	2,000	3/3	–
Bignoniaceae	650	2/3	–
Canellaceae	16	2/2	–
Caryophyllaceae	1,750	0/0	15
Chrysobalanaceae	400	4/4	–
Combretaceae	600	8/12	2
Compositae	13,000	13/26	19
Cucurbitaceae	640	4/6	21
Euphorbiaceae	5,000	27/46	2
Hippocastanaceae	15	0/0	3
Lauraceae	2,500	3/3	–
Leguminosae	12,000	76/94	56
Meliaceae	1,400	4/5	1
Myrtaceae	3,000	11/17	–
Olacaceae	250	1/1	3
Phytolaccaceae	100	5/8	5
Piperaceae	2,000	3/8	–
Polygalaceae	800	1/4	3
Polygonaceae	800	12/16	–
Primulaceae	1,000	0/0	6
Proteaceae	1,050	0/0	11
Ranunculaceae	1,900	3/3	12
Rhamnaceae	2,000	2/4	3
Rosaceae	2,000	1/2	3
Rubiaceae	6,000	16/32	5
Sapindaceae	2,000	4/7	6
Sapotaceae	800	1/1	3
Scrophulariaceae	3,000	3/3	7
Simaroubaceae	120	1/1	5
Solanaceae	2,000	9/16	ca. 114
Verbenaceae	3,000	7/14	–
Zygophyllaceae	240	5/5	1

reported tested for molluscicidal activity, with 288 showing some activity and 276 being reported as inactive (Table 6). An additional 435 species not previously tested for molluscicidal activity would be predicted active, based on the reported presence in them of one or more active molluscicidal agents.

Table 6. Summary of Plants Tested for Molluscicidal Activity[a]

	Total	+	±	−	Not Tested but Predicted Active
Monocots					
Species Tested	54	5	14	35	50
Genera Represented	37				
Families Represented	18				
Dicots					
Species Tested	512	108	161	241	384
Genera Represented	321				
Families Represented	91				
Totals					
Species Tested	566	113	175	276	435
Genera Represented	358				
Families Represented	109				

[a]Data compiled from Tables 2 and 4.

In the group of 276 species of angiosperms that were inactive when tested for molluscicidal activity, 16 species contain chemical compounds with known molluscicidal activity, and thus extracts from these plants should have shown molluscicidal activity. As indicated previously, plant parts were not taken into consideration in Tables 2 and 4 and most of the unexpected negative data might be due to the absence of the active compound or compounds in the plant part tested. On the other hand, of the 288 species showing molluscicidal activity, 46 could have been predicted as active, since they contained one or more active molluscicides.

The 435 predicted active species were identified on the basis of incomplete knowledge of the distribution and occurrence in plants of monodesmosidic triterpene and spirostanol saponins; monodesmosidic spirosolane saponins; the sesquiterpenes ambrosin, damsin, muzigadial, and warburganal; the aliphatic amide affinin (spilanthol); and alkenyl phenols (anacardic acids). Data were unavailable that would allow consideration of all sesquiterpene lactones or aliphatic amides, for example, as having molluscicidal activity. If such data were available, hundreds of additional predicted active plants could be identified on the basis of published reports of their occurrence.

Additional probable predictive characters were not used to identify mollusicidal plants in the present exercise but should be used to provide as complete a list as possible. These predictive characters could be obtained from published reports on the piscicidal, insecticidal, and anthelmintic activity of plants, including ethnomedical and experimental data, as well as from the results of tests said to be screening plants for the presence of saponins.

A literature search on the occurrence of rotenone, bergapten, chalepensin, isopimpinellin, xanthotoxin (8-methoxypsoralen), and ent-kaur-16-en-19-oic acid (kaurenoic acid), which are reported present in hundreds of plant species, would provide a large number of additional predicted active plants. For example, rotenone most likely is present in most species of *Derris*, *Entada*, *Millettia*, *Mundulea*, *Piscidia*, and *Tephrosia*, of the Leguminosae, but not in other taxa. The molluscicidal furanocoumarins bergapten, chale-pensin, isopimpinellin, and xanthotoxin are widespread in the Rutaceae, especially *Citrus*, *Orixa*, *Poncirus*, *Ptelea*, and *Ruta*; in the Umbelliferae, especially species of *Ammi*; and in the Leguminosae, *Psoralea* species. The anacardic acids seem to be ubiquitous in members of the Anacardiaceae (*Gluta*, *Holigarna*, *Melanorrhea*, *Metopium*, *Pentaspadon*, *Semecarpus*, *Smo-dingium*, and *Toxicodendron* species) and the Proteaceae (*Cardwellia*, *Gre-villea*, *Hakea*, *Opisthiolepis*, *Persoonia*, and *Petrophilia* species). Aliphatic amides such as affinin (spilanthol) are to be anticipated in selected taxa of the Compositae (*Heliopsis*, *Spilanthes*, and *Wedelia* species) and especially of the Piperaceae (*Piper* species). The molluscicidal activity of triacontan-1-ol, the flavonoids, and the rotenoids (Table 1) should be confirmed before considering them as predictive characters to identify potential molluscicidal plants.

Even though we have not been able to make full use of information on the occurrence of all molluscicidal principles in plants as predictive characters, it has been possible to identify 435 predicted active species. That is about one and one-half as many species as have shown molluscicidal activity on the basis of actual experimentation.

POTENTIAL TOXICOLOGIC PROBLEMS

Because certain groups of plants appear more promising than others and the number of candidates that might eventually be subjected to molluscicidal bioassay is almost endless, a few comments regarding the general toxicologic potential of some taxa seem appropriate. If it can be reasonably predicted that the molluscicidal activity of a plant is due to the known or predicted presence of a known toxic principle, such a plant might be eliminated as a practical means for snail control with a saving of time and money.

A mixture of alkenyl phenols, referred to as anacardic acids, has been shown to be the active molluscicide in *Anacardium occidentale* (Anacar-diaceae) (Sullivan et al., 1982). Alkenyl phenols, in general, are highly unstable compounds and are known to induce allergenic reactions or vesicant effects in persons exposed to plants containing them. For example, these compounds are responsible for the toxic effects of poison-ivy, poison-oak, and poison-sumac dermatitis (Baer, 1979). Hence, plants of the Anacar-diaceae may well be unsuitable for general use as molluscicides.

The furanocoumarins are reported to be active as molluscicides at levels as low as 2 ppm (Schonberg and Latif, 1954). However, since the furanocoumarins are considered phototoxic to humans and the coumarins are generally believed to be hepatotoxic, it would not be advisable to expose people to plants containing high levels of compounds with this skeletal structure (Musajo and Rodighiero, 1962; Pathak and Fitzpatrick, 1959). Furanocoumarins are generally found not only in most species of the Umbelliferae but also in the Leguminosae, Moraceae, and Pittosporaceae, and less frequently in other taxa (Hegnauer, 1971).

The potential toxicity of plants whose molluscicidal activity may be due to rotenone or rotenoids must be recognized. Rotenoids are generally restricted to the Leguminosae and specifically species of *Amorpha, Dalbergia, Derris, Entada, Lonchocarpus, Millettia, Mundulea, Neorautanenia, Pachyrrhizus, Piscidia, Swartzia,* and *Tephrosia* (Hegnauer, 1971). A report of rotenone in *Verbascum thapsus* (Scrophulariaceae) is unexpected and requires confirmation. Rotenone might be considered rather innocuous if ingested, since the oral LD_{50} in rats is 1.5 g/kg. Studies have shown, however, that a single dose as low as 10 mg/kg, administered to pregnant rats, produces adverse effects on the offspring (Haley, 1978; Khera et al., 1982).

Species of the Euphorbiaceae and Thymelaeaceae that have molluscicidal activity (see Table 4) most likely contain phorbol esters that initiate or promote tumors or are irritants. In the Euphorbiaceae, phorbol esters are frequently encountered in species of *Aleurites, Baliospermum, Croton, Elaeophorbia, Euphorbia, Excoecaria, Hippomane, Hura, Micrandra, Cunuria, Ostodes, Pedilanthus, Sapium, Stillingia,* and *Synadenium.* In the Thymelaeaceae, they appear to occur only in the genera *Aquilaria, Daphne, Daphnopsis, Dirca, Gnidia, Lasiosiphon, Pimelia,* and *Synaptolepis* (Hecker and Schmidt, 1974; Kinghorn, 1979). The molluscicidal use of these plants or others known to contain this type of phorbol esters obviously should not be considered, even though they may be highly active.

There may, however, be taxa of the Euphorbiaceae that show molluscicidal activity but do not contain these compounds. We have not encountered phorbol esters in species classified in tribes lower than 39 of this plant family, and literature reports seem to confirm that the phorbol esters do not occur in the lower tribes of this taxon. They are frequently found, and are the major toxic constituent, only in certain species classified in tribes 39 to 52. The classification system is that of Webster (1975). Of the 17 genera of Euphorbiaceae listed in Table 4, nine – *Acalypha, Alchornea, Antidesma, Bridelia, Hevea, Manihot, Phyllanthus, Putranjiva,* and *Ricinus* – are classified in tribes lower than 39. Accordingly, one would not expect toxicity or molluscicidal activity for species of these genera to be due to phorbol esters. However, if the species representing tribes 39 to 52 are dismissed, only *Bridelia atroviridis* and *Putranjiva roxburghii* appear to be of interest as potential molluscicides.

In a major program designed to identify highly effective molluscicides, in which the intent is to employ plants in powdered form or as extracts, a knowledge of chemotaxonomic and biochemotaxonomic relationships of the plant kingdom will prove invaluable.

One of the most interesting molluscicidal plants that most likely could be considered for practical use is *Phytolacca dodecandra*. But related species, especially *P. americana,* contain highly active mitogenic lectins (McPherson, 1979; Stobo, 1980). To the best of our knowledge, *P. dodecandra* has not been studied for mitogenic activity.

When consideration is being given to placing plant material into water supplies that people may use, its potential for producing contact irritant or allergenic effects must be considered. Several classes of chemical compounds are capable of producing allergic contact dermatitis in humans. The most important are the sesquiterpene lactones, which occur in hundreds of species of plants. It has been shown that almost any sesquiterpene having an α, β-unsaturated lactone moiety (Mitchell and Dupuis, 1971; Mitchell and Rook, 1979; Mitchell et al., 1970; Towers, 1979) will be allergenic for humans. This moiety is also an essential feature for cytotoxicity in this group of compounds (Kupchan et al., 1971; Lee et al., 1971). The general subject of dermatologic problems caused by exposure to plants and plant-contained principles has been thoroughly reviewed by Mitchell and Rook (1979). The sesquiterpene lactones, which are widespread in nature, are found primarily in plants of the Compositae, one of the largest of the plant families. They are also found in the Aristolochiaceae, Labiatae, Lauraceae, Magnoliaceae, and Umbelliferae of the dicots and in one species of gymnosperms but appear to be absent from the monocots (Fischer et al., 1979).

CONCLUSIONS

The literature to date indicates that only about 566 species of flowering plants have been tested for molluscicidal activity, and about half of them are reported to produce some killing of snails. One might presume, then, that in a random-selection screening program about half of all plants would show some molluscicidal effect. However, this generalization would be valid only if the test sample were truly random. Available data suggest that the sample we have discussed does not represent the worldwide distribution of plants. The vast majority of species reported tested were collected in Brazil, Egypt, Nigeria, Puerto Rico, and the Sudan (Chapter 3). Furthermore, it has been suggested that most of the plants tested were selected on the basis of local use in traditional medicine. If so, selection appears to have been made on nonlogical grounds, most likely on the basis of ready availability.

If we consider that there are about 53 monocot families, that only 18 of them are represented by the 54 species tested for molluscicidal activity, and that only 19 of the species have shown molluscicidal activity, we must conclude that the sample size is exceedingly small. Similarly, of the estimated 292 dicot families, only 91 are represented by the 512 species tested, of which 269 were active. Thus, only about one third of the angiosperm families (109 of 345) have been sampled; of 566 species tested, 288 showed activity.

Forty-four plant-derived chemical compounds of known structure have been reported to have molluscicidal activity (Table 1). Our preliminary study has attempted to extrapolate data concerning the occurrence in plants of a few of these, including the monodesmosidic triterpene and spirostanol saponins and the monodesmosidic spirostanol (steroid alkaloid) glycosides, whose presence would seem to predict that the species would have mol- ᵢ luscicidal activity. In this way we have identified (Tables 2 and 4) 50 monocot and 385 dicot species which, although they have not been tested, should be considered predicted active molluscicidal plants. Toxicologic considerations, based initially on data from the literature, could determine whether further studies on any of these species should be undertaken.

No justification or need exists at this time for more blind screening of plants to identify new taxa for consideration as practical molluscicides. It might be profitable, however, to extend the list of predicted active species by carrying out a systematic search of the literature and compiling data already published on results from screening plants for saponins and experimental and ethnomedical information on piscicidal and insecticidal plants. These data could be analyzed and a number of species selected that appear most promising on the basis of geographic location, potential human toxicity, ready availability in endemic areas, and other factors. From this list, species could be selected for comparative bioassay procedures to isolate those having maximal molluscicidal and minimal piscicidal activity and therefore likely candidates for toxicologic and field studies.

Acknowledgments

Certain aspects of the data collection required to compile this report were supported, in part, by a contract from the National Cancer Institute (USA, CM-97259), as well as by funds from the World Health Organization Special Programme for Research Development and Research Training in Human Reproduction (78135) and the UNDP/WORLD BANK/WHO Special Programme for Research and Training in Tropical Diseases. We would like to thank the following persons for their assistance in collecting and organizing the data: Professor A. S. Bingel, Dr. W. D. Loub, Ms. C. Lewandowski, Ms. M. L. Quinn, Mrs. P. Canas, and Mrs. P. Tai. Special thanks are extended to Dr. H. Kloos for providing valuable references and to Dr. K. E. Mott for his interest and encouragement. We also acknowledge Ms. D. L. Guilty and Ms. D. M. Lattyak for typing the manuscript on short notice.

REFERENCES

Agarwal, S. K., and R. P. Rastogi. Triterpenoid saponins and their genins. *Phytochemistry* 13:2623–2645 (1974).

Alzérreca, A., and G. Hart. Molluscicidal steroid glycoalkaloids possessing stereoisomeric spirosolane structures. *Toxicol. Lett.* 12:151–155 (1982).

Baer, H. The poisonous Anacardiaceae. In *Toxic Plants*, ed. A. D. Kinghorn, 161–170. New York: Columbia University Press (1979).

Basu, N., and R. P. Rastogi. Triterpenoid saponins and sapogenins. *Phytochemistry* 6:1249–1270 (1967).

Boiteau, P., B. Pasich, and A. Rakoto Ratsimamanga. *Les triterpenoides en physiologie végétale et animale*. Paris: Gauthier-Villars (1964).

Chandel, R. S., and R. P. Rastogi. Triterpenoid saponins and sapogenins: 1973–1978. *Phytochemistry* 19:1889–1908 (1980).

Engler, A. *Syllabus der Pflanzenfamilien*. 12th ed. Vol. 2. Berlin-Nikolassee: Gebrüder Borntraeger (1964).

Farnsworth, N. R. Biological and phytochemical screening of plants. *J. Pharm. Sci.* 55:225–276 (1966).

Farnsworth, N. R., W. D. Loub, D. D. Soejarto, and G. A. Cordell. Computer services for research on plants for fertility regulation. *Korean J. Pharmacognosy* 12:98–110 (1981).

Fischer, N. H., E. J. Olivier, and H. D. Fischer. The biogenesis and chemistry of sesquiterpene lactones. *Fortschr. Chem. Org. Naturst.* 38:47–390 (1979).

Haley, T. J. A review of the literature of rotenone. *J. Environ. Pathol. Toxicol.* 1:315–337 (1978).

Hecker, E., and R. Schmidt. Phorbolesters – the irritants and cocarcinogens of *Croton tiglium* L. *Fortschr. Chem. Org. Naturst.* 31:377–467 (1974).

Hegnauer, R. Chemical patterns and relationships of Umbelliferae. In *The Biology and Chemistry of the Umbelliferae*, ed. V. H. Heywood, 267–277. London: Academic Press (1971).

Hostettmann, K. Saponins with molluscicidal activity from *Hedera helix* L. *Helv. Chim. Acta* 63:606–609 (1980).

Hostettmann, K., M. Hostettmann-Kaldas, and K. Nakanishi. Molluscicidal saponins from *Cornus florida* L. *Helv. Chim. Acta* 61:1990–1995 (1978).

Hostettmann, K., H. Kizu, and T. Tomimori. Molluscicidal properties of various saponins. *Planta Med.* 44:34–35 (1982).

Khera, K. S., C. Whalen, and G. Angers. Teratogenicity study on pyrethrum and rotenone (natural origin) and Ronnel in pregnant rats. *J. Toxicol. Environ. Health* 10:111–119 (1982).

Kinghorn, A. D. Cocarcinogenic irritant Euphorbiaceae. In *Toxic Plants*, ed. A. D. Kinghorn, 137–159. New York: Columbia University Press (1979).

Kupchan, S. M., M. A. Eakin, and A. M. Thomas. Tumor inhibitors: 69. Structure-cytotoxicity relationships among the sesquiterpene lactones. *J. Med. Chem.* 14:2105–2126 (1971).

Lee, K.-H., E.-S. Huang, C. Piantadosi, J. S. Pagano, and T. A. Geissman. Cytotoxicity of sesquiterpene lactones. *Cancer Res.* 31:1649–1654 (1971).

Mahato, S. B., A. N. Ganguly, and N. P. Sahu. Steroid saponins. *Phytochemistry* 21:959–978 (1982).

McPherson, A. Pokeweed and other lymphocytic mitogens. In *Toxic Plants*, ed. A. D. Kinghorn, 83–102. New York: Columbia University Press (1979).

Mitchell, J. C., and G. Dupuis. Allergic contact dermatitis from sesquiterpenoids of the Compositae family of plants. *Br. J. Dermatol.* 84:139–150 (1971).

Mitchell, J., and A. Rook. *Botanical Dermatology – Plants and Plant Products Injurious to the Skin.* Vancouver: Greengrass (1979).

Mitchell, J. C., B. Fritig, B. Singh, and G. H. N. Towers. Allergic contact dermatitis from *Frullania* and Compositae. *J. Invest. Dermatol.* 54:233–239 (1970).

Musajo, L., and G. Rodighiero. The skin-photosensitizing furocoumarins. *Experientia* 18:153–160 (1962).

Pathak, M. A., and T. B. Fitzpatrick. Relationship of molecular configuration to the activity of furocoumarins which increase the cutaneous responses following long-wave ultraviolet radiation. *J. Invest. Dermatol.* 32:255–262 (1959).

Roddick, J. G. The steroidal glycoalkaloid α-tomatine. *Phytochemistry* 13:1–25 (1974).

Schonberg, A., and N. Latif. Furochromones and coumarins: XI. The molluscicidal activity of bergapten, isopimpinellin and xanthotoxin. *J.A.C.S.* 76:6208 (1954).

Schreiber, K. Steroid alkaloids: The *Solanum* group. *Alkaloids* (London) 10:1–192 (1968).

Stobo, J. D. Mitogens. *Clin. Immunobiol.* 4:55–77 (1980).

Sullivan, J. T., C. S. Richards, H. A. Lloyd, and G. Krishna. Anacardic acid: Molluscicide in cashew nut shell liquid. *Planta Med.* 44:175–177 (1982).

Towers, G. H. N. Contact hypersensitivity and photodermatitis evoked by Compositae. In *Toxic Plants,* ed. A. D. Kinghorn, 171–183. New York: Columbia University Press (1979).

Tschesche, R., and G. Wulff. Chemie und biologie der Saponine. *Fortschr. Chem. Org. Naturst.* 30:461–606 (1973).

Webster, G. L. Conspectus of a new classification of the Euphorbiaceae. *Taxon* 24:593–614 (1975).

Willis, J. C., and H. K. Airy Shaw. *A Dictionary of the Flowering Plants and Ferns.* 8th ed. Cambridge: The University Press (1973).

6

ECOSYSTEM DATA BASE AND GEOGRAPHIC DISTRIBUTION OF MOLLUSCICIDAL PLANTS

G. E. Wickens

Royal Botanic Gardens
Kew, Richmond, Surrey, England

Human survival depends on plants as a source of food, fuel, timber, fiber, fodder, thatching, medicines – even poisons. To use plants efficiently, they need to be identified and classified, and the large array of information about them needs to be communicated to all concerned. This chapter investigates the sources of information available to help in the search for molluscicidal plants.

TAXONOMY

Although no accurate figures are available, the total number of higher plant species (flowering plants and ferns) is probably around 500000, ranging in form and size from minute duckweeds to giant forest trees (Brenan, 1981a). Obviously, some form of internationally recognized system for naming and classifying them is essential.

The International Code of Botanical Nomenclature has been developed and continuously modified for this purpose. It enables us to pigeonhole the taxa (a taxonomic group of any rank) or groups of taxa into families, genera, species, subspecies, varieties, and so forth. The names, authors, and places of publication of these taxa are listed in the *Index Kewensis* (Jackson, 1893–1895) and its quinquennial supplements (subspecies, varieties, etc. are only included in the latest supplement, Brenan, 1981b). References to taxonomic monographs, revisions, name changes, etc. are published annually in *The Kew Record of Taxonomic Literature* (Royal Botanic Gardens, Kew, 1972–).

Species within a geographic area may be identified using various floras at regional, national, and district levels of varying standards of completeness and reliability. Unfortunately, the compilation of regional floras is exceedingly time consuming because of taxonomists' desire for perfection. On the other hand, local district floras may be produced hurriedly to fulfill a need, with less attention paid to taxonomic niceties.

Provided the species described in the flora has been accurately determined, the use of a synonym instead of the correct nomenclature should not unduly confuse taxonomic botanists. The misapplication of a name, however, can cause considerable confusion when herbarium specimens are not cited for future reference. The same is true of plants named in works covering such fields as ecology, anatomy, physiology, cytology, and biochemistry.

Experimental work should always be supported by properly preserved and annotated herbarium specimens safely deposited in a herbarium whose location is recorded in the published work. These are called voucher specimens. They provide necessary checks when similar experiments with allegedly the same species produce divergent results. For example, critical taxonomic research carried out at Kew by Charles Jeffrey (personal communication, 1984) has shown that *Ambrosia senegalensis* is a distinct species and should not be regarded as a synonym of *A. maritima*, a species under investigation as a molluscicide in Egypt (El-Magdoub, 1980). Voucher specimens received from Senegal and Burkina Faso confirmed that *A. senegalensis* was being used in their trials and not *A. maritima*; both species appear to be equally efficacious (Vassiliades and Diaw, 1980).

The taxonomic treatment of plants through their hierarchical classification is very uneven. Some families are neglected because they are too obscure, uninteresting, or unimportant, while others may be popular and overworked. Still others contain taxa so difficult that they require a lifetime's devotion.

Taxonomic botanists vary in their ability, standards, and training and are influenced by the availability of comprehensive herbarium collections and reference libraries. The herbaria officially recognized around the world are cited in the *Index Herbariorum* (Holmgren et al., 1981). The size and scope of their collections are also listed. The larger herbaria, such as those in Brussels, Geneva, Kew, St. Louis, and Paris, generally have more comprehensive collections of plants and literature and are consequently more likely to provide a correct name. On the other hand, the smaller local herbaria are often able to provide faster identifications but may not have the facilities for revising their collections.

Since relatively few species are involved in the search for plant molluscicides, I would recommend the duplication of specimens for identification; should the identifications differ, they can be queried.

Major herbaria are listed in Table 1.

SOURCES OF INFORMATION

The two basic sources of information that can be used in the search for effective plant molluscicides – the herbarium and the literature – should supplement each other, not be used in isolation. Unfortunately, for both sources the information is often scattered and therefore time consuming to collate.

The herbaria of the world are ever increasing in size. The volume of material within an institution is such that the manual retrieval of information can be exceedingly difficult; information is being buried in a data crypt (Hall, 1974). More efficient cataloging of collections is needed to make the information they contain more readily available.

Although many herbaria have the information filed away in card systems, a small but growing number have computer data banks; they are listed in Table 2. They are generally small, relatively new herbaria, rather parochial in content and mainly confined to the New World. The older and larger herbaria cannot afford to undertake the expensive and time-consuming exercise of data capture.   ·

There are also a few specialized plant data banks, some of which can be useful sources of information; they are listed in Table 3.

A computerized herbarium data bank would appear to be a useful system for storing and extracting information on plants, their distribution, habitat, uses, and so forth. However, experience at the Botanical Research Institute, Pretoria, suggests otherwise. It is the largest herbarium to have computerized its entire stock, the progress of which has been well documented (Morris and Glen, 1978; Morris, 1980; Morris and Manders, 1981). A statistical analysis of the frequency of the various descriptors assigned to a taxon revealed that a great deal of the computer storage space was being wasted because of lack of data (Morris and Manders, 1981). Thus only 1% of the specimens had information on economic use. Similarly, a survey of the 2.5 million specimens at Harvard University produced only 5000 new or little-known species used for food, medicine, and other biological properties (Altschul, 1973, 1977).

One reason for this shortage of information is that collectors fail to appreciate that their specimens, which to them are merely a convenient way of obtaining identifications, are regarded by taxonomists as permanent records; herbaria destroy specimens only if they are too poor to identify. Unfortunately, collectors sometimes believe that publication elsewhere of the use of a plant will eliminate the need for such information on the herbarium sheet (that is, a voucher specimen). In time, name changes, more critical taxonomic judgment, and so forth make the necessary correlation between specimens and publication impossible; the information can no longer be verified.

Table 1. Major Herbaria (More Than 500,000 Specimens) with Specialist Interests in the Tropics and Subtropics

Herbarium	No. of Specimens	Interests
NORTH AMERICA		
Herbarium, Department of Botany, University of California, Berkeley, California 94720, USA	1,500,000	N. & S. Am., Pac. Basin
Herbaria of Rancho Santa Ana Botanic Garden and Pomona College, 1500 N. College Avenue, Claremont, California 94711, USA	606,000	Calif., Mex.
Herbarium, Department of Botany, California Academy of Sciences, Golden Gate Park, San Francisco, California 94118, USA	1,400,000	S.W. U.S., Mex.
Dudley Herbarium of Stanford University, Department of Botany, California Academy of Sciences, Golden Gate Park, San Francisco, California 94118 USA	850,000	S.W. U.S., Mex.
John G. Searle Herbarium, Field Museum of Natural History, Roosevelt Road at Lake Shore Drive, Chicago, Illinois 60605, USA	2,358,300	N. & S. Am.
Herbarium, Arnold Arboretum of Harvard University, 22 Divinity Avenue, Cambridge, Massachusetts 02138, USA	1,092,000	Malaysia, E. As.
Gray Herbarium of Harvard University, 22 Divinity Avenue, Cambridge, Massachusetts 02138, USA	1,791,000	N. Am., Mex.
Herbarium of the University of Michigan, North University Building, Ann Arbor, Michigan 48109, USA	1,420,000	N. & trop. Am., W.I., S.W. Pac., S.E. As.
Herbarium, Missouri Botanic Garden, P. O. Box 299, St. Louis, Missouri 63166, USA	2,850,000	C. & trop. S. Am., Af.
Herbarium, New York Botanical Garden, Bronx, New York 10458, USA	4,286,000	N. & S. Am.
University of Texas Herbarium and C.L. Lundell Herbarium, Department of Botany, Plant Resources Center, University of Texas, Austin, Texas 78712, USA	900,000	Tex., Mex.
National Herbarium, Department of Botany, Smithsonian Institution, Washington, D.C. 20560, USA	4,110,000	C. & S. Am.
SOUTH AMERICA		
Herbario, Museo de Botánica y Farmacologia Juan A. Dominguez, Junín 954, 1er. Piso, 1113 Buenos Aires, Argentina	750,000	Argent.

Table 1 *cont'd.*

Herbarium	No. of Specimens	Interests
Herbario, Departmento de Botánica Agricola, Instituto Nacionale de Tecnologia, Agropecuaria, 1712 Castelar, Buenos Aires, Argentina	500,000	S. Am.
Herbario, Instituto Miguel Lillo de la Fundación Miguel Lillo, Calle Miguel Lillo 251, Casilla de Correo 91, 4000 San Miguel de Tucumán. Tucumán, Argentina	700,000	S. Am.
EUROPE		
Natural Historisches Museum Botanische Abteilung, Burgring 7, Postfach 417, A-1014, Wien, Austria	3,500,000	Near & Mid. E.
Institut für Botanik und Botanische Garten der Universität, Rennweg 14, A-1030 Wien, Austria	over 1,000,000	Medit., Near E.
Herbarium, Jardin Botanique National de Belgique, Domein van Bouchout, B-1860 Meise, Belgium	over 2,000,000	Trop. Afr.
Botanical Museum and Herbarium, Gothersgade 130, DK-1123 Copenhagen K, Denmark	2,000,000	S. Afr., S. Am., S.E. As.'
Institut de Botanique, 163 Rue Auguste Broussonnet, 34000 Montpellier, France	4,000,000	Medit.
Muséum National d'Histoire Naturelle, Laboratoire de Phanérogamie, 16 Rue de Buffon, 75005 Paris, France	6,500,000	Global
Botanischer Garten und Botanisches Museum Berlin-Dahlem, Königin-Luise-Strasse 6-8 D-1000 Berlin 33, Federal Republic of Germany	2,000,000	S.W. As., Togo, Hawaii
Systematisch-Geobotanisches Institut der Universität Göttingen, Untere Karspüle 2, D-3400 Göttingen, Federal Republic of Germany	750,000	Near East
Herbarium, Institut für Allgemeine Botanik und Botanischer Garten, Jungiusstrasse 6–8 D-2000 Hamburg 13, Federal Republic of Germany	800,000	Global
Herbarium Botanische Staatssammlung, Menzinger Strasse 67, D-8000 München, Federal Republic of Germany	2,000,000	S.W. Afr., Afghan.
Herbarium, Universitatis Florentinae, Museo Botanico, Via Giorgio La Pira 4, I-50121 Firenze, Italy	3,500,000	Medit., N.E. Afr.

Table 1 *cont'd.*

Herbarium	No. of Specimens	Interests
Rijksherbarium, Schelpenkade 6, 2313 ZT Leiden, Netherlands	2,500,000	S.E. As., Australia
Institute of Systematic Botany, Heidelberglaan 2, P. O Box 80.102, 3508 TC Utrecht, Netherlands	600,000	Trop. Am.
Botanical Institute of the University of Coimbra, 3049 Coimbra, Portugal	760,000	Medit., Afr., Micrones.
Museu Laboratório e Jardim Botanico, Rua da Escola Politécnica, 1294 Lisboa, Portugal	500,000	Angola
Herbarium, Botanical Museum, Carl Skottsbergs Gata 22, S-413 10 Göteborg, Sweden	1,250,000	S. Am.
Botanical Museum, Ö. Vallgaten 18, S-223 61 Lund, Sweden	2,300,000	Medit., S. Afr.
Herbarium, Swedish Museum of Natural History, Roslagvägen 106, P. O. Box 50007, S-10405 Stockholm, Sweden	4,000,000	S. Am., S. Afr.
Herbarium, University of Uppsala, P. O. Box 541, S-761-21 Uppsala, Sweden	2,200,000	Trop. Afr.
Herbarium, Conservatoire et Jardin botanique de la Ville de Genève, C. P. 60, CH-1292 Chambésy, Switzerland	5,000,000	Global
Herbarium, Royal Botanic Garden, Inverleith Row, Edinburgh, EH3 5LR, Scotland, UK	1,700,000	As.
Herbarium, Royal Botanic Gardens, Kew, Richmond, Surrey TW9 3AB, UK	over 5,000,000	Global
Herbarium, British Museum (Natural History), Cromwell Road, London SW7 5BD, UK	4,000,000	Global
AFRICA National Herbarium, Botanical Research Institute, Private Bag X101, Pretoria, 0001 South Africa	800,000	S. Afr.
ASIA Herbarium, Institute of Botany, Academia Sinica, Beijing, People's Republic of China	1,200,000	China
Herbarium, Botanical Institute, Academia Sinica, Guangzhou, Guangdong, People's Republic of China	520,000	China
Herbarium, Jiangsu Botanical Institute, 210014 Nanjing, Jiangsu, People's Republic of China	500,000	China

Table 1 *cont'd.*

Herbarium	No. of Specimens	Interests
Herbarium, Northwest Botanical Institute, Academia Sinica, Wugong, Shaanxi, People's Republic of China	500,000	China
Herbarium of Kunming Institute of Botany, Academia Sinica, Helongtan, Kunming, Yunnan, People's Republic of China	692,000	China
Central National Herbarium, Botanical Survey of India, P. O. Botanic Garden, Howrah-711103, India	1,250,000	S.E. As.
Herbarium Borgorense, Jalan Raya Juanda 22–24, Bogor, Indonesia	1,600,000	Malaysia
Herbarium, Department of Botany, The Hebrew University, Givath Ram, Jerusalem, Israel	600,000	Mid. E.
Herbarium, Department of Botany, Faculty of Science, Kyoto University, Sakyo-ku, Kyoto, Japan	950,000	E. & S.E. As.
Herbarium, Botanic Gardens Koishikawa, Hakusan 3-7-1. Bunkyo-ku, Tokyo 112, Japan	1,300,000	As.
Herbarium, Natural Sciences Museum, Department of Botany, 3-23-1 Hyakunin-cho, Shinjuku-ku, Tokyo 160, Japan	540,000	E. As., New Guinea
Herbarium, Botanical Institution of the Academy of Sciences of the Georgia SSR, Kodjorskoje Highway, Tibilis 380007, USSR	1,000,000	Turkey, Mid. E., Iran
AUSTRALIA National Herbarium of New South Wales, Royal Botanic Garden, Mrs. Maguaries Road, Sydney, NSW 2000 Australia	1,000,000	Australia
National Herbarium of Victoria Royal Botanic Gardens, Birdwood Avenue, South Yarra, Victoria 3141, Australia	1,000,000	Australia

Clearly, the development of such relevant information as plant uses has to be a separate exercise from that of a general herbarium data bank. The Arid Vegetation Information Service (AVIS) at the University of Arizona (Holland, 1979) has been developed for this purpose. At Kew, the Survey of Economic Plants for Arid and Semi-Arid Lands (SEPASAL) functions in a similar way (Wickens, 1984).

In pursuing the second basic source of information in the search for plant molluscicides, literature surveys may be hastened by using the various

Table 2. Herbarium Data Banks

Herbarium	No. of Specimens	Economic Uses Recorded	Reference
NORTH AMERICA			
University of British Columbia, Vancouver 8, British Columbia, Canada			
Rapid Access Plant Information Center, Colorado State University, Fort Collins, Colorado 80521, USA	152,000	Yes	Adams et al., 1975
Florida Technological University Herbarium, Orlando, Florida 32816 USA	10,000	Yes	Sweet and Poppleton, 1977
Edward Lee Greene Herbarium, University of Notre Dame, Notre Dame, Indiana 46556, USA	65,000	Yes	Crovello, 1972
University of Michigan Herbarium, Ann Arbor, Michigan 48109, USA	Type specimens only	No	Estabrook, 1979
New York Botanical Garden, Bronx, New York 10458, USA	Selective data capture	Proposed	Morris, 1974
Department of Botany and Plant Pathology, Oklahoma State University, Stillwater, Oklahoma 74074, USA	125,000		
National Museum of Natural History, Smithsonian Institution, Washington, D.C. 20002, USA	Selective data capture	?	Mello, 1975
Instituto Nacional de Investigaciones sobre Recursos Bióticas, Apartado Postal 63, Xalapa, Veracruz, Mexico	?	Yes	Gomez-Pompa and Nevling, 1973
SOUTH AMERICA			
Instituto Nacional de Pesquisas da Amazónia, Caixa Postal 478, Manaus, Amazonas, Brazil	24,000	?	Forero and Pereira, 1976
Museu Paraense Emilio Goeldi, Caixa Postal 399, Belém, Pará, Brazil	40,000	?	Forero and Pereira, 1976
Herbario Nacional Colombiano, Instituto de Ciencias Naturales de la Universidad Nacional, Apartado Aéreo 7495, Bogotá, Colombia	150,000	Yes	Forero and Pereira, 1976

Table 2. *cont'd.*

Herbarium	No. of Specimens	Economic Uses Recorded	Reference
Facultad de Ciencias Forestales, Universidad de los Andes, Mérida, Venezuela	30,000	?	Verbal information
AFRICA			
Botanical Research Institute, Private Bag X101, 0001 Pretoria, Republic of South Africa	500,000	No	Morris and Manders, 1981
AUSTRALIA			
Queensland Herbarium, Meiers Road, Indooroopilly, Queensland 4068, Australia	300,000	No	Morris, 1974
EUROPE			
British Antarctic Survey Botanical Section, University of Birmingham, Edgbaston, Birmingham, UK	30,000	No	Greene and Greene, 1975
British Museum (Natural History), Cromwell Road, London, SW7 5BD, UK	Selective data capture	No	Verbal information

abstracting journals. The time required to search for the latter can be reduced by means of on-line literature search services, such as those listed in Table 4 (see also Hall and Brown, 1981, and Bisby, 1985).

Nor should the botanists be forgotten; many have the field knowledge of plants they collected and, equally important, understand their systematic treatment and relationships – vital matters in the organized search for more effective plant molluscicides. And, as another source of information, a list of selected references on plant insecticides and piscicides follows the list of references to this chapter.

DATA BANK INPUT

Species that have been scanned for molluscicidal properties were reviewed by Kloos and McCullough (1981). On checking these species for additional toxic and medicinal attributes, it was found that all were reputed to have some medicinal property. In some instances, that was to be expected because the researchers had selected medicinal plants for their study (for example,

Table 3. Specialist Plant Data Banks

Data Bank	Subject	Economic Uses Recorded	Reference
Natural Products ALERT (NAPRALERT), Medical Center, University of Illinois, Chicago, Illinois 60680, USA	Ethnomedical uses of plants	No	Farnsworth et al., 1981
School of Renewable Natural Resources, University of Arizona, Tucson, Arizona 85721, USA	Comprehensive information on native and cultivated plants	Yes	Holland, 1979
Economic Botany Laboratory, USDA, Beltsville, Maryland 20705, USA	Ecological data on economic plants	Yes	Duke, 1983
American Horticultural Society Plant Sciences Data Center, Mount Vernon, Virginia 22121, USA	American records of cultivated or ornamental plants in North America	No	Brown, 1975
Herbarium, Royal Botanic Gardens, Kew, Richmond, Surrey TW9 3AB, UK	Survey of economic plants for the arid and semiarid lands	Yes	Wickens, 1984
Tropical Products Institute, 56/62 Gray's Inn Road, London, WC1X 8LU, UK	Economic plants of the world (bibliographic)	Yes	Dutta, 1974

El-Kheir and El-Tohami, 1979). The rationale behind such selection would appear to be that if the plant was known to have some toxic medicinal property, it might also prove to be molluscicidal. However, I believe there should and could be a more scientific approach to the selection of species for evaluation.

The active ingredients of many traditional folk medicines are often deliberately obscured by the presence of nonactive materials in order to conceal the identity of the former and preserve the reputation of the medicine man or witch doctor. In some cases the use of a plant is based on wishful thinking, the doctrine of signatures, or some other fantasy. In other cases the plants, whose chemical constituents have been analyzed and documented, have

Table 4. Useful On-Line Literature Search Services[a]

Subject	Coverage From					Data Systems			
		BLAISE	BRS	DIALTECH	DIMDI	IRS (ESD)	LOCKHEED DIALOG	PERGAMON INFOLINE	SDC-ORBIT
Agriculture	1970 1972/3 1973 1977	AGRICOLA		AGRIS	?AGRICOLA CAB ?PASCAL ?AGRIS	CAB PASCAL AGRIS	AGRICOLA CAB		AGRICOLA
Chemistry	1965 1967 1972	CHEMLINE	CAS	CAS	CHEMLINE	CAS	CAS CHEMNAME	CAS	
Life Sciences	1969 1970 1970	BIOSIS AGRICOLA		BIOSIS	?AGRICOLA	BIOSIS	BIOSIS AGRICOLA AQUACULTURE ASFA	BIOSIS	AGRICOLA
	1978				?ASFA				
Medicine	1966 1974	MEDLINE	MEDLINE		MEDLINE EXCERPTA MEDICA		EXCERPTA MEDICA		

[a]Data from Hall and Brown, 1981; ?denoted data base services planned for 1980.

real pharmaceutical properties. Of the thousands of plants whose economic properties have been recorded, an educated guess is that 50% are reputed to have some medicinal property and that of these, perhaps 10% to 15% are of any value.

Toxicity can, of course, vary according to the plant organ sampled, the age of the plant, the season, and even the soil in which the plant grew. If volatile substances are involved, storage conditions and the time stored also have to be considered. The effectiveness of the agent depends not only on its concentration but also on the species of snail and the possible variables in its habitat.

The disadvantages of a molluscicide that also acts as a general piscicide are obvious. Surprisingly, this aspect does not appear to have been systematically investigated by the majority of researchers included in the survey of Kloos and McCullough (1981).

In Table 5, all the species mentioned in the Kloos and McCullough (1981) survey (with later additions by Kloos, personal communication, 1982) are listed, together with their other reputed and generally proven toxic uses as insecticides, herbicides, piscicides, homicides, rodenticides, arrow and ordeal poisons, and mammalian toxins that could affect grazing livestock. If a species appears to have no alternative uses, that is more likely to mean that the search has been too rapid and superficial than that a good molluscicide with no side effects has been found.

The geographic distribution, habit, and habitat are also shown in general terms. The geographic distribution of biota can usually be presented in a more meaningful manner by distribution maps. The effectiveness of their presentation depends on such factors as map projection, scale, method of plotting, symbols, and so forth. For an informative introduction to the subject, see Stearn (1951) and Royal Geographical Society and Systematics Association (1954). The value of such maps is obviously related to both the standard of taxonomy and the number of specimens plotted.

Information regarding existing plant distribution maps for Africa can be obtained from Lebrun and Stork (1977) and Stork and Lebrun (1981). The more comprehensive world index edited by Tralau (1969–1981) is incomplete; the final volumes dealing with the dicotyledonous genera D–Z await publication.

The active ingredients – mainly alkaloids, rotenones, saponins, and tannins – of molluscicidal plants are often recorded. A screening of other known biologically active species such as those used for insecticides and piscicides doubtlessly would result in the discovery of additional molluscicides and certainly would be more cost-effective than screening medicinal plants. Alternatively, the screening of families with known chemical attributes could also be considered, using such guides as Darnley Gibbs (1974).

Investigation of related species should not be forgotten. In Africa,

Ambrosia maritima and *A. senegalensis* have been found effective as molluscicides, with so far no piscicidal effects. *Ambrosia* is essentially a widespread American genus, and I would certainly recommend screening Central and South American species for similar or better molluscicidal properties. It is also possible that screening of relatives of *Tephrosia vogelii*, for example, would reveal species with less drastic piscicidal action.

ECOLOGIC DATA

There are two sources of ecologic information: ecologic publications and notes found on herbarium specimens. The publications usually lack traceable voucher specimens, so that such matters as misidentification cannot be checked, whereas in the notes the ecologic status may be difficult to evaluate without local knowledge and access to other specimens from the same collecting area.

Unfortunately, there is no universally accepted standard among ecologists for naming plant communities. Thus in tropical Africa *savanna* implies grassland with an open canopy of trees stretching as far as the eye can see; in Asia and America the same term refers to relatively small clearings within a forest. Even *forest* can be variously interpreted to mean tropical rainforest, tropical or temperate woodlands, or even savanna. There is clearly a need for a standardized terminology for use within any one data bank, ideally providing a thesaurus for users.

Some plants have wide tolerances of habitat type, while others have very narrow ones. The latter can usually be fairly readily recognized and described; the former are less easy to classify, especially if they are very far ranging and accompanied by changes from mesophytic to xerophytic habitats with decreasing rainfall. Regional information on such plants can be compiled from ecologic reports, local floras, and herbarium specimens.

A slight bias can arise with the use of herbarium specimens because of the general tendency to collect in areas readily accessible by motorized transport and to collect the first specimens in flower at the start of a season, which usually means that they are from moister and generally uncharacteristic sites. Another foible of the collector is to ignore well-known dominants or conspicuous species.

Soil descriptions need to be recorded. There are a number of internationally accepted systems of classification, including the U.S. Department of Agriculture Seventh Approximation; the FAO/UNESCO, Soviet, French, and Australian systems; and other minor ones (Kalpage, 1976).

Climatic considerations, too, need to be taken into account. The basic data on temperature, rainfall, and other such factors are relatively straightforward, but for countries with few recording stations the data may need to be interpolated.

Table 5. Molluscicidal Plants – Distribution, Habit, Habitat, and Toxic Use.

Plant Family and Species	Distribution (Introductions)	Habit[a]	Habitat	Toxic Use[b]
ACANTHACEAE				
Brillantaisa vogeliana	W. Afr.	H	Forest	m
Crossandra flava	W. Afr.	p H	Forest	m
Lankesteria elegans	Afr.	S	Forest	m
AGAVACEAE				
Agave sisalana	C. Am.	p H	Cult.	he, i, m, p
Sanseviera liberica	W. Afr.	p H	Savanna	m
Sanseviera trifasciata	W. Afr.	p H	Savanna	m
Yucca aloifolia	N. & C. Am.	p H	Savanna	m
Yucca pallida	s. N. Am.	p H	Desert	m
Yucca schidigera	s. N. Am.	p H	Desert	m
AMARYLLIDACEAE				
Amaryllis vittata	S. Am. (Eur.)	b H	Cult.	m
ANACARDIACEAE				
Anacardium occidentale	C. & S. Am. (pantrop.)	T	Cult.	a, i, m, p
Astronium fraxinifolium	S. Am.	T	Forest	m
Spondias mombin (= *S. lutea*)	C. & S. Am. (pantrop.)	T	Forest/cult.	m
Tapirira guianensis	S. Am., W. I.	T	Forest	m
ANNONACEAE				
Annona senegalensis	Afr., Madag.	S/T	Savanna	a, i, m
Annona squamosa	W. I. (pantrop.)	S/T	Cult.	i, m, p
Cleistopholis patens	Afr.	T	Forest	i, m
ANTONIACEAE				
Antonia ovata	S. Am.	S/T	Savanna/forest	m, p
APOCYNACEAE				
Hunteria umbellata	W. Afr.	T	Forest	m
Himatanthus bracteata (=*Plumeria bracteata*)	S. Am. (pantrop.)	T	Savanna/cult.	m
Rauvolfia caffra	S. & E. Afr.	T	Riverine forest	m, r
Rauvolfia ternifolia	W. I., S. Am.	S	Riverine forest	m
Voacanga africana	Afr.	S/T	Forest	m
Peschiera affinis	S. Am.	S/T	Savanna/forest	m
ARACEAE				
Pistia stratiotes	pantrop.	H	Aquatic	m
ARALIACEAE				
Hedera helix	Eur., S. Afr., As.	cl S	Woodland	i, m, ma
Polyscias guilfoylei	S. As. (Pac.)	S	Cult.	m

Table 5 *cont'd.*

Plant Family and Species	Distribution (Introductions)	Habit[a]	Habitat	Toxic Use[b]
ASCLEPIADACEAE				
Asclepias curassavica	C. & S. Am.	p H	Weed	m, ma
Cryptostegia grandiflora	Madag., Australia	cl S	Riverine savanna	ho, m, ma
BALANITACEAE				
Balanites aegyptiaca	Afr.	T	Savanna, riverine	m, p
Balanites maughamii	S. Afr.	T	Savanna, riverine	m, p
BIGNONIACEAE				
Crescentia cujete	C. & S. Am., W. I. (Afr., As.)	T	Cult.	m
Kigelia africana	Afr.	T	Savanna	m
Tabebuia caraiba	S. Am.	T	Savanna	m
BOMBACACEAE				
Bombax costatum	W. Afr.	T	Savanna	m
BURSERACEAE				
Bursera simaruba (=*Elaphrium simaruba*)	C. Am., W. I.	T	Savanna	m
Dacryodes edulis	Afr.	T	Forest (cult.)	m
Protium heptaphyllum	S. Am.	T	Forest, savanna	m
CANELLACEAE				
Warburgia ugandensis	E. & S. Afr.	T	Forest	m
CANNABACEAE				
Cannabis sativa (=*C. indica*)	As. (cosmop.)	a H	Cult.	m
CANNACEAE				
Canna indica	S. Am.	p H	Cult.	i, m, o
CASUARINACEAE				
Casuarina equisetifolia	Pac. (pantrop.)	T	Cult.	m
CHENOPODIACEAE				
Chenopodium ambrosioides	pantrop.	a H	Weed	i, m, ?ma
CHRYSOBALANACEAE				
Acioa barteri	W. Afr.	cl S	Forest, cult.	m
Acioa rudatisii (=*A. lehmbachii*)	W. Afr.	T	Forest	m
Hirtella racemosa var. (=*H. americana*)	C. & S. Am.	S	Savanna/ forest	m
COCHLOSPERMACEAE				
Cochlospermum insigne	S. Am.	S/T	Savanna	m
Cochlospermum regium	S. Am.	S/T	Savanna	m

Cont'd.

Table 5 *cont'd.*

Plant Family and Species	Distribution (Introductions)	Habit[a]	Habitat	Toxic Use[b]
COMBRETACEAE				
Combretum fragrans (=*C. ghasalense*)	Afr.	T	Savanna	m
Terminalia mollis	Afr.	T	Savanna	ho, m
COMPOSITAE				
Ambrosia maritima	Medit.	H	Maritime	m
Ambrosia senegalensis	Afr.	H	Savanna	m
Artemisia sp.		H	m	
Baccharis genistelloides var. *trimera* (=*B. trimera*)	S. Am.	p H	Savanna	m
Eupatorium odoratum	C. & S. Am. (Afr., S. E. As.)	S	Forest, savanna	m, p
Heliopsis longipes	C. Am.	H	Savanna	i, m
Spilanthes oleracea	S. Am., W. I., (Afr., India)	H	Weed, swamp	m
Wedelia caracasana (=*W. scaberrima*)	S. Am.	cl S	Savanna	m
Wedelia parviceps	C. & S. Am.	H	Savanna	m
CORNACEAE				
Cornus florida	N. Am.	S/T	Woodland	m
CUCURBITACEAE				
Cucurbita foetidissima	s. N. Am.	p H	Desert	m
Luffa operculata	C. & S. Am.	a H	Savanna	m
Momordica charantia	pantrop.	a H	Forest, cult.	a, i, m
DILLENIACEAE				
Curatella americana	C. & S. Am.	S/T	Savanna	m
DIOSCOREACEAE				
Dioscorea cayenensis sp. *rotundata* (=*D. rotundata*)	W. Afr. (W. I.)	cl H	Forest, cult.	m
EUPHORBIACEAE				
Acalypha ornata	Afr.	S	Savanna	m
Bridelia atroviridis	Afr.	S/T	Forest	m
Croton macrostachys	Afr.	T	Savanna	m, p
Croton tiglium	India, S. E. As.	S/T		a, ho, i, m, p
Cyrtogonone argentea	W. Afr.	T	Forest	m
Euphorbia candelabrum	N. E. Afr.	T	Savanna	a, ho, m
Euphorbia cotinifolia	C. & n. S. Am.	S		m, o, p
Euphorbia pulcherrima	C. Am. (pantrop.)	S/T	Cult.	m, p
Euphorbia tirucalli	Afr., As.	S	Savanna, cult.	i, p

Table 5 *cont'd.*

Plant Family and Species	Distribution (Introductions)	Habit[a]	Habitat	Toxic Use[b]
Jatropha aceroides	N. E. Afr.	S	Savanna	m
Jatropha aethiopica	N. E. Afr.	H	Savanna	m
Jatropha curcas	C. Am. (S. Am., Afr., India)	S/T	Cult.	a, ho, m, p
Jatropha gossypiifolia	S. Am. (pantrop.)	S	Cult.	m
Jatropha podagrica	C. Am. (Afr.)		Cult.	m
Manihot glaziovii	S. Am. (pantrop.)	T	Cult.	m
FLACOURTIACEAE				
Casearia guianensis	S. Am.	S/T	Savanna	m
HYDROPHYLLACEAE				
Hydrolea spinosa	C. & S. Am. (As.)	p H	Swamp	m
LABIATAE				
Hyptis pectinata	Am. (pantrop.)	H	weed	m
Ocimum basilicum (=*O. canum*)	pantrop.	p H	Cult.	i, m
Ocimum gratissimum	pantrop.	p H	Cult.	m
LAURACEAE				
Persea americana (= *P. gratissima*)	C. Am.	T	Cult.	m
LEGUMINOSAE-CAESALPINIOIDEAE				
Caesalpinia coriaria	C. & S. Am., W. I.	S/T	Savanna	m
Caesalpinia ferrea	S. Am. forest	T	Savanna, riverine	m
Cassia singueana	Afr.	S/T	Savanna	m
Delonix regia	Madag. (pantrop.)	T	Cult.	m
Detarium senegalense	W. Afr.	T	Forest	m
Dialium guianensis	W. Afr.	T	Forest	m
Swartzia madagascariensis	Afr.	T	Savanna	i, m, p
Tamarindus indica	Afr., As.	T	Savanna	m
LEGUMINOSAE-MIMOSOIDEAE				
Acacia albida	Afr.	T	Savanna	m
Acacia dudgeoni	W. Afr.	T	Savanna	m
Acacia nilotica	Afr., Arabia, India	T	Savanna, riverine	m

Cont'd.

Table 5 *cont'd.*

Plant Family and Species	Distribution (Introductions)	Habit[a]	Habitat	Toxic Use[b]
Anadenanthera colubrina var. *cebil* (=*Piptadenia macrocarpa*)	S. Am.	S/T	Savanna, riverine	m
Calliandra portoricensis	C. Am., W. I., W. Afr.	S/T	Savanna, woodland	m
Dichrostachys cinerea (=*D. glomerata*)	Afr., Arabia, As.	S/T	Savanna	m
Entada phaseoloides	S. E. As.	cl S	Forest, riverine	m, p
Chloroleucum foliolosum (=*Pithecellobium foliolosum*)	S. Am.	S/T	Savanna	m
Pithecellobium multiflorum (=*Albizia polyantha*)	S. Am.	S	Forest margin	m, p
Samanea saman (=*Albizia saman*) biunifera)	S. Am. (pantrop.)	T	Savanna, cult.	m
Stryphnodendron coriaceum	S. Am.	S/T	Savanna	m
Tetrapleura tetraptera	Afr.	T	Forest	m, p
LEGUMINOSAE-PAPILIONOIDEAE				
Calopogonium velutinum (=*Stenolobium velutinum*)	S. Am.	p H	?Cult.	m
Derris elliptica	India, E. I.	cl p H	Forest, cult.	a, he, i, m, p
Dioclea reflexa	pantrop.	cl S	Forest	i, m
Indigofera kerstingii	W. Afr.	H	Weed	m
Indigofera secundiflora	Afr.	a H	Savanna, swamp	m
Indigofera spicata	Afr., Madag., S. E. As. (Am.)	p H	Savanna	m, ma
Indigofera suffruticosa	S. Am.	p H	Cult.	i, m
Neorautanenia mitis (=*N. pseudopachyrhiza*)	Afr.	cl H	Savanna	i, m, p
Sesbania sesban	Afr., India	T	riverine, lakes	i, m
Stylosanthes viscosa	C. & S. Am., W. I.	p H	Savanna	i, m
Tephrosia sinapou (=*T. toxicara*)	C. & S. Am., W. I.	S	Savanna	a, i, m, p
Tephrosia vogelii	Afr.	p H	Savanna, cult.	a, i, m, p
LILIACEAE				
Dipcadi fesoghlense	N. E. Afr.	b H	Savanna	m
Eriospermum abyssinicum	Afr.	b H	Savanna	m

Table 5 *cont'd.*

Plant Family and Species	Distribution (Introductions)	Habit[a]	Habitat	Toxic Use[b]
LOGANIACEAE				
Strychnos dendron				m
Strychnos parviflora	S. Am.	cl S	Forest	m
MALPIGHIACEAE				
Byrsonima sericea	S. Am.	S/T	Riverine, savanna	m
MELIACEAE				
Azadirachta indica	India (pantrop.)	T	Cult.	i, m
Ekebergia senegalensis	Afr.	T	Savanna	m
Guarea trichilioides	W. I.	T	Riverine, forest	m
MORACEAE				
Dorstenia cayapia	S. Am.	p H	Forest	m
Ficus glumosa	Afr.	T	Savanna	m
MYRTACEAE				
Eucalyptus acmenoides (=*E. triantha*)	Australia	T	Woodland, cult.	m
Eucalyptus alba	Australia, New Guinea	T	Forest (cult.)	m
Eucalyptus citriodora	Australia (pantrop.)	T	Cult.	i, m
Eucalyptus globulus	Australia (pantrop.)	T	Woodland, cult.	i, m
Eucalyptus largiflorens (=*E. bicolor*)	Australia	T	Riverine, woodland, cult.	m
Eucalyptus pilularis	Australia	T	Forest, cult.	m
Eucalyptus robusta	Australia	T	Swamp, forest, cult.	m
Eucalyptus saligna	Australia	T	Cult.	m
Syzygium cumini	S. E. As. (pantrop.)	T	Cult.	m
(=*S. jambolana*)				
OCHNACEAE				
Lophira alata	W. Afr.	T	Forest	i, m
Ouratea fieldingiana	S. Am.	S	Savanna	m
OLACACEAE				
Ximenia americana	pantrop.	S	Savanna	ho, i, m, p
ONAGRACEAE				
Ludwigia leptocarpa	Afr., C. & S. Am.	p H	Riverine	m

Cont'd.

Table 5 *cont'd.*

Plant Family and Species	Distribution (Introductions)	Habit[a]	Habitat	Toxic Use[b]
Ludwigia octonervis subsp. *brevisepala* (=*L. angusti folia*)	Afr.	p H	Riverine	m
OPILIACEAE				
Agonandra brasiliensis	S. Am.	T	Forest	m
ORCHIDACEAE				
Eulophia guineensis	Afr.	p H	Forest	m
PAPAVERACEAE				
Argemone mexicana	C. & S. Am.	a H	Weed	i, m, ma
PHYTOLACCACEAE				
Phytolacca americana	N. Am.	p H	Wayside	m, ma
Phytolacca dodecandra	Afr., Madag.	cl S	Riverine, forest, savanna	ho, m, p
Phytolacca icosandra	C. & S. Am., W. I.	p H	Savanna	m
Phytolacca octandra	C. & S. Am. (pantrop.)	p H	Weed	m, ma
Phytolacca rivinoides	S. Am., W. I.	p H	Weed	m, p
PIPERACEAE				
Piper marginatum	C. & S. Am.	S	Savanna	m
Piper tuberculatum	C. Am., W. I.	S/T	Riverine forest	m, p
POLYGALACEAE				
Securidaca longepedunculata	Afr.	T	Savanna	a, ho, m, o, p
POLYGONACEAE				
Polygonum senegalense (*P. glabrum, P. sambesicum*)	Afr., Madag.	H	Semi-aquatic	m
Polygonum meisnerianum	W. I., C. & S. Am., S. Afr.	p H	Semi-aquatic	m
Polygonum mite	Eur., W. As.	a H	Semi-aquatic	m
Polygonum pulchrum (=*P. tomentosum*)	Afr., As.	p H	Semi-aquatic	?ho, m
PONTEDERIACEAE				
Eichhornia natans	Afr., S. Am.	p H	Aquatic	m
PUNICACEAE				
Punica granatum	As. (pantrop.)	S	Cult.	ho, i, m
RHAMNACEAE				
Maesopsis eminii	Afr.	T	Forest, riverine	m

Table 5 *cont'd.*

Plant Family and Species	Distribution (Introductions)	Habit[a]	Habitat	Toxic Use[b]
Ziziphus joazeiro	S. Am.	T	Savanna	m
Ziziphus undulata	S. Afr.	T	?Savanna	m
ROSACEAE				
Quillaja sp.	S. Am.	T		m
RUBIACEAE				
Catunaregam nilotica (=*Randia nilotica,* *Lachnosiphonium niloticum,* *Xeromophis nilotica*)	Afr.	S	Savanna	i, m, p
Canthium subcordatum	W. Afr.	T	Forest	m
Coffea spathicalyx	Afr.	S/T	Savanna	m
Gardenia ternifolia	Afr.	S	Savanna	m, p
Gonzalagunia spicata (=*Duggena hirsuta*)	S. Am., W. I.	S		m
Gardenia vogelii	Afr.	S/T	Forest	m
Morinda lucida	Afr.	T	Forest	m
Oldenlandia affinis	Afr., Madag., As.	p	Savanna, forest	m
Rothmannia urcelliformis	Afr.	S/T	Forest	m
Rothmannia whitfieldii	Afr.	S/T	Forest	m
RUTACEAE				
Zanthoxylum rhoifolium (=*Fagara rhoifolia*)	S. Am.	T	Forest	m
Pilocarpus sp.	Am.			m
Ruta chalepensis	S. Eur., N. Afr. (pantrop.)	p H	Cult.	m
SAPINDACEAE				
Magonia pubescens	S. Am.	T	Savanna	m, p
Paullinia pinnata	Afr., Madag., Am.	cl S	Forest, riverine, savanna	a, i, m, p
Sapindus saponaria	C. & s. N. Am.	T	Woodland streams	m, p
SAPOTACEAE				
Butyrospermum paradoxum	W. Afr.	T	Savanna	m
SCROPHULARIACEAE				
Capraria biflora	Am. (W. Afr.)	S	Weed	m
Scoparia dulcis	pantrop.	p H	Weed	i, m
SIMAROUBACEAE				
Simarouba versicolor	S. Am.	T	Savanna	m

Cont'd.

Table 5 *cont'd.*

Plant Family and Species	Distribution (Introductions)	Habit[a]	Habitat	Toxic Use[b]
SOLANACEAE				
Capsicum frutescens	C. Am. (pantrop.)	p H	Cult.	m
Cestrum laurifolium	W. I.	S	Savanna	m
Cestrum macrophyllum	W. I. (S. Am.)	S	Savanna	m
Nicotiana tabacum	C. Am. (pantrop.)	a /p H	Cult.	a, i, m, p
Solanum americanum (=*S. nodiflorum*)	cosmop.	a H	Woodland waste	i, m
Solanum mammosum	W. I., C. & n. S. Am.	p H	Savanna, cult.	i, m, r
Lycopersicon esculentum	cosmop.	H	Cult.	i, m
STERCULIACEAE				
Guazuma ulmifolia	C. & S. Am.	S/T	Cult.	m
STYRAXACEAE				
Styrax officinalis	S. Eur. (N. & C. Am.)	S/T	Woodland	i, m, p
THEACEAE				
Camellia sinensis (=*Thea oleosa*)	E. As.	S	Montane forest, cult.	m, p
Schima wallichii ssp. *noronhae* var. *superba* (=*S. argentea*)	E. As.	T	Montane forest	m, p
THYMELAEACEAE				
Gnidia kraussiana	Afr.	p H	Savanna	a, ho, m, ma, p
TURNERACEAE				
Turnera sp.				m
ULMACEAE				
Trema guineensis	Afr., As.	S/T	Forest, savanna	m
UMBELLIFERAE				
Ammi majus	Medit., Mid. E., N. E. Afr.	a H	Waste places	m
Eryngium foetidum	Am. (pantrop.)	p H	Forest	m
VERBENACEAE				
Vitex oxycuspis	W. Afr.	T	Forest	m
XYRIDACEAE				
Xyris anceps	Afr.	H	Marshes, rice fields	m

Table 5 *cont'd.*

Plant Family and Species	Distribution (Introductions)	Habit[a]	Habitat	Toxic Use[b]
ZINGIBERACEAE				
Hedychium coronarium	pantrop.	p H	Forest, cult.	m
Zingiber officinale	As. (pantrop.)	p H	Forest, cult.	m
ZYGOPHYLLACEAE				
Peganum harmala	N. Afr., Mid. E., E. As. (Australia)	S	Savanna, desert	m

[a]H = herb, T = tree, S = shrub, a = annual, b = bulbous, cl = climbing, p = perennial.
[b]a = arrow poison, he = herbicide, ho = homicide, i = insecticide, m = molluscicide, ma = mammalicide, o = ordeal poison, p = piscicide, r = rodenticide.

An attempt to computerize soil and climatic data for a selection of economic plants has been undertaken in the United States (Duke, 1979). It is not yet perfected; because of insufficient data it is possible to obtain slightly misleading information about some species.

The complexity of the problems that can be expected from including ecologic information in a data bank depends largely on whether it is a local territorial data bank, such as that at Pretoria for South Africa, or one covering several countries or even continents. In the former, some degree of uniformity of treatment is to be expected; with increasing coverage involving many countries, languages, and cultures, the problems inevitably increase but are not insuperable.

REFERENCES

Adams, R. P., D. H. Wilken, W. M. Klein, G. Bryang, and R. G. Walter. RAPIC: The missing link? *Bioscience* 25(7):433–437 (1975).

Altschul, S. *Drugs and Foods from Little-Known Plants: Notes in Harvard University Herbaria.* Cambridge: Harvard University Press (1973).

——. Exploring the herbarium. *Sci. Am.* 236:96–104 (1977).

Bisby, F. A. Plant information services for economic plants of arid lands. In *Plants for Arid Lands*, ed. G. E. Wickens, J. R. Goodin, and D. V. Field, 413–425. London: George Allen and Unwin (1985).

Brenan, J. M. P. Plants for man – Their diversity, codification and exploitation. *Int. Relations* 7(1):1005–1020 (1981a).

——. *Index Kewensis Plantarum Phanerogamarum.* Suppl. 16. New York: Oxford University Press (1981b).

Brown, R. A. American Horticultural Society Plant Records Center. In *Computers in Botanical Collections*, ed. J. M. P. Brenan, R. Ross, and J. T. Williams, 125–138. London: Plenum Press (1975).

Crovello, J. J. Computerization of specimen data from the Edward Lee Greene Herbarium (ND-G) at Notre Dame. *Brittonia* 24:131–141 (1972).

Darnley Gibbs, R. *Chemotaxonomy of Flowering Plants*. 4 vols. Montreal and London: McGill and Queen's University (1974).

Duke, J. A. Ecosystematic data on economic plants. *Q. J. Crude Drug Res.* 17(3–4):91–110 (1979).

——. The USDA Economic Botany Laboratory's database on minor economic plant species. In *Plants: The Potentials for Extracting Protein, Medicines and Other Useful Chemicals*. Workshop Proceedings, 196–214. Washington, D. C.: U.S. Congress, Office of Technology Assessment (OTA-BP-F-23 Sept. 1983).

Dutta, V. K. The Technical Index of the Tropical Products Institute. *Trop. Sci.* 16(4):237–245 (1974).

El-Kheir, Y. M., and M. S. El-Tohami. Investigation of molluscicidal activity of certain Sudanese plants used in folk medicine. *J. Trop. Med. Hyg.* 82:237–241 (1979).

El-Magdoub, A. A. I., M. F. El-Sawy, H. K. Basiouny, I. A. El-Sayed, R. A. Galil, and E. M. Hassan. An evaluation of the plant *Ambrosia maritima* as a molluscicide. *Ir. Vet. J.* 34:157–159 (1980).

Estabrook, G. F. A TAXIR data bank of seed plant types at the University of Michigan Herbarium. *Taxon* 28(12/3):197–203 (1979).

Farnsworth, N. R., W. D. Loub, D. D. Soejarto, G. A. Cordell, M. L. Quinn, and K. Mulholland. Computer services for research on plants for fertility regulation. *Korean J. Pharmacognosy* 12(2):98–110 (1981).

Forero, E., and F. J. Pereira. EDP-IR in the National Herbarium of Colombia. *Taxon* 25(1):85–94 (1976).

Gomez-Pompa, A., and L. I. Nevling. The use of electronic data processing methods in the flora of Veracruz program. *Contrib. Gray Herb.* 203:49–64 (1973).

Greene, D. M., and S. W. Greene. The data bank of the British Antarctic Survey's Botanical Section. In *Computers in Botanical Collections*, 79-87. See Brown, R. A.

Hall, A. V. Museum specimen record data storage and retrieval. *Taxon* 23(1):23–28 (1974).

Hall, J. L., and J. J. Brown. *Online Bibliographic Databases: An International Directory*. 2d ed. London: ASLIB (1981).

Holland, M. AVIS: A prototype Arid Vegetation Information System. *Desert Plants* 1(2):71–76 (1979).

Holmgren, P. K., W. Keuken, and E. K. Schofield. *Index Herbariorum, Part 1: The Herbaria of the World*. 7th ed. The Hague and Boston: Dr. W. Junk (1981).

Jackson, B. D. *Index Kewensis Plantarum Phanerogamarum*. 4 vols. Oxford: Clarendon Press (1893–1895, with 16 supplements, 1901–1981).

Kalpage, F. S. C. P. *Tropical Soils: Classification, Fertility and Management*. London: Macmillan (1976).

Kloos, H., and F. McCullough. Plant molluscicides: A review. WHO Unpublished Document WHO/VBC/81.834; WHO/SCHISTO/81.59 (1981).

Lebrun, J. P., and A. L. Stork. *Index 1935–1976 des Cartes de Répartition des Plantes Vasculaires d'Afrique*. Geneva: Conservatoire et Jardin Botaniques (1977).

Mello, J. F. The use of SELGEM system in support of systematics. In *Computers in Botanical Collections*, 125–138. See Brown, R. A.

Morris, J. W. Progress in the computerization of herbarium procedures. *Bothalia* 11(3):349–353 (1974).

——. Encoding the National Herbarium (PRE) for computerized information retrieval. *Bothalia* 13(12):149–160 (1980).

Morris, J. W., and H. F. Glen. PRECIS, the National Herbarium of South Africa (PRE) computerized information system. *Taxon* 27(56):449–462 (1978).

Morris, J. W., and R. Manders. Information available with the PRECIS data bank of the National Herbarium, Pretoria, with examples of uses to which it may be put. *Bothalia* 13(34):473–485 (1981).

Royal Botanic Gardens, Kew. *The Kew Record of Taxonomic Literature*. London: HMSO (Issued annually from 1972).

Royal Geographical Society and Systematics Association. *The Cartographical Presentation of Biological Distributions*. London: Royal Geographical Society (1954).

Stearn, W. T. Mapping the distribution of species. In *The Study of the Distribution of British Plants*, ed. J. E. Louseley, 48–64. Arbroath: Buncle (1951).

Stork, A. L., and J. P. Lebrun. *Index des Cartes de Répartition des Plantes Vasculaires d'Afrique: Complément 1935–1976, Supplément 1977–1981 (avec Addendum A-Z)*. Maisons-Alfort: Institut de l'Elevage et de Médecine Vétérinaire des. Pays Tropicaux, Etude Botanique No. 8 (1981).

Sweet, H. C., and J. E. Poppleton. An EDP technique designed for the study of a local flora. *Taxon* 26(2/3):181–190 (1977).

Tralau, H., ed. *Index Holmensis. A World Index of Plant Distribution Maps*. Vol. 1, Equisetales-Gymnospermae (1969); Vol. 2, Monocotyledoneae A-I (1972); Vol. 3, Monocotyledoneae J-Z (1973); Vol. 4, Dicotyledoneae A-B (1974); Vol. 5, Dicotyledoneae C (1981). Zurich: Scientific Publications.

Vassiliades, G., and O. T. Diaw. Action molluscicide d'une souche sénégalaise d'*Ambrosia maritima*. Essais en laboratoire. *Rev. Elev. Méd. Vét. Pays Trop.* 33(4):401–406 (1980).

Wickens, G. E. Plants for man – The Kew data bank of economic plants. *Int. Relations* 8(1):73–80 (1984).

SELECTED REFERENCES ON PLANT INSECTICIDES AND PISCICIDES

Cox, P. A. Use of indigenous plants as fish poisons in Samoa. *Econ. Bot.* 33:397–399 (1979).

Drake, N. L., and J. R. Spiers. The toxicity of certain plant extracts to goldfish, I. *J. Econ. Entomol.* 25:129–133 (1933).

Fanshaw, D. B. Fish poisons and insecticides. *Br. Guiana Forest. Bull.* 2(NS):47–50 (1948).

——. Fish poisons of British Guiana. *Kew Bull.* 2:239–240 (1953).

Gatty, H. The use of fish poison plants in the Pacific. *Trans. Proc. Fiji Soc. Sci. Ind.* 3:152–159 (1945).

Holman, H. J. *A Survey of Insecticide Materials of Vegetable Origin*. London: Imperial Institute, Plant and Animal Products Department (1940).

Howes, F. N. Fish poison plants. *Bull. Misc. Inf. Kew* 1930:129–152 (1930).

——. *Tephrosia macropoda* as a possible insecticidal plant. *Bull. Misc. Inf. Kew* 1937:510–513 (1937).

Jacobson, M. Plants, insects and man – Their interrelationships. *Econ. Bot.* 36(13):346–354 (1982).

Kawazu, K. The piscicidal plants in southern Asia and their active constituents. *Japan Agric. Res. Q.* 3(2):20–24 (1968).

Killip, E. P., and A. C. Smith. The use of fish poisons in South America. *Annual Rep. Smithsonian Inst.*, 401–408 (1930).

Lamba, S. S. Indian piscicidal plants. *Econ. Bot.* 24:134–136 (1970).

Moretti, C., and P. Grenard. Les nivrée ou plantes ichtytoxiques de la Guyane Française. *J. Ethnopharmacol.* 6(2):139–160 (1982).

Nishimoto, S. K. Plants used as fish poisons. *Newslett. Hawaiian Bot. Soc.* 8(3):20–23 (1969).

Pennington, C. W. Tarahumar fish stupefaction plants. *Econ. Bot.* 12: 95–102 (1958).

Schultes, R. E., and J. Cuatercasas. A new species of ichthyotoxic plant from the Amazon. *Bot. Mus. Leaflets Harvard Univ.* 23(3):129–136 (1972).

Spies, J. R. The toxicity of certain plant extracts to goldfish, II. *J. Econ. Entomol.* 26:285–288 (1933).

Tattersfield, F., and C. T. Giminghan. The insecticidal properties of *Tephrosia macropoda* Harv. and other tropical plants. *Ann. Appl. Biol.* 19(2):253–262 (1932).

Tattersfield, F., J. Martin, and F. N. Howes. Some fish poison plants and their insecticidal properties. *Bull. Misc. Inf. Kew* 1940:169–180 (1940).

Tattersfield, F., C. Potter, K. A. Lord, E. M. Gillham, M. J. Way, and R. I. Stoker. Insecticides derived from plants. *Kew Bull.* 3:329–349 (1948).

Vellard, J. A. Poisons de pêche et poissons de chasse en Amerique du Sud. *Bol. Mus. Nac. Rio de Janeiro* 14–17:345–362 (1938–1941).

——. Les poisons de pêche de l'Amerique du Sud. *Bull. Mus. Hist. Nat.* II(6):497–507 (1939).

Worsley, R. The insecticidal properties of some East African plants, I. *Ann. Appl. Biol.* 21:649–669 (1934).

7

FEASIBILITY OF GROWTH AND PRODUCTION OF MOLLUSCICIDAL PLANTS

C. B. Lugt

Agronomist/Phytochemist/Plant Pathologist
The Hague, The Netherlands

The feasibility of growing and producing molluscicidal plants depends on several factors: their potency; the nature of the plant material to be harvested; the suitability of the plants for cropping; their yields, prices, costs, and benefits; their availability for use when needed; and the difficulties of harvesting, handling, and storing them. Although these factors are interlinked, this chapter stresses their separate roles in the agricultural process. And, in assessing feasibility, it is assumed that the molluscicidal products will be applied in self-help programs in villages and farming communities and that cropping on a large scale will not take place in the near future.

MOLLUSCICIDAL POTENCY

With potential plant molluscicides, as with plants in general, growth depends on environment – climate, day length, altitude, soil. Thus it is important to see that circumstances will allow the plants to develop the desired type and amount of constituents.

For example, the seeds of *Amni visnaga* growing wild in the Mediterranean contain such desirable constituents as coumarins and chromones; but cultivated in Arizona its seeds, though plentiful, are almost devoid of these constituents (Claus et al., 1970). With *Digitalis purpurea*, the availability of manganese in the soil appears to influence the quantity of glycosides in the

231

plant; manganese uptake by roots is negatively influenced by a high pH and an excess of water in the soil (Lugt, unpublished observations, 1973). Furthermore, *D. purpurea* from various sites in Europe showed differences in content and composition of glycosides when planted at random on a small plot of soil (Lugt, 1976). And *Nicotiana tabacum* possesses chemical variability in such a way that nicotin, nor-nicotin, and anabasin types have been determined (Steinegger and Hansel, 1972).

Many more examples are available in crop production, and it is likely that such variations will also occur in growing molluscicidal plants. These plants need to be looked at carefully to take into account how the environment influences blossoming, fruit setting, and quantity of desired compounds.

Phytolacca dodecandra, a plant with molluscicidal potency, is limited in growth and berry production by certain unknown environmental circumstances (Lugt, unpublished observations, 1980). Plants originated from one mother plant through cuttings did not blossom and performed badly in Bati, a town 400 km north of Addis Ababa at an altitude of 1560 m. However, at 2500 m in Addis Ababa and at 1500 m on the Wonji/Shoa Sugar Estate some 90 km to the south, the plants performed well and produced berries abundantly. Since altitude obviously did not influence growth, other factors were sought – in vain. With respect to soil, indications were that it had no influence on growth performance, either (pH KCl at the three sites was 6, 5, and 6, respectively). As for day length, its influence was questionable, since results in Bati were bad throughout the year. One is inclined to conclude that this species occurs as a so-called ecotype, a phenomenon also known with sorghum and sweet potatoes.

One should start with seeds from a plant with known molluscicidal potency; a search for valuable types can proceed from there. Moreover, the start must be performed under various environmental conditions to be able to select types that can overcome the new circumstances to which they will be exposed.

The strength of the plant material's snail kill also must be looked at. The higher the molluscicidal potency, the better – it means that a smaller amount of material will need to be applied and a smaller crop will need to be grown. A 100% snail kill under field conditions at concentrations around 50 mgL^{-1} is desirable.

NATURE OF THE MATERIAL

The nature of the plant material (plant organ) to be harvested for molluscicidal purposes is important for two reasons. First, leaves, wood or bark, and seeds differ considerably in water content and consequently in percentage of valuable and wanted compounds. Second, plant organs that are not juicy or bulky are easier to handle and store.

Percentage of Compounds

Plant parts such as leaves, inflorescences, and juicy fruits contain more than 90% water, while woody parts and bark contain up to 50% and seeds less than 10% (Steinegger and Hansel, 1972). The quantity of chemical compounds in the organs differs proportionately.

The leaves of *Atropa belladonna*, for example, contain 0.1% to 1.2% of the tropine alkaloids for which they are propagated or collected in the wild, and the leaves of *D. purpurea* contain a mixture of cardiac glycosides varying between 0.09% and 0.23%, depending on place of growth and climate. But the bark of *Cinchona ledgeriana*, a medicinal material, can contain up to around 12% of cinchona alkaloids, and the seeds of *Foeniculum vulgare* contain 4% to 6% of volatile and 12% to 18% of fixed oils (Karsten et al., 1962). Thus it is important that seeds in particular be given special attention.

Among the well-known plants that have shown molluscicidal potential, only *P. dodecandra* has been extensively screened for its content of molluscicidal compounds. While the leaves and other parts were found to possess some potency, the berries provided a biologically active light tan powder, representing 20% to 25% of the initially ground berries containing a mixture of related molluscicidal saponins of which the aglycon is oleanolic acid (Lemma et al., 1972). This finding means that labor-intensive and costly extraction of the berries can be avoided.

Research has also been carried out with *Hedera helix* berries (Hostettmann, 1980). Other plants already characterized for their molluscicidal potential – such as *Ambrosia maritima* (El-Sawy et al., 1978), which as a complete plant was used in snail-control tests; *Polygonum senegalense* forma *senegalense* (Dossaji et al., 1977); and the plants preliminarily screened by Adewunmi and Sofowora (1980) – should be checked extensively for the molluscicidal potential of their seeds or small fruits.

Handling and Storing

The larger the volume of plant materials collected, the bigger the problems of harvesting and storing them. Thus seeds, berries, and small fruits are to be preferred not only because they generally contain a higher percentage of valuable compounds but also because they are less bulky and contain less water.

Another consideration is whether the plant material is available at the time when it is most likely to be used to control snails. If so, the plant organs, after a short processing period, can be applied at once and no storage facilities will be needed.

If complete plants are to be harvested or if, to avoid destruction of the crop, only part of the leaves, branches, or roots are to be collected, then they must be of high molluscicidal potency. Using the fully grown plant shortens the period of availability of the desired material and means that snail control has to be carried out over a shorter period of time. Storage facilities may also be needed, and true-to-type multiplication will be necessary to preserve the plants' molluscicidal potency.

The use of perennial plant species would of course give the advantage of constant production of the material wanted.

SUITABILITY FOR CROPPING

Once a plant species has been recognized as valuable for its molluscicidal potential, the next step is to investigate whether it is suitable for cropping. That involves such matters as ecology, varieties, propagation, and field establishment and maintenance (Acland, 1975).

Ecology

While water, temperature, shade, and soil will be the decisive factors in the success or failure of any new crop, the potential plant molluscicide is likely to receive the best of none of them. The plant will almost certainly be a wild one to begin with, and it will be subjected to a complete process of domestication. The environment will not be friendly, and special care must be taken in the process of adaptation to new circumstances. Neither will the new crop be regarded as the principal one on farms where it is grown, and it will not be planted in spots where conditions are optimal. The farmer will not give it priority, for it will be of little economic importance compared with the main crop. The same is true if the crop is planted for self-help purposes in villages – unless the occurrence of bilharzia weighs very heavily in the balance with economic considerations.

These plants must be modest in their demands for water, since it may not always be available and watering is labor intensive. *Anacardium occidentale* (cashew), which requires little water and is more or less drought resistant, would be in a strong position to meet the primitive requirements of cropping at the farm or village level. But not *Coffea* spp. (coffee), which requires a constant supply of soil moisture plus a short dry period for flowering.

A potential molluscicidal plant should be studied for sensitivity to high and low altitudes. For example, the promising species *P. dodecandra*, which occurs naturally at altitudes above 1500 m, should be tested at sea level, too, where bilharzia is endemic.

Temperature requirements are also important. Mountains and deserts are subject to relatively low night temperatures that may cause frost damage and affect fruit setting.

It would be better, too, if the molluscicidal crop required no shade, for shade must be specially provided and the extra cost will make the farmer reluctant to grow it.

Finally, the promising plant should not make high demands on the soil. It may be forced to thrive on wasteland.

Varieties

Not only chemical variety but also morphological variety may occur within each plant species. Both affect productivity, molluscicidal potency, resistance to pests, growth performance, and other factors – all of which may in turn be influenced by certain environmental circumstances.

A plant taken out of its natural habitat may lose its specific characteristics. Thus a thorough testing of the species selected is necessary to meet all possible negative influences on crop growth. Tests should be undertaken in various bilharzia endemic areas to determine as early as possible the plant's demands in response to a new environment.

Propagation

Any potential molluscicidal plant that is to be cropped must propagate a true-to-type offspring, whose product is at least as potent as that of the mother plant. The best way to achieve that is through vegetative means. But, while vegetative propagation results in new plants that are similar to the mother plant, it makes the supplying of sufficient plants difficult. It also requires skilled workers to carry out multiplication procedures in special centers, which in turn must be responsible for distributing the crop in endemic areas.

If vegetative multiplication is not possible, the only other solution is multiplication via seeds. With this method, it is the seeds that must be provided by a special center. This center must also check the seeds for purity and uniformity. An advantage of this method is that it opens up the possibility of improving the quality and molluscicidal potency of the crop.

Research into propagation is of utmost importance. Two examples will show why. First, *Olea europea* (olive), which occurs in numerous varieties, gives varying results in multiplication via cuttings. It depends largely on the variety being propagated. Culture variety (CV) Shemlaili, for instance, can be propagated via seeds, after which the offspring is more or less homozygotic; but vegetative multiplication via cuttings results in only 5% to 10% rooting. In contrast, CV Leccino can be multiplied via cuttings with 100% success. Second, *P. dodecandra*, whose types sometimes differ so much that one tends to speak of subspecies, must also be multiplied via cuttings; it is not possible to get true-to-type offspring via seeds. Furthermore, among *P.*

dodecandra types in Ethiopia, success of propagation via cuttings appears to depend on the type selected: Type 17 was easy and Type 3 difficult. (Specimens of these types are in the Herbarium Vadense of the Laboratory for Plantsystematics and Geography, P. O. Box 8010, NL-6700 ED Wageningen, The Netherlands.)

From the foregoing, it becomes clear that propagation must be done in special breeding centers that can also carry out research on improving the crop and organize the distribution of the plant material.

Field Establishment and Maintenance

Once a potential molluscicidal crop has been established or is under maintenance in the field, such factors as spacing and pruning, fertilizing, and controlling weeds, pests, and diseases need to be looked at. These factors will be most relevant where large-scale farming is carried out but of minor importance at the village level. In both cases, of course, the less labor and money expended the better.

Spacing and Pruning. In generative propagation of the crop, the space allotted for individual plants in the field will be related to the cost of planting versus higher yields, which in turn will be related to competition among plants in the field. In vegetative propagation, the plant must be perennial, since cropping would be too costly. Spacing will be closely related to pruning. With the perennial *P. dodecandra*, which is propagated through cuttings, a yearly pruning after the rainy season in October was carried out in Ethiopia and a spacing of one meter on the row and two meters between the rows was found satisfactory. Moreover, pruning favored the formation of new shoots, the only parts out of which the inflorescences emerged. Thus spacing is largely a matter of experience and research.

Fertilizers. Fertilizers should not play an important role. The only reason for applying them would be to bring about a dramatic increase in the end product and, as is known with other crops, that is not likely to be the case. Because of the extra cost, a potential molluscicidal plant would be less attractive for cropping if fertilizers had to be applied.

Weed Control. For both large- and small-scale farming, weed control can be time consuming. But it is absolutely necessary, since weeds compete with crops for moisture, nutrients, and light and, if successful, lower the crop yields. That is especially true when the roots of the crop are close to the topsoil. Weeds might be suppressed by spacing the plants so that the soil is completely covered, thus not allowing weeds to interfere. If the plants have an open structure, allowing the sun to reach the soil, weeds can be controlled by regular cultivation of the soil or by slashing and mulching the weeds.

Pest and Disease Control. Any plant species being considered for cropping as a molluscicide must be subjected to thorough research to determine its sensitivity to pests and diseases. Tests must be carried out both where the species originates and where it will be planted and cropped. Measures must be taken not only against pests or diseases that might be introduced into the new environment but also against those that can be expected to challenge any plant species taken out of its original environment. If, for instance, seeds are the plant organs that are wanted, birds may be a serious problem, as is the case with sunflower and sorghum (Acland, 1975).

Suitability for cropping of a particular plant species thus depends on numerous factors that need serious consideration. A newly discovered and developed molluscicidal plant should be regarded as a new cash crop and treated accordingly. But also, since time may be needed for the new crop to prove its benefits, the development of a simple means of cropping it is important. Nevertheless, the means is not likely to be as simple as the one suggested for *P. dodecandra* of planting the bushes along canals and streams populated with snails and simply allowing the berries to drop in the water and thus to do the work of killing the snails. Unfortunately, birds were keen on the berries, so that most of them disappeared before they could drop in the water. Besides, as was later discovered (Lugt, 1980), even had they fallen as planned they would not have reduced the snail population, since the ripe berries possess little molluscicidal potency.

YIELDS, PRICES, COSTS, AND BENEFITS

As with any crop, a yield projection is required to determine the feasibility of growing and producing the desired molluscicidal plant organs. Land use is assessed from an economic point of view by means of yields, prices, costs, and benefits. In the case of plants to be cropped for molluscicidal purposes, absolutely nothing is known about these essential factors.

In this discussion, a farm is considered to consist of at least 1000 hectares on which a few principal crops are grown; such a farm will grow the molluscicidal plant crop to control snails in its irrigation system. Cropping at the village level starts with small settlements of a few square meters. There the cost of growing and applying the molluscicidal plant can be paid by a community health center.

Yields

Water availability, which has a decisive influence on yields, may differ considerably at the village and farm levels. Normally, farmers can be expected to have better water supply systems than villagers do, although such may not be the case with villages located close to rivers or streams or in completely irrigated places such as the Nile Delta.

Assuming that sufficient water is available and that the crop is not over-watered, thus allowing for sufficient aeration of the soil (Acland, 1975), differences in yield will be determined by whether the land is rain-fed or artificially supplied with water (Table 1).

Drought-resistant plants are of course to be preferred. But some plants such as cashew that require little water will increase their productivity considerably when supplied with more water than is strictly necessary for survival.

Yields will also be influenced by the amount of attention the farmer gives the crop. If proper use is not made of the results of research into weed control, pruning, and water, yields will undoubtedly decrease proportionately. Both care and yield may improve if the plant organs are bought by an established center that guarantees a price based on quality.

Prices

Left to itself, it is unlikely that the molluscicidal crop will bring a price that will tempt farmers to grow it. But if a health center or a molluscicidal research center purchases the plant organs from the farmers at a competitive price while paying higher prices for higher quality materials, the farmers would be more likely to grow the crop and to look after it to get the best possible price. The center could then have a standardized product whose snail-killing potency was known and could oversee its application. Although such a system would need to be subsidized initially, if the material proves effective it might eventually be able to support itself.

Table 1. Yields of Rain-Fed and Irrigated Crops (Tons per Hectare)

	Rain-Fed	Irrigated
Cashew	1.00–1.10	2.20
Coffee (dried beans)	0.63–1.25	2.50
Sesame	0.22–0.33	0.45–0.55
Sorghum	1.30–2.50	4.00–5.00
Sunflower	0.90–1.10	2.00
Tea	0.50–0.70	1.50–2.50

Costs and Benefits

If the molluscicidal plant material were produced as described, costs and benefits would have to be kept in balance as much as possible. Production and application costs must be as low as possible if the material is to compete with commercial molluscicides.

In the case of self-help production on farms and in villages, costs and benefits will be difficult to determine. Benefits can only be expressed in terms of a healthier population resulting in higher productivity in areas where the crop is grown and used against snails. That requires not just molluscicides but also treatment and cure of infected persons. Farmers and villagers must see that they can get a reasonable price for their crop and that at the same time medical treatment will follow. No matter how much people care about their health, if growing a molluscicidal crop is not profitable they will not grow it and will continue to be sick.

While at first the whole program may require a fully supported system, as soon as the importance of the product becomes evident to the producers in terms of health and other benefits, the crop will be grown more enthusiastically. And in countries that can produce and process the crop for export, the material will bring in foreign exchange while making it unnecessary to buy artificially produced molluscicides. At that point the plant molluscicide will be on an economically sound basis.

AVAILABILITY

It would be most advantageous if the material could be applied immediately after harvesting. No expensive storage facilities would be needed. But such an ideal situation rarely occurs.

The productive period for some potential plant molluscicides may fall in the rainy season, when snails are apt to be controlled simply by being washed away, and when the molluscicide, too, would be easily washed away. Ideally, a plant molluscicide should be available during the dry season. There are places, however, with permanent irrigation where the material could be applied at any time of the year if the crop were producing its wanted plant organs.

The search for molluscicidal plant sources should therefore be carried out among plant species such as *A. maritima* (El-Sawy et al., 1978) and *P. senegalense* forma *senegalense* (Dossaji et al., 1977) that are already growing in bilharzia endemic areas. If the plant source does not occur in these areas, a period of adaption must be added to the period of research to counter all possible negative environmental influences on the newly introduced species.

An ideal situation exists in the case of *P. dodecandra* (Lugt, unpublished observations, 1978). The berries that provide the molluscicide are ready for harvesting shortly after the rainy season ends. After drying and grinding, the end product can be applied. The period of berry production almost exactly coincides with the dry period, giving – if regular irrigation is carried out – nearly seven months of a constant supply of molluscicidal material at just the right time for treating the snails.

A less than ideal situation occurs when, for example, complete plants are the source of the molluscicide and require a period of growth that shortens the productivity period and prevents snail treatment during the most suitable period of the year. The situation is also less than ideal if the harvest must be done in a short, labor-intensive period.

HARVESTING

Harvesting on farms or in villages will almost certainly be done by hand or with simple tools. Fruits, seeds, leaves, and branches will be hand picked and collected in baskets. Complete plants will be cut by sickle or machete or rooted up and transported in bags or bundles. Roots or other soil-borne organs and bark are more difficult and thus more costly to harvest.

The right time for harvesting will be determined by when the plant or plant part reaches maximum molluscicidal potency and when the largest amount of end product is available. To know when and how to harvest, phytochemical and basic agricultural research must be carried out.

PROCESSING

Processing the harvested plant materials involves drying, cleaning, grinding, and packing. If the end product is not readily soluble in water, aids must be added to increase solubility. Nonvoluminous materials such as seeds and berries are easier and less expensive to process than bulky leaves and roots and juicy fruits.

Drying

To preserve the active molluscicidal compounds for a long period, drying should yield material that contains no more than 5% moisture. Most farmers will know from their experience with other crops just when the material is at the desired state of preservation.

The cheapest and simplest drying method is to spread the plant parts on clean ground and let the sun dry them. Drying on a larger scale is done in simple shade houses that let the wind do the work while protecting the material from deterioration by fungi.

Quick drying usually is best to preserve the active ingredients. The material should be given enough space to dry; otherwise it might get too hot and develop fungi or burn down enzymes and other valuable compounds.

Cleaning

After optimal drying, soil particles are sifted out and unwanted plant parts and weeds taken away. A clean product gives optimal activity and brings the best price, but the law of diminishing returns will determine how much labor to expend on cleaning.

Grinding

The grinding that follows should be based on the principle that the smaller the particles the easier it is for the active molluscicidal compounds to go into solution. But fine grinding presents some problems.

In a primitive village, grinding will be less efficient than in a village or farm that has electricity and a grinding mill. The difficulty may come in getting the owner of the mill to let it be used to grind a molluscicide, since he uses it primarily for food crops. The possible dangers involved must be elucidated by toxicologic studies.

Another possible hazard comes from the molluscicidal plant's dust, which may induce allergic reactions. *P. dodecandra* berries, when ground to a fine powder, cause such reactions (Lugt, unpublished observations, 1978). The grinding of the berries must be done so as to allow the air to flow away from the person doing the grinding.

Packing

The material, if it is to be stored, must be packed in such a way that it does not lose its potency. Plastic bags, sealed tins, or plastic containers with screw tops work fine. Dust can also present a problem at the packing stage, and care should be taken as in the grinding stage.

Solubility

To play its snail-killing role, the material must be soluble in water. Grinding to fine particles sometimes keeps water from penetrating the dried plant material, and thus the plant tissues and the water do not mix fully. Substances must be added to increase solubility. Grinding fatty seeds can result in an amorphous product that is almost insoluble in water; this material must be defatted.

It is unlikely, however, that these suggestions for improving solubility will be carried out, since they add to production costs and make the cropping of these plants even less attractive. Nevertheless, if these materials are found to be competitive with artificially produced molluscicides, phytochemical research may be able to find cheaper ways to get better solubility.

In the case of *P. senegalense* forma *senegalense*, *Croton macrostachys*, *Sesbania sesban*, and *A. maritima*, which are only marginally molluscicidal and whose usable parts are extremely difficult to dissolve in water, it is questionable whether research should continue on these plants.

CONCLUSIONS

Potential molluscicidal plants should first be sought in areas where schistosomiasis is endemic, since any plants found there will already be adapted to the right circumstances. Once such a plant is discovered, extensive research must be carried out to determine the influence of growing conditions in other endemic areas. Limitations caused by such environmental circumstances as altitude, day length, soil, and climate can then be evaluated.

For chemical and practical reasons, the most desirable plant organs for harvesting and use are seeds or small fruits. These materials are likely to possess relatively high proportions of molluscicidal compounds and are easy to handle in the field under primitive conditions.

The plant must be modest in its demands for water and for soil quality and must not interfere with crops already being grown for other purposes. It would be preferable if propagation and multiplication of the plant could be carried out via seeds that produce true-to-type offspring. If not, a simple means of multiplying the plant vegetatively must be developed.

The possibility of the occurrence of varieties among plant species with promising molluscicidal potency should be considered. If it happens that the plant must be vegetatively multiplied to preserve a constant quality in offspring, further selection of these plants for more valuable types should be carried out with wild varieties. Such an undertaking is in conflict with the idea of self-help, however, since the research can be carried out only in qualified centers.

In cropping, the plant must also be modest in its demands for space and for control of weeds, diseases, and pests. Field establishment and maintenance will be hindered by any extra labor costs. The interest of the farmer in growing the crop will be determined by the price of the molluscicidal end product.

One can only speculate about the yields, prices, costs, and benefits of a crop when the plant to be grown is not known. However, as soon as the end product has shown itself to be of evident importance, resulting in a significant control of snails and leading to a healthier and more productive population, the plant undoubtedly can be produced on an economically sound basis.

For effective control of snails, availability of the molluscicidal plant organs at the right moment is important. If the harvest period is short, requiring considerable storage facilities and thus extra costs, the plant will not be cropped enthusiastically by the farmer. But if harvesting can be done throughout the year and the end product applied instantly, the crop will be more attractive.

A relatively long period of harvesting would also be beneficial for proper processing of the plant organs. Less space and fewer facilities for drying would be required, and the whole process could be carried out more efficiently. A peak harvest period of, say, one month when a large amount of material becomes available presents problems of loss of molluscicidal potency through improper management of drying and grinding. The drying needs to be carried out as swiftly as possible, without overheating the material.

Finally, the end product must be readily soluble in water and have its highest snail-killing potency when made into a watery suspension. Solubility must be investigated at the very beginning of the search for a plant that is suitable for use in controlling snails.

Acknowledgments

The author expresses his deep gratitude to Dr. Th. Visser of the Institute for Horticultural Plant Breeding in Wageningen for his willingness to render assistance in the preparation of this manuscript and to Ms. Barbara Reese, Cairo American College, for reading it.

REFERENCES

Acland, J. D. *East African Crops*. 3d ed., 29–32, 58–89, 170–172, 186–191, 202–203, 208–221. London: FAO/Longman (1975).

Adewunmi, C. O., and E. A. Sofowora. Preliminary screening of some plant extracts for molluscicidal activity. *Planta Med*. 39:57–65 (1980).

Claus, E. P., V. E. Tyler, and L. R. Brady. *Pharmacognosy*. 6th ed., 7. London: Lea and Febiger (1970).

Dossaji, S. F., M. G. Kairu, A. T. Gondwe, and J. H. Ouma. On the evaluation of the molluscicidal properties of *Polygonum senegalense* forma *senegalense*. *Lloydia* 40:290–293 (1977).

El-Sawy, M. F., H. K. Bassiouny, A. Rashwan, and A. I. El-Magdoub. *Ambrosia maritima* (Damsissa) a safe and effective molluscicide in the field. *Bull. High Inst. Public Health Alexandria* 8:307–317 (1978).

Hostettmann, K. Saponins with molluscicidal activity from *Hedera helix* L. *Helv. Chim. Acta* 63:606–609 (1980).

Karsten, G., U. Weber, and E. Stahl. *Lehrbuch der Pharmacognosie*. 9th ed., 193, 237, 286, 487, 504. Stuttgart: Gutav Fischer Verlag (1962).

Lemma, A., G. Brody, C. W. Newell, R. M. Parkhurst, and W. A. Skinner. Endod (*Phytolacca dodecandra*), a natural product molluscicide, increased potency with butanol extraction. *J. Parasitol*. 58:104–107 (1972).

Lugt, C. B. The cardiac-glycoside composition of different *Digitalis purpurea* populations. *Pharm. Weekbl.* 111:441–455 (1976).

——. Development of molluscicidal potency in short and long staminate racemes of *Phytolacca dodecandra*. *Planta Med.* 38:68–72 (1980).

Steinegger, E., and R. Hänsel. *Lehrbuch der Pharmacognosie.* 3d ed., 11–12, 309, 311. Berlin: Springer-Verlag (1972).

8

TOXICOLOGIC SCREENING OF MOLLUSCICIDAL PLANT PRODUCTS

J. H. Koeman

Department of Toxicology
Agricultural University
Wageningen, The Netherlands

Plants of many kinds have been found to be molluscicidal and thus of possible use in the control of the snail vectors of schistosomiasis. The introduction of these plants or their parts into the environment requires prior investigation of their possible toxic effects on mammals and certain other nonmollusc groups of organisms.

The molluscicidal and other chemical constituents of these plants are in principle not inherently different from conventional pesticides, unless it can be proven that they are biorational pesticides, which include microbial and biochemical or pheromonal pest-control agents. The United States guidelines for registering pesticides state that the most important differences between biorational and conventional pesticides are that the former have target species specificity, a generally nontoxic mode of action, and natural occurrence (USA-EPA, 1982). These factors are the basis for the expectation that many classes of biorational pest-control agents pose a lower potential hazard than do conventional pesticides. Hence the data requirements for the biorational agents in a number of cases would be different and occasionally less extensive than those for conventional pesticides. However, there is no evidence that the molluscicidal plant compounds identified thus far (saponins, tannins, and so on) can be placed under the heading of biorational pesticides despite their natural origin. Thus they should be considered as comparable to conventional pesticides.

Toxicologic assessment is complicated by the fact that the exact nature of the active ingredients in molluscicidal plants is unknown or at least only partially known. Moreover, the plant materials may contain nonmolluscicidal constituents with toxic properties that should be checked. Consequently, the toxicity of molluscicidal plants or extracts can be investigated only by bioassay procedures that, from a toxicologic point of view, are difficult to interpret.

This chapter presents some criteria for toxicologic screening that may help in developing an adequate approach to these problems.

AVAILABILITY OF PLANTS WITH CONSTANT PROPERTIES

A major problem in toxicologic screening of plant products is that the concentrations of molluscicidal constituents may vary according to genetics, conditions of growth or breeding, and season. Molluscicidal potential may also vary considerably by plant type or variety, and so may the possible toxic potential of other constituents. To adequately characterize the toxic potential of a molluscicidal plant therefore implies that one should undertake studies with all possible varieties that could be used. The costs and efforts required for such an exercise would of course be prohibitive. One may conclude that the availability of a plant product with more or less constant properties is an essential prerequisite.

TOXICOLOGIC RESTRAINTS

When molluscicidal preparations are applied to an aquatic environment, at least three potential hazards should be taken into consideration: The preparation may pollute the drinking water; compounds present in the preparation may accumulate in food products of aquatic origin, such as fish and molluscs; and species of nontarget aquatic organisms may be exposed either directly or through the food web.

In the case of a new chemical pesticide, studies can usually be carried out to assess the fate of the chemical and its possible metabolites in the environment in order to learn whether human or other organisms will be at risk. In the case of molluscicidal plant materials, a reliable assessment cannot be made because one does not know just what to look for. For the same reason, no adequate assessment can be made of the toxicity of other potentially hazardous constituents. Assessing all the possible hazards of molluscicidal plants is comparable to the assessment of impure chemicals.

ASSESSMENT OF HAZARDS

Tests should be carried out on the molluscicidal preparation derived from the plant material – for instance, water or organic solvent extracts.

An attempt should first be made to assess the toxic properties of the material, following a stepwise approach. Preliminary screening could include tests of acute oral toxicity in rats; a 30-day diet study with rats; a few short-term mutagenicity tests, such as the Ames tests and tests in eukaryotic systems (for example, point mutation tests in mammalian cells or yeast with and without metabolic activation and the sex-linked recessive lethal test in *Drosophila melanogaster*); and, finally, tests of toxicity to fish, *Daphnia*, and algae.

Depending on the outcome of these preliminary trials, additional tests may be required. These are described in various guidelines for toxicity testing, including the OECD (1981) guidelines, the FAO (1981) criteria for registration of pesticides, the WHO (1978) guidelines for evaluating the toxicity of chemicals, the United Kingdom (1979) pesticides safety precautions, and the Netherlands (1980) report on carcinogenicity.

The ultimate approval of a molluscicidal preparation cannot be based solely on the outcome of the toxicity trials. Additional information is required on the environmental fate of the toxic properties.

If the molluscicidal preparation is based on a water extract of the plant material, the conclusion can be drawn that the possible toxic constituents are water soluble and will very probably not accumulate substantially in fish, molluscs, and other organisms that may serve as food for human and other nontarget organisms. However, direct exposure of people through drinking water and of aquatic organisms cannot be excluded. In that case only the water extracts should be submitted to the toxicity tests, but an investigation should also be made into whether and at what rate any toxic properties will disappear under circumstances likely to prevail in the field. For this purpose both bioassay tests in the laboratory and limited field trials (for example, assessment of the effects on nontarget organisms and of the toxic properties of water samples collected in the field) may provide sufficient information.

When the toxic constituents are water soluble there is still a chance that the substances will give rise to the release of compounds with lipophilic properties, such as the aglycon moieties of glycosides. This possibility can be investigated by examining the toxicity of the extract after hydrolytic treatment in one or more relatively simple tests (short-term mutagenicity tests or bioassay experiments with fish). If these tests prove positive, the evaluation becomes more complicated because there is a possibility that toxic constituents with accumulative properties are present. This possibility should also be taken into consideration when the molluscicidal preparation is based on an organic solvent extract.

Thus one cannot rely on toxicity trials with the extract itself. When the extract is toxic and the toxic properties show some persistence, trials should be performed to find out whether the toxic properties can be transferred into aquatic organisms that may serve as food for man and other organisms. In principle, one should then conduct toxicity trials with the contaminated products. However, such a situation presents a number of technical difficulties and uncertainties. It will, for instance, be extremely difficult to accomplish balanced long-term feeding trials with raw fish products. Tests with extracts of fish and other food products also pose problems because one does not know whether the toxic compounds or possible toxic metabolites formed in the product are extracted from the tissues. Therefore it seems unavoidable that one needs to know the identity of the toxic principles of a lipophilic nature before any meaningful toxicity study can be carried out.

SUMMARY

The following criteria should be adopted with regard to the toxicologic screening of plant products:

• Plant material should be available with more or less constant properties (molluscicidal and otherwise), (see page 323).

• In case the water extract contains the molluscicidal principles, one should assess the toxicity by using the preliminary tests proposed in this chapter; examine the persistence of the toxicity in the aquatic environment, either in laboratory trials or limited field experiments; investigate whether hydrolysates have toxic properties of a lipophilic nature that may accumulate in food products; and, if preliminary toxicity tests prove insufficient, carry out additional toxicity trials (semichronic, chronic, reproduction) according to guidelines developed by OECD, WHO, FAO, and national authorities.

• In the case of the possible accumulation of toxic properties of a lipophilic nature in food products and when organic solvent extracts contain the molluscicidal principles, one should assess the toxicity by using the preliminary tests proposed in this chapter; examine the persistence of the toxicity in the aquatic environment, either in laboratory trials or limited field experiments; and carry out additional toxicity tests as required. When there is persistence of toxicity and likelihood of contamination of food products, the chemical nature of the toxic compound should be known before any meaningful hazard assessment can be made.

Acknowledgments

The suggestions made by Dr. M. Vandekar, of the Division of Vector Biology and Control at the World Health Organization in Geneva, Dr. W. N. Aldridge, of the Medical Research Council Toxicology Unit in Carshalton, United Kingdom, and Dr. J. P. Brown, of the Zoecon Corporation in Palo Alto, California, were highly appreciated and fully considered in the preparation of this chapter.

REFERENCES

FAO. *Environmental Criteria for Registration of Pesticides*. FAO Plant Production and Protection Paper 28. Rome: Food and Agricultural Organization (1981).

The Netherlands. *Report on the Evaluation of the Carcinogenicity of Chemical Substances*. The Hague: Health Council of the Netherlands (1980).

OECD. *OECD Guidelines for Testing of Chemicals*. Paris: Organization for Economic Cooperation and Development (1981).

United Kingdom. *The United Kingdom Pesticides Safety Precautions Scheme*. Agreed between Government Departments and Industrial Associations. Revised. London: Ministry of Agriculture, Fisheries and Food and HMSO (1979).

USA-EPA. *Guidelines for Registering Pesticides in the United States Federal Register*. Vol. 47, No. 227, 53202–53218 (Sections 158.65–158.165). Washington: Environmental Protection Agency, Department of Health Education and Welfare (1982).

WHO. *Principles and Methods for Evaluating the Toxicity of Chemicals*, Part I. Environmental Health Criteria Document No. 6. Geneva: World Health Organization (1978).

9

LABORATORY EVALUATION OF POTENTIAL PLANT MOLLUSCICIDES

J. Duncan

Center for Overseas Pest Research

R. F. Sturrock

London School of Hygiene and Tropical Medicine

London, England

A standardized procedure has been devised for the laboratory screening and evaluation of synthetic chemical molluscicides (WHO, 1965). It recommends that candidate compounds pass through three stages: preliminary screening, definitive screening, and comprehensive evaluation.

In preliminary screening, the aim is to separate materials into those without any molluscicidal activity and those with some, however weak. No guidance was given on how that should be done.

For definitive screening, a procedure was provided, including methods for both wholly aquatic and amphibious snail intermediate hosts of schistosomiasis. It allows for the determination of the LC_{50} and LC_{90} values of the molluscicide (the concentration in water that kills 50% or 90% of a target snail population) and the comparison of results from different laboratories.

Any further characterization of molluscicidal activity comes at stage three, comprehensive evaluation, in which factors that affect molluscicidal performance under field conditions can be investigated by laboratory simulation. The variables most commonly examined are the effects of pH, temperature, light, and absorption onto organic matter.

This procedure emphasizes a logical progression, both in the selection of the more active molluscicides and in the increasing complexity of the tests applied to them. The approach is relatively straightforward in that most of the materials being evaluated are physically or chemically pure and are

251

either already formulated (usually as emulsifiable concentrates or wettable powders) or readily formulable by the investigator.

The screening procedure initially considered compounds that were essentially acutely toxic to snails. Exposure and recovery times were therefore relatively short, 24 and 48 hours respectively, although 48-hour exposures were advocated in the case of *Oncomelania* snails. The method was modified when, for example, it was found that the completed response to organometal poisoning was evident only after several days (Hopf et al., 1967) so that longer recovery periods were necessary.

More recently, attempts have been made to develop slow-release formulations for molluscicides in order to extend the exposure time of the snails, limit the amounts of chemical entering the environment, and reduce application costs. A standard methodology for examining such formulations is not available.

Plant products present a larger number of variables than do synthetic molluscicides. The plant materials come in a wide variety of forms and physical states. The molluscicidally active ingredients are often more complex molecules than those found in synthetic molluscicides. They may be present as precursors, requiring a biological or chemical agent to release them, and other components of the plant may synergize or inhibit the molluscicidal activity. Thus laboratory evaluation work should be done mainly with whole plant material (see page 323). The isolation of active ingredients is a more specialized task; if it is necessary, it should come in the later stages of comprehensive evaluation, that is, in studies on mode of action and in some aspects of toxicology.

Furthermore, phytoconstituents may vary with the strain of the plant, the part of the plant being examined, and spatial (geographical), temporal (seasonal), and even diurnal influences. Plant material may release active ingredients slowly so that the effect on snail populations is delayed, that is, the plant may act as a slow-release matrix. In this mode, there may also be effects on feeding and oviposition.

Other plant substances affect the orientation and feeding behavior of snails (Thomas and Assefa, 1979), and although they could be used in attracting snails to baits or slow-release formulations, we will not be concerned with such compounds in this chapter.

The major problem in assaying plant molluscicides therefore is to construct a sequential testing plan that takes all these factors into account. It is hoped that the methods outlined here will provide guidelines to newcomers to this work and to those formulating research proposals and projects. Variations in plant material doubtlessly will necessitate modifications of the general scheme. However, the aim is to provide a systematic approach that will help in assessing both the inherent molluscicidal activity of the plant material and its potential for use in the field to control snail populations.

FEASIBILITY STUDY

A prerequisite for investigating plant molluscicides is to conduct a feasibility study of local requirements for and practical constraints on the use of material in an area where schistosomiasis is endemic. The conclusions from such a study not only will determine the usefulness of the plant but also will assist in choosing candidates and determining the bioassay tests to be emphasized.

The investigator should therefore consider the importance of schistosomiasis in his area of interest, the control methods in use or likely to be used, and the role plant molluscicides might play, either alone or in an integrated control strategy. It is well to know, even at this early stage, whether local or central health authorities are likely to encourage or give moral or financial support to the development and use of plant molluscicides. Knowledge of transmission patterns (seasonal or throughout the year, focal or of a wider distribution) will affect such planning decisions as how much plant material might be required; the need to collect or cultivate these amounts; the availability of the plant at the right time of year or the need for storage; whether the plant grows or could be grown near transmission sites; and transportation requirements.

Possible application methods should be reviewed: whether the plant will be applied as dry or fresh material or as pastes or slurries of dried, whole, or ground material and whether it will be done by government health workers or by village or farming communities. In the latter case, the acceptability of the method to local people and the likelihood of its sustained use should be assessed. Knowing how toxic the material is to organisms other than snails is important. Risks to nontarget organisms will have to be minimized, for example, by formulation or by timing or method of application. Any regulatory requirements of the governments concerned, with respect to toxicology, will also have to be adhered to. While it may be possible to obtain preliminary acute toxicologic data relatively cheaply, the cost of longer-term studies on chronic toxicity, should these be required, may mitigate against even initial or exploratory work on plant molluscicides.

The selection of candidate plants can be guided by the known molluscicidal effects of related species, genera, or larger groupings. Kloos and McCullough (1981) provided a list of plants already studied (see also Chapter 3). The correct identification of plant specimens is important, and voucher specimens should always be lodged with a herbarium (see Chapter 6). Advice should be sought from a competent botanist on the preparation of museum voucher plant specimens (see also Womersly, 1981).

Plants that already have some other use will be that much easier to develop for molluscicidal purposes. Plants that are known to be toxic to mammals should be avoided unless the toxicity can be easily attenuated. Gloves and possibly protective glasses or goggles should be worn when handling plant material whose properties are unknown. Grinding, particularly of large quantities of dried plant samples, should be carried out under controlled and supervised conditions.

A feasibility study should be aimed at obtaining information on the epidemiology of schistosomiasis in the area where the plant molluscicide will be used; the control methods already in use; the likelihood that plant molluscicides are needed, are available, and will be used; and the requirements for developing and using a plant molluscicide.

The information on epidemiology will tell which species of *Schistosoma* is present, the disease's prevalence and intensity, the kinds of transmission sites involved, and whether transmission is focal or general, seasonal or more or less continuous. The answers will show whether schistosomiasis requires control and what quantities of plant molluscicide might be required.

The information on control methods will tell what methods are being used already, what molluscicides are being used or have been used, how they were applied and whether by a government agency or privately, how a plant molluscicide could be used either alone or with other control measures, whether the government is likely to give moral or financial support to the use of a plant molluscicide, and whether any regulatory requirement or legislation would affect the use of such a plant substance. With this information, a prediction can be made about the eventual use of a promising candidate.

The information on plant development and use will tell whether the plant is available at or near transmission sites, whether it could be cultivated at such sites, who would cultivate it (local health authorities or the community), whether it would be more economical to grow the plant on a large scale and whether any horticultural/agricultural support is required, what the opportunity costs are in large-scale growing, who will be responsible for the growing (the government or private concerns), whether the plant is available early in the transmission season, whether storage is required and where (villages or government-controlled centers), whether processing (drying, extracting, formulation) is required, and whether transport is required to transmission sites and who will provide and pay for it. The scale of operations and the infrastructure required to support the organization for providing the plant should now begin to emerge.

LABORATORY BIOASSAY

The procedure outlined for preliminary screening, definitive screening, and comprehensive evaluation of synthetic molluscicides can also be used for plant molluscicides (Fig. 1).

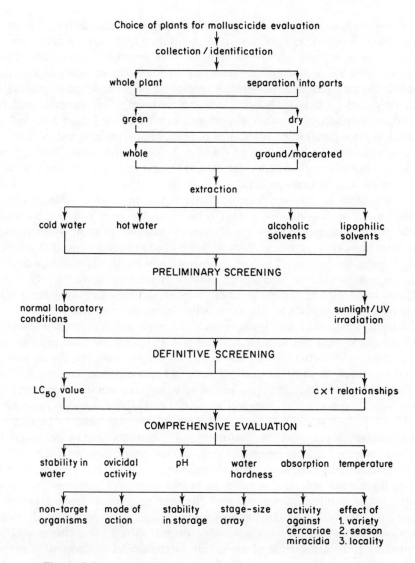

Fig. 1. Scheme for laboratory evaluation of a plant molluscicide.

Preliminary Screening

Plant material will be collected whole or in part in the fresh state. If fresh material will be used in snail control, then there is clearly a case for screening the material in this condition. But this means that it will be necessary to determine the weight of the plant material at the time of collection in order to know the amount of fresh plant that will kill snails when put into a measured volume of water. The weighing of plant material during collecting expeditions will not be easy, and though ultimately it may be necessary to experiment accurately with fresh material, this may be better reserved for later detailed assay of candidate materials of marked molluscicidal activity.

Initially, and particularly while screening a large number of plants, it will be easier to deal with dried, ground material. This material will be homogeneous and of constant weight and can be studied over a period of time. It is known that some substances of plant origin (for example, alkaloids) can be decomposed by milling of dried materials (Farnsworth, 1966). That is not likely to be a major consideration in the present context, but it is well to keep such exceptions in mind.

Some system of cross-checking of activities should be used throughout the evaluation. For example, even though the work has been conducted mainly with solutions prepared from dried, ground material, at some stage similar solutions should be prepared from fresh plant sample to ensure that activity is not being lost or gained by the various treatments that the plant material has been subjected to. Obviously, that will become progressively more difficult to do with increasing numbers of plant samples. Cross-checking may thus have to be limited to the more active candidate materials.

Temperatures used for drying should be high enough to stop enzyme activity quickly but not so high as to destroy the plant material or drive off volatile compounds that may be molluscicidally important. The drying period should be kept as short as possible to avoid fermentation or rotting. The berries of endod (*Phytolacca dodecandra*) have been sun-dried for three to four days (Lemma, 1970). Sherif et al. (1962) oven-dried *Ambrosia maritima* at 30° C. Larger succulent fruits or storage organs may need to be divided into smaller pieces prior to desiccation. Precautions should be taken in storing the dried material to avoid pest infestation or fungal attack.

The plant molluscicide may act on snails by ingestion, but more likely the active ingredients will need to be brought into aqueous solution so as to pass through the external membranes of the snails. Thus the plant material must be extracted, and the solvents employed may be aqueous, alcoholic (methanol, ethanol), or lipophilic (chloroform, ether). The choice will be guided by prior knowledge of any likely molluscicidal compounds present. Selection of an appropriate solvent obviously presents a problem when the nature of the chemical involved is unknown. A further complication is that some phytoconstituents can affect the solubility of others. Saponins can improve the solubility of some compounds, possibly by acting as a wetting agent and aiding the formation of micelles. Plant lipids may interfere by retaining solutes. Farnsworth (1966) cited a case where a group of alkaloids could be extracted from a plant material using hexane, whereas the isolated, purified alkaloids were insoluble in this solvent. It was assumed that lipid present in the original plant material made the alkaloids more available to solution in hexane.

No ready answer can be offered for solubility problems other than that crude plant material should be examined with a variety of solvents. However, since any plant molluscicide is destined for use in water and since organic solvents may not be readily available at field stations, cold- and hot-water extractions should be attempted first. *Cold water* here means water used under ambient conditions. The temperature or temperature range employed should be noted. Hot-water extraction (decoction, infusion) may be more efficient than cold when the active ingredient is sparingly soluble. A range of extraction times and weight of plant material to water volume ratios should be investigated to determine the conditions under which maximal molluscicidal activity or loss of activity is attained.

Adewunmi and Sofowora (1980) used methanolic extracts in examining a wide range of plant species for molluscicidal activity. The molluscicidal agent of endod exists in the plant in an inactive form as oleanolic acid with a carboxy group present as part of a sugar ester. When Jewers and King (1972) used ethanol as a solvent in the extraction of the saponin fraction of endod, only a molluscicidally inactive material resulted. Subsequent alkaline hydrolysis of the ethanol extract was required to release a free -COOH group from the ester and thereby produce the molluscicidally active molecule. The hydrolysis occurs in aqueous extracts of the endod berries through the action of a naturally occurring enzyme. Endod saponin has also been prepared by initial extraction of powdered berries by light petroleum ether, which removes waxes. The defatted powder is soaked in water and the mixture extracted with butanol. The butanol layer is removed and evaporated to dryness to give a fraction that is molluscicidally superior to a straight-forward water extract (Lemma et al., 1972). A similar improvement in activity is claimed by Lemma and Lemma (1977) if an aqueous suspension of powdered berries is fermented for a few days, although prolonged fermentation destroys the molluscicidal effect. These authors suggested that both butanol extraction and fermentation remove substances such as lipids that the molluscicide may partition into or otherwise react with instead of passing into the water phase.

When alcoholic or lipophilic solvents are employed, Soxhlet or repeated small-volume extractions may be used for dried plant material. Fresh plant material may need to be broken down under the solvent in a pestle and mortar or a blender (Odebiyi and Sofowora, 1978). The supernatant is removed and evaporated, usually with a rotary evaporator. It may be possible to take up the residue in water, before or after trituration, to provide a stock solution of molluscicide. Otherwise, a stock solution may be provided by taking up the residue in as small a volume as possible (up to 1 mL) of a solvent such as dimethyl sulfoxide and dispersing this solution by stirring into water.

The water used in bioassay of aqueous or organic solvent extracts should be from the snail habitat where molluscicidal treatment is to take place; if that is not readily available, dechlorinated tap water can be used. The pH should be noted. Although snails will survive for quite some time in distilled water, its use is not recommended because evidence suggests that the amount of molluscicide taken up by snails is significantly reduced in distilled water media (J. Duncan, unpublished data, 1975).

The molluscicidally active ingredient may be affected by ultraviolet light. Synthetic molluscicides such as pentachlorophenol are degraded by sunlight. The cercaricidal activity of α-terthienyl from the roots of the marigold *Tagetes patula* (Compositae) is increased 30-fold by UV irradiation (Graham et al., 1980). Polyacetylenes of plant origin are also phototoxic to bacteria, fungi, and human fibroblast cells (Wat et al., 1979) and to mosquito larvae (Arnason et al., 1981).

When a large number of plant samples are to be examined in a preliminary screening, it is suggested that the following conditions be used: a concentration of up to 1000 mgL^{-1} (ppm), an exposure period of three to five days, two to five adult *Biomphalaria* or *Bulinus* snails in a solution volume of 200 to 500 mL or five *Oncomelania* adults in 100 mL in a glass container, and water temperatures of 25 to 27° C. The precise experimental conditions should be reported. Care should be taken when using higher concentrations of plant materials to judge whether snail deaths are due to molluscicidal compounds or a general pollution of the water. One gram of plant material in a liter of water provides a nominal 1000 mgL^{-1}. One treatment should be kept continuously under laboratory conditions, while a replicate solution is exposed daily to natural sunlight to detect any activating or deactivating photochemical effect. UV light will penetrate water only to a depth of a centimeter or so, and thus a shallow container should be used. Precautions should be taken to prevent a rise in the water temperature by placing the container of molluscicidal solution in a much larger volume or in a flowing body of water. It is not essential to put up control experiments with every preliminary screening, especially if large numbers of candidate plants are being examined; but initial experiments should be made to check, for example, the viability of snails under the experimental conditions or the stability of water temperatures in outdoor tests. Control experiments can then be included from time to time.

The number of snails dead and alive on each day of the test should be recorded. Dead snails are immobile and either retracted well into or hanging out of the shell, with the body and shell discolored. Death is confirmed by lack of reaction to prodding of the body with a blunt seeker or failure to see any heart activity under a dissecting microscope. Dead snails should be removed as soon as possible.

The use of up to 1000 mgL^{-1} of solution for up to five days should demonstrate molluscicidal activity even if the available sample is of low potency, the activity resides in just one plant part or is diluted by grinding (especially of whole plants), the active ingredient is sparingly soluble in water or the plant material in cold water acts like a slow-release formulation, the test conditions are not optimal for the active ingredient, a fermentation process is necessary (it can still proceed in water at ambient temperatures), or the mortality response of the snail to the molluscicide is slow.

Definitive Screening

Plants that have shown molluscicidal activity in any of the preliminary tests may be considered for definitive screening. The necessary level of activity depends on operational considerations. A plant showing only moderate levels of activity may still be usable if it is or can be made available in adequate quantities at or near schistosomiasis transmission sites.

Synthetic molluscicides are commonly compared on the basis of their LC$_{50}$ values computed from a dosage-mortality response curve (usually a log.dosage-probit mortality curve) with a 24-hour exposure and a 48-hour recovery period. If a plant material appears to be fast acting (that is, activity is demonstrable within 24 hours or so), then it should be possible to determine the LC$_{50}$. Plant material, however, may be slow acting; the active ingredient may be somewhat insoluble, taking time to leach out, or unstable, or both. Yasuraoka (personal communication, 1982) pointed out that in the typical stagnant-water habitats of *Oncomelania*, molluscicides tend to linger. In such cases, where time is a common variable, LC$_{50}$ values may not be appropriate; concentration x time (c x t) susceptibility relationships may be more relevant.

If the whole plant is to be used in control, the first step in definitive screening is to identify more closely the concentration range in which the activity lies. Solutions of the samples that were active in preliminary screening should be prepared anew and diluted in the same water used to prepare the stock solution. A typical concentration range of dosages might be 1000, 500, 200, 100, 50, and 25 mgL^{-1}. If activity was demonstrated only in hot-water extracts, the 1000 mgL^{-1} of stock solution should be diluted with hot water and the resulting solution allowed to cool to the bioassay temperature. Two to five snails should be added to 200 mL of each dosage and the numbers alive and dead noted each day for three to five days. If the survival of the test species is good under the conditions used for preliminary screening, replicate dosages are not necessary at this stage. If only a small number of plant samples are being dealt with, dilution and bioassay of the stock solution can be done at the preliminary screening stage.

If only the more active part of the plant is to be used in control, the plant sample must be divided (best done soon after collection) into flowering heads, fruit, leaves, stem, roots, or any other appropriate subdivision, then dried. The bioassay procedure and concentration range given in the previous paragraph can then be followed.

A bioassay of whole plants or parts in which snails are killed within 24 hours at a dosage below 100 mgL^{-1} of dried material indicates that the molluscicide is released quickly and the material may be a candidate for LC$_{50}$ determination. Further dosage series must be prepared with narrower ranges until the investigator is confident that the LC$_{50}$ value is known within fairly narrow limits.

The LC$_{50}$ or LC$_{90}$ may then be determined. Materials that are affected by UV irradiation must be tested under the appropriate lighting conditions. Between five and ten dosages or treatment levels might be employed. Adult snails can be exposed singly (to keep dying snails from affecting the survival of others) in 200-mL volumes for 24 hours with at least ten replicates of each dosage. The number of replicates depends on the variation in susceptibility of the snails. Inbred laboratory colonies should have less variability than field-collected specimens. A uniform shell size or height should be used. Snails brought in from the field for bioassay must be acclimatized to the laboratory conditions for a few days prior to experimentation. The number of replicates, snail size, and acclimatization time depend on the particular laboratory's experience in working with the biological material. The experimental conditions should be reported.

At the end of the exposure period, the exposure solution is decanted off and the container rinsed twice and filled with untreated water – the same as that used in preparing the original molluscicidal solutions. The numbers of snails dead and alive are noted after a 48-hour recovery period or whenever a response endpoint is achieved. LC$_{50}$ and LC$_{90}$ values may then be calculated, using log.dosage-probit mortality graph paper and following the method of Litchfield and Wilcoxon (1949), or using the appropriate computer software.

Concentration x time (c x t) relationships can be worked out using the methods given by WHO (1965). Dosage ranges are prepared suitable for computing LC$_{50}$ and LC$_{90}$ values. The dosage-mortality response is determined for l-, 6-, and 24-hour exposure in the manner described. For plant molluscicides, longer exposure times may be required.

An alternative approach to c x t relationships is that commonly used in fish toxicity testing (Muirhead-Thomson, 1971). Batches of test animals are exposed to a range of dosages and the time at which any individual animal dies is noted. The log.dosage is plotted against log.median survival time (time at which 50% mortality was recorded), from which the LC_{50} at any particular time interval may be determined. A disadvantage of this method is that the experiments need observation at regular intervals, which may mean night duty at least in the early part of the study. The results of c x t studies may yield very useful information in deciding field application regimes.

Comprehensive Evaluation

The purpose of comprehensive evaluation is to examine more carefully in the laboratory those candidate materials that appear to have potential for use in the field. The tests that may be carried out at this stage tend to fall into two groups.

The first group includes factors of immediate interest: the effect of pH, temperature, absorption onto surfaces, stability in water, and ovicidal activity. It is not always easy to fix all these variables so that each may be examined alone. It does, of course, make for a more efficient experiment if a number of variables are examined simultaneously with the appropriate statistical design and analysis for main effects and interactions of treatments.

The second group includes effects against nontarget organisms; activity against miracidia and cercariae; the effect on molluscicidal potency of variety, season, time of day, method of drying, length of storage, and locality; the differential susceptibility of the size-range (stage-size array) of the snail; and mode of action, that is, identification of the main active ingredients and the way they kill snails or control the snail population.

Stability in Water. The stability of synthetic molluscicides is tested by preparing dosages expected to give a 90% snail kill and letting them stand for 0.25, 1, 2, 4, and 8 days before adding the snails for a 24-hour exposure (WHO, 1965; Ritchie, 1969). Mortalities are recorded after allowing a 48-hour recovery period in molluscicide-free water.

Water instability could be the result of several factors acting singly or in combination: hydrolysis, which may be pH dependent; UV irradiation; initial concentration; and bacterial decomposition. There may also be spurious effects due, for example, to absorption of active ingredient on the walls of the solution container. The test conditions must therefore be carefully considered and reported.

Preparation of plant molluscicide solutions giving LC_{90} values within 24 hours may mean starting with hot water or alcohol extracts or detailed information on the rate of cold-water extraction. The pH of solutions may be altered by small additions of dilute sulfuric acid or caustic soda solutions. The pH should be monitored at 24-hour intervals. Duncan and Pavlik (1970) used 0.005 M Sorensen's phosphate buffer fortified with 0.104 g of $CaCl_2$ and 0.26 g of $MgSO_4.7H_2O$ per liter. That provided a pH range of 5.5 to 8.0. The relationship between pH, conductivity, and water hardness should be borne in mind if attempts are made to study the effects of the latter on molluscicidal activity.

Any effects of UV irradiation will be known from the preliminary screening stage. It may be necessary to confirm this finding by using UV lamps with a range of known emission spectra and comparing exposed molluscicidal solutions with those kept in darkness. Of course, even a light-sensitive molluscicide could be applied in turbid water or in the field in the evenings.

Etges et al. (1965) showed that the molluscicidal action of three synthetic molluscicides (niclosamide, trifenmorph, and isobutytriphenylmethylamine) was reduced by bacterial action – a result that may not be totally undesirable, since it means that such compounds may not persist in the environment.

Temperature. With synthetic molluscicides, an increase in temperature usually results in increased susceptibility of snails. The same trend is seen with endod (Lemma, 1970). The temperature range employed in laboratory experiments is usually large. Care must be taken to interpret the results in terms of the temperature range experienced in the natural habitat. The length of time snails are acclimatized to the temperature conditions prior to exposure to the molluscicide may have an effect on the magnitude of the response.

Ovicidal Activity. The susceptibility of snail eggs can be affected by the stage of development of the embryo. Egg masses at four days old or showing the trochophore stage are convenient for preliminary experiments. Adult snails will oviposit on a polythene sheet, and small circles of polythene with one egg mass attached can be cut out. A range of dosages is prepared as already described, and one or two egg masses are placed in each dosage. Volumes of 100 mL are suitable. Larger volumes are preferred to smaller ones to avoid any reduction of the nominal concentration of the molluscicide by absorption onto the polythene. A final check for absorption effects can be carried out using eggs laid on a water weed normally used by snails as a substrate for oviposition. If sufficient numbers of egg masses are available, the testing for ovicidal activity can be included in the first stages of definitive screening. LC_{50} values are not commonly computed for egg masses; it is usually sufficient to know that the susceptibility of the eggs is less or in excess of that of the adults. Even if dosages approximating the levels lethal to adult snails do not kill eggs immediately, it is still possible that these or lower treatment levels will either prevent hatching or will kill over a prolonged period of time, and this possibility should be investigated.

Ritchie et al. (1963) examined the susceptibility of the various developmental stages of ova and the snail to molluscicides, thus extending the tests done on egg masses alone. This approach against a stage-size array was designed to provide information on the dosage range needed in the field to control all stages. The effect of various exposure times was also studied. Although susceptibility might be expected to decrease with increasing shell size, Boyce et al. (1967) showed for trifenmorph that the 3-mm *Biomphalaria glabrata* was the most sensitive of the size range.

Absorption. It might be expected that the mud substratum, macrophytes, or algae would be mainly responsible for absorption of molluscicides from solution. However, these materials are not easy to define or standardize; nor is it easy to say what proportions of each to water should be used in laboratory tests as a model of a natural habitat. WHO (1965) suggested that feces of rats fed on a standard diet could represent the various contaminants of natural water. A standard concentration of 50 mgL^{-1} was recommended. Paulini and Pereira de Souza (1970), using colloidal clay, calcium carbonate, and yeast cells, showed that various molluscicides had differing affinities for these materials. They also discussed the various types of bonding involved in absorption phenomena. Meyling and Pitchford (1966) showed that the amounts of certain molluscicides lost by precipitation depended upon pH and that the amounts lost by absorption could depend on pH and dissolved solids.

NONTARGET SPECIES

Laboratory tests of toxicity to nontarget organisms have often been carried out with small fish such as the guppy. The tests can be grouped into static-water methods, periodic replacement tests, and flowing-water systems (Muirhead-Thomson, 1971). The cladoceran *Daphnia* is generally more sensitive than other microcrustacea to chemical pollution. It is relatively easily maintained in the laboratory and has been widely used in preliminary screening. The ciliate protozoan *Spirostomum ambiguum* has been used for the bioassay of certain synthetic molluscicides (Meredith and Meredith, 1972).

The choice of animals or plants for toxicity tests will clearly be influenced by local economic and environmental considerations, such as what crops are growing in irrigated areas and what kind of commercial fishing is being done. In view of the possible diverse effects of molluscicidal action, it is likely that the most useful information will be obtained by monitoring wild populations of ecologically relevant species during field trials of candidate plant molluscicides. That will also allow assessment of chronic effects, for example, on reproductive rates or growth. Such studies require considerable time and expertise.

While molluscicides can also be lethal to cercariae and miracidia, the chemical or its formulation would almost certainly need to have a residual or chronic effect in order to have an impact on schistosomiasis transmission. Broberg (1980) recorded the effect of extracts of *Glinus lotoides* on miracidia.

CONCLUSION

Few plants have been examined in any depth for their molluscicidal properties. In preparing a plan for laboratory evaluation, we have had to rely on experience gained from assay of commercially available molluscicides. However, it is hoped that a systematic and unifying approach to the problems brought about by the variable structure and chemistry of plant materials has been evolved and that workers in this area will find this chapter to be a useful guide in processing their own materials, in comparing data with other laboratories, and in reporting their results.

Acknowledgments

The authors gratefully acknowledge comments made by Professor C. F. A. Bruijning, University of Leiden, The Netherlands; Professor K. Yasuraoka, University of Tsukuba, Japan; and Dr. F. S. McCullough, WHO, Geneva.

REFERENCES

Adewunmi, C. O., and E. A. Sofowora. Preliminary screening of some plant extracts for molluscicidal activity. *Planta Med.* 39:57–65 (1980).

Arnason, T., T. Swain, C.-K. Wat, E. A. Graham, S. Partington, and G.H.N. Towers. Mosquito larvicidal activity of polyacetylenes from species of the Asteraceae. *Biochem. Systematics Ecol.* 9:63–68 (1981).

Boyce, C. B. C., J. W. Tieze-Dagevos, and V. N. Larman. The susceptibility of *Biomphalaria glabrata* throughout its life-history to N-tritylmorpholine. *Bull. WHO* 37:13–21 (1967).

Broberg, G. Observations on the mode of action of extracts of *Glinus lotoides* on the miracidia of *Fasciola gigantica* and *Schistosoma mansoni*. *Suomen Elainlaakarilehti* 86:146–147 (1980).

Duncan, J., and J. W. Pavlik. Effects of structural rearrangement on the molluscicidal activity of certain fluorinated aromatic compounds. *Bull. WHO* 42:820–825 (1970).

Etges, F. J., E. J. Bell, and D. E. Gilbertson. Bacterial degradation of some molluscicidal chemicals. *Am. J. Trop. Med. Hyg.* 14(5):846–851 (1965).

Farnsworth, N. R. Biological and phytochemical screening of plants. *J. Pharm. Sci.* 55:225–276 (1966).

Graham, K., E. A. Graham, and G. H. N. Towers. Cercaricidal activity of phenylhepatatriyne and α-terthienyl, naturally occurring compounds in species of Asteraceae (Compositae). *Can. J. Zool.* 58(11):1955–1958 (1980).

Hopf, H. S., J. Duncan, J. S. S. Beesley, D. J. Webley, and R. F. Sturrock. Molluscicidal properties of organotin and organolead compounds with particular reference to triphenyllead acetate. *Bull. WHO* 36:955–961 (1967).

Jewers, K., and T. A. King. *Improvements Relating to Molluscicides*. Patent Specifications 1277417. London: The Patent Office (1972).

Kloos, H., and F. S. McCullough. Plant molluscicides: A review. WHO Unpublished Document WHO/VBC/81.834; WHO/SCHISTO/81.59 (1981).

Lemma, A. Laboratory and field evaluation of the molluscicidal properties of *Phytolacca dodecandra*. *Bull. WHO* 42:597–617 (1970).

Lemma, T., and A. Lemma. New approaches to endod (*Phytolacca dodecandra*) extraction: A comparative study. Unpublished Research Note 1. Addis Ababa: Institute of Pathobiology (1977).

Lemma, A., G. Brody, G. W. Newell, R. M. Parkhurst, and W. A. Skinner. Studies on the molluscicidal properties of endod (*Phytolacca dodecandra*):1. Increased potency with butanol extraction. *J. Parasitol.* 58(1):104–107 (1972).

Litchfield, J. T., and F. Wilcoxon. A simplified method of evaluating dose-effect experiments. *J. Pharmacol. Exp. Ther.* 96:99–113 (1949).

Meredith, R., and G. C. Meredith. The assay of the molluscicide niclosamide, using the ciliate protozoan, *Spirostomum ambiguum*. *Bull. WHO* 46:404–407 (1972).

Meyling, A. H., and R. J. Pitchford. Physico-chemical properties of substances used as molluscicides: Preciciptation and absorption. *Bull. WHO* 34:141–146 (1966).

Muirhead-Thomson, R. C. *Pesticides and Freshwater Fauna*. London: Academic Press (1971).

Odebiyi, O. O., and E. A. Sofowora. Phytochemical screening of Nigerian medicinal plants II. *Lloydia* 41(3):234–246 (1978).

Paulini, E., and C. Pereira de Souza. Influence of different suspended solids in the water upon molluscicidal activity. WHO Unpublished Document PD/MOL/70.12 (1970).

Ritchie, L. S. Chemical stability of molluscicidal compounds in water. *Bull. WHO* 40:471–473 (1969).

Ritchie, L. S., L. P. Frick, L. A. Berrios-Duran, and I. Fox. Molluscicidal qualities of sodium pentachlorophenate (NaPCP) revealed by 6-hour and 24-hour exposures against representative stages and sizes of *Australorbis glabratus*. *Bull. WHO* 29:421–424 (1963).

Sherif, A. F., A. H. Abdou, and M. F. El-Sawy. Laboratory trials of the molluscicidal action of Egyptian herbs: *Ambrosia maritima* (damsissa). *Proceedings of the First International Symposium on Bilharziasis*, Cairo, 689–694 (1962).

Thomas, J. D., and B. Assefa. Behavioral responses to amino acids by juvenile *Biomphalaria glabrata*, a snail host of *Schistosoma mansoni*. *Comp. Biochem. Physiol.* 63:99–108 (1979).

Wat, C.-K., R. K. Biswas, E. A. Graham, L. Bohm, and G. H. N. Towers. Ultraviolet-mediated cytoxic activity of phenyheptatriyne from *Bidens pilosa* L. *J. Nat. Prod.* 42(1):103–111 (1979).

WHO. Molluscicide screening and evaluation. *Bull. WHO* 33:567–581 (1965).

Womersley, J. S. *Plant Collecting and Herbarium Development*. Rome: Food and Agriculture Organization (1981).

10

FIELD EVALUATION OF PLANT MOLLUSCICIDES

R. F. Sturrock

London School of Hygiene and Tropical Medicine

J. Duncan

Center for Overseas Pest Research

London, England

A logical progression of trials is needed to demonstrate that a candidate plant molluscicide can kill snails under local field conditions; can kill enough snails, when applied by simple means, to reduce transmission significantly; can be used in a simple and effective self-help program to control transmission; and can be cost-effective. These trials include testing the molluscicide's effect on snails in simulated and in isolated field conditions; testing its effect on transmission, using snails, other animals, and especially humans, in the area where it will be used; and devising a plan for its use in self-help programs. Methods used in the development of synthetic molluscicides for field use will, in many ways, be equally appropriate for plant molluscicides.

So far, systematic field testing of plant molluscicides has been confined to endod (*Phytolacca dodecandra*) (Lemma et al., 1978), damsissa (*Ambrosia maritima*) (El Sawy et al., 1983, 1984), and to a lesser extent Arian (*Tetrapleura tetraptera*) (Adewunmi et al., 1982).

Comprehensive laboratory testing (Chapter 9) should have preceded field testing to provide a substantial body of technical information: how rapidly the potential plant molluscicide acts, how much is needed, and how to apply it. Can it be applied directly, or is an extraction process necessary? If extraction is necessary, can it be done on site or must it be done elsewhere in advance? How are such factors as temperature, silt, turbidity, sunlight,

PROVISIONAL FLOWCHART FOR FIELD TESTING

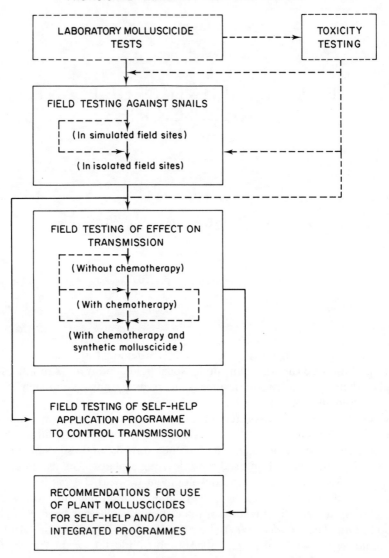

and pH likely to affect the molluscicide's performance in the field? The answers to all these questions should be known in advance.

Field evaluators should also have information on the molluscicide's acute and chronic mammalian toxicity so that they can assess its safety for persons handling it, persons who may be exposed to it in test sites, and domestic and wild animals that may come into contact with treated water. The molluscicide's possible effect on irrigated crops should also be known. If such information is not available, caution must be exercised; preliminary tests should be confined to isolated and possibly to simulated field sites. Simulated sites may, in any case, provide a useful halfway house to full field testing and allow a more accurate assessment of snail mortalities than would be possible in the field.

In addition to all this information, field evaluators must have sufficient plant material available for the initial field trials. The material may be either fresh or dried and either part or all of the plant.

APPLICATION TECHNIQUES

Techniques for applying a plant molluscicide range from very simple to quite complicated. No method at all is required for a plant that grows within the transmission site and does its work by simply being there; such a plant needs only to be propagated. At the other extreme, complicated extraction techniques may be needed to provide powders, granules, or emulsifiable concentrates that are applied by the methods developed for synthetic molluscicides (WHO, 1965a, b, 1973; Ansari, 1973; Sturrock and Barnish, 1973; Chu, 1978).

Attempts should be made to develop effective, simple, and cheap application methods that can be used by villagers and farmers in a self-help program or by impoverished public health departments. The simplest method is to throw all or part of a freshly cut plant, intact or chopped up, into the transmission site. Sun drying between cutting and application can be done relatively easily. Measuring the volume of material applied is done easily and cheaply with a plastic scoop. A possible disadvantage of using dried material is that there is some delay while it becomes wet and sinks; in large habitats such as lakes and dams, this delay may allow currents and waves to carry it away from the snails' habitat. Wind and wave action may also be a problem if the plant – dried or not – is applied to habitats with intermittent or permanent flows. It is possible to anchor the plant material in bags or nets weighted with stones, but then the active ingredient may be lost as it is leached out of the bags.

If the active ingredient must be extracted by hot- or cold-water infusions, possibly with some degree of fermentation, the molluscicide will be in liquid form and may contain suspended plant material. This liquid can be poured directly into the site or, if the suspended material can be removed easily by filtration or decanting, the liquid can be sprayed onto the site. Measuring the volume of liquid is easy, but precision is doubtful unless standardized extraction methods are used. A disadvantage of using liquids is that slow-acting molluscicides or those that require some time to be leached from the plant material may be washed out of flowing sites before they have time to act, unless they are sorbed onto suitable materials, such as vegetation or the substrate, that remain in the site. Continuous application, even over relatively short periods, requires some form of dispenser, adding to the cost and complexity of the operation.

Hard rules applicable to all plant molluscicides cannot be laid down. The choice of application method depends on the properties of the plant molluscicide to be applied and the characteristics of the sites to be treated. Wherever possible, though, the aim should be simplicity.

EFFECT ON SNAILS

The biggest problem in moving from laboratory to field tests is to assess the effect of the molluscicide on the snails. In the laboratory, known numbers of snails of known size (age) are exposed under controlled conditions and can be examined individually at predetermined times to measure mortality or changes in their biochemistry, physiology, or behavior.

The quantitative measurement of field snail populations is beset by a multiplicity of sampling problems. These problems have been discussed by many workers (Hairston et al., 1958; Pesigan et al., 1958; Yeo, 1962; Shiff, 1964; Webbe, 1965a; WHO, 1965a, b; Sturrock, 1973b, 1975; and others). Measurement is difficult in relatively uniform environments such as man-made canals and drains because of the discontinuous, aggregated distribution of the snails (Yeo, 1962; Hairston, 1965; Sturrock, 1975). The difficulties are magnified in the nonuniform conditions that are the rule in natural field sites.

In mollusciciding trials a further complication arises: The initial snail population is large and can be estimated by a relatively insensitive sampling technique, whereas successful treatment leaves small, residual populations that may be detectable only by more sensitive sampling techniques. Several authors have discussed ways to solve this problem (Hairston, 1965; Sturrock et al., 1974; El-Sawy et al., 1983).

Simulated Field Testing

A helpful step between laboratory tests and full field trials is the use of artificial, simulated field sites. Simulated field tests of plant molluscicides can provide accurate quantitative data about the effect on snails. They can assess the molluscicide's performance under local climatic conditions in local water. The effect f specific physicochemical factors can be investigated under partially controlled conditions, and other assay techniques that may be useful in evaluating full field trials can be tested and their results compared with the effect of the molluscicide on free snail populations. An additional advantage is that substances of unknown or suspect toxicity can be tested under local conditions without exposing the local population (Hopf et al., 1967; Sturrock et al., 1974).

The exact design of simulated field tests depends on characteristics of the plant molluscicide to be used. These characteristics determine how the molluscicide is applied, how long the test must last, and which of the prevailing local physicochemical factors should be investigated, alone or in combination.

Simulated sites are constructed outdoors, where they are subject to local climatic conditions. They are filled with local field water so that its quality (pH, dissolved salts, turbidity, and so forth) is tested. Other factors such as silt, aquatic vegetation, organic matter, shade, and other biota can be added and their individual or combined effects tested.

Test snails are added to the simulated sites and recovered after mollusciciding. Starting with a known number of snails and exhaustively recovering both the survivors and the dead snails avoids the subsampling problems experienced in field sites and gives accurate data to assess the effect of the molluscicide.

Simple standing water pools are easily simulated, and a series of pools can be made to allow replicated tests with a range of doses, including untreated controls. A convenient site is about 1 m diameter or square and 0.25 m deep. Holes may be dug in suitable soils, or the ponds may be made of plywood, fiberglass, or concrete. Lining the ponds with heavy-duty polythene prevents leaks and avoids problems associated with the sorption of the molluscicide onto the substrate; the lining can be replaced after each trial if it absorbs the molluscicide. Various workers have used such pools for molluscicide and other tests (Hopf et al., 1967; Upatham, 1972).

Flowing sites are more difficult to make. An open concrete channel used for infection studies by Webbe (1966) required enormous amounts of water, which ran to waste, even for periods as short as one hour. Berrios-Duran et al. (1968) used a converted section of a natural stream. More realistic is a closed system in which the water is recycled by a pump. Careful design with varied cross sections and widths can provide a range of flows within a single system. The effect of various flows can be tested by separating off sections with plastic or fiberglass mosquito netting. Sorption of molluscicides onto the substrate may still present problems.

A desirable though not essential feature of both standing and flowing water systems is a screened drainage device to facilitate the recovery of snails stranded when the system is emptied.

Snails of standardized size or sizes can be added to the simulated field site and left for a sufficient period to acclimatize and move freely throughout the system before the molluscicide is added – probably not more than one day later. Feeding is unnecessary for short-term exposures up to 24 hours.

The free snails may move to the surface for respiration and will behave generally in the same way as field snails. Free snails may be left *in situ* for several weeks or months to produce a breeding population with a more normal age-structure. However, minor variations between the ponds can produce substantially different snail populations, necessitating randomized and replicated experiments to assess the effect of various molluscicidal treatments.

The use of snails restrained in suitable cages, usually of plastic or fiberglass mosquito netting over a wood, metal, or plastic frame, allows the recovery of the snails added to the test system. Obvious disadvantages include the snails' inability to move throughout the system and the difficulties if long-term exposure is necessary. Uneven distribution of the molluscicide within the system can result in the caged snails' exposure to higher or lower doses than free snails receive. Caged snails cannot avoid exposure by moving from higher to lower concentrations or by crawling out of the water. They are also unable to feed, unless food is added to the cage, or to move to the surface for respiration. Nevertheless, the use of caged snails can provide useful information, especially for quick-acting molluscicides. A comparison of their response with that of free snails is valuable in determining their suitability for monitoring the effect of field applications (Crossland, 1967). Floating, open cages developed for studying snail bionomics in the field (J. O'Keefe, personal communication, 1982) might be worth testing in this role. (Similar cages may be used simultaneously for exposing other animals, such as fish, shrimp, and tadpoles, to see how they respond to the chemical.)

Various chemical tests can be tried out during simulated field testing to evaluate their suitability for monitoring field trials. Such tests, which must be sensitive enough to detect the low concentrations expected in the field, depend on the successful identification of the active ingredient or ingredients in the laboratory testing program. Relatively sophisticated chemical analyses can be used to examine water samples taken from the simulated field habitats and to investigate the effect locally of relevant physicochemical factors. However, simplified chemical tests will probably be required for field trials. It may be optimistic to expect them to be developed successfully; the field tests for synthetic molluscicides still leave much to be desired.

Simple bioassays, using the target snails as the test organism, are probably easier than chemical tests to develop, based on the initial laboratory testing techniques (WHO, 1965a; and see Chapter 9). Water samples taken at appropriate intervals from the simulated field sites can be used for standardized laboratory tests with fixed exposure and recovery times if the molluscicide is quick acting. For slow-acting chemicals, it may be necessary to leave test snails in the water sample indefinitely and to observe mortality after an appropriate time interval. Bioassays can be developed using other, more sensitive species than the target snail (although the existence of such species would not augur well for the environmental safety of the plant molluscicide).

Isolated Field Testing

Field testing of any potentially toxic molluscicides must be carried out in field sites to which the general public has no access. For man-made habitats such as irrigation canals, it may be possible to construct special blocks in which there is good control of water management (Crossland, 1962; WHO, 1965a).

The principal assessment of the effect of the molluscicide is achieved by measuring the snail population before and after treatment and calculating the percentage mortality produced. As already discussed, subsampling problems will be encountered. These problems apply to both natural and man-made habitats. The choice of habitats depends on which ones are important locally in transmission and therefore presupposes that this information is available.

In theory, direct, exhaustive subsampling methods should be more reliable than fractional sampling techniques, but that is not always the case (El-Sawy et al., 1983). Exhaustive sampling techniques are often unsuitable for natural habitats (Sturrock, 1973a). As a general principle, a large number of small samples should be taken throughout the habitat to avoid the risk of being misled where snails are not distributed randomly. The efficiency of the sampling method can be checked by methods involving the release and recapture of marked snails (Sturrock, 1973a) or the recovery of objects of the same size and shape (Jobin, 1970).

It is common practice to take samples repeatedly from a series of fixed sampling points chosen to ensure complete coverage of a habitat. However, repeated sampling may alter conditions at these sampling stations, usually unfavorably for the snails, and artificially low snail recoveries may be obtained. Thus it is good practice to make periodic qualitative searches between sampling stations. Such searches sometimes reveal quite large snail colonies not detected by routine sampling.

Supplementary information for assessing the effect of the molluscicide may be obtained by exposure of caged snails, chemical assays, and bioassays.

Immediate, Short-Term Mortality. The effect of quick-acting molluscicides should be detectable within a few days. Mortality (M) is calculated by the formula

$$M = 100 \ (1 - P_2/P_1)$$

where P_1 and P_2 are the snail populations before and after treatment. Because estimates of the total snail populations introduce errors, it is better to use actual snail counts. These counts must be normalized by some suitable transformation, and some constant such as 1 must be added to the raw count to allow the inclusion of zero counts before statistical analysis (Yeo, 1962; Hairston, 1965; Sturrock, 1975; El-Sawy et al., 1983).

With slow-acting molluscicides, the delay between estimating P_1 and P_2 may be considerable. Other extraneous factors may also affect the surviving snail populations, either adversely or favorably. The use of untreated control sites, if they are available, is desirable to allow for such effects. However, snail populations in apparently similar habitats often behave differently, even in the absence of an imposed factor such as mollusciciding.

In general, the period between P_1 and P_2 estimates should be kept as short as possible. If an extended period is unavoidable, evidence from caged snails, chemical assays, and bioassays may assist in interpreting the field population sampling data.

Long-Term Mortality. Eradication (100% mortality) is rarely achieved by mollusciciding in the field. In some cases, that may be a result of differential susceptibility between eggs and snails of different sizes. Sometimes the molluscicide fails to cover the habitat completely because of poor application techniques or degradation of the molluscicide by physicochemical or biological factors. Sometimes the snails take successful evasive action. Whatever the explanation, a residual snail population – perhaps as small as 0.1% or less of the original population – can be expected, especially when a quick-acting, nonresidual molluscicide is used.

Many of the snails that transmit schistosomes are adapted to survive in unstable habitats. Natural factors such as floods or droughts can produce a catastrophic population decline of a magnitude resembling that caused by molluscicides. To compensate for catastrophes, the survivors have a breeding capacity that allows them to repopulate habitats explosively when favorable conditions return. Populations can reach their original levels within weeks or months of a natural catastrophe (Webbe, 1962a, b; Dazo et al., 1966; Shiff, 1964; Jobin and Michelson, 1967; Sturrock, 1973 a, b; Klumpp and Chu, 1977; and others). Repopulation rates after mollusciciding have been calculated for several snail species (Hairston, 1965; Jordan and Webbe, 1969, 1982; Sturrock, 1973b).

Immigration can be expected if the treated site is not completely isolated from snail-containing untreated sites in the vicinity. Even isolated sites are not immune from immigration. Snails reportedly have been transported by birds, water beetles, large mammals, and people into isolated, snail-free habitats (Chernin and Adler, 1967). If conditions are favorable, the immigrant snails repopulate the site or augment the residual population.

Thus it is of practical importance to establish how long a plant molluscicide can be expected to keep a site either entirely or minimally free of snails and to find out whether repopulation is due to the survivors or immigrants. A rapid reappearance of large numbers of adult snails suggests immigration, whereas a slow build-up, starting mainly with small snails, points to breeding by survivors; such evidence, however, is equivocal. A persistent, long-acting plant molluscicide released into the habitat for prolonged periods should be more effective in preventing repopulation than a quick-acting but short-lived molluscicide.

A study of long-term snail mortality cannot be lightly embarked upon. Care is needed in its execution and recording and in analyzing the results. Nevertheless, it can provide valuable experience in the use of the plant molluscicide and make possible the testing of various application techniques and dosage schedules. Untreated areas provide useful information on the seasonal behavior of the snail populations, if it is not already available. Information may be obtained on any seasonal pattern of transmission by using the field-testing techniques to be described in this chapter.

A long-term study is essentially the same as a short-term study except that the snail populations are sampled at regular intervals (two to six weeks) after treatment to give estimates P_2, P_3 ... P_n. How long the study continues depends largely on how rapidly repopulation takes place. If information about seasonality in the natural snail populations is available, the effects of applying the molluscicide at favorable and unfavorable periods can be compared. Information covering the complete seasonal cycle is desirable. The behavior of an untreated snail population should also be monitored for comparative purposes. If possible, several treated and untreated populations should be studied to avoid being misled by aberrant results from a single site.

It is not necessary at this stage, however, to cover all snail habitats and transmission sites in a specific area. If the molluscicide proves capable of controlling snail populations within a limited number of sites, the way will be open to proceed to the next stage of development.

EFFECT ON TRANSMISSION

Demonstrating that transmission can be controlled is considerably more difficult than showing that snail populations can be reduced. An attempt must be made to treat all the transmission sites within the trial area. While the ultimate role of the plant molluscicide may be to supplement or follow up chemotherapy, it is important at this stage to demonstrate that the molluscicide, preferably used in a manner suitable for a self-help program, can control transmission. If it cannot, further development is pointless.

Doubt about the ability of molluscicides to control transmission (Gilles et al., 1973) has been shown to be groundless (Sturrock et al., 1974; Jordan, 1977). Thus it may seem unnecessary to repeat studies on transmission control. Previous studies, however, involved 'imposed' mollusciciding with relatively sophisticated equipment by supervised teams of trained field workers. Our present aim is different: to find a plant molluscicide that can be used by villagers and farmers with only the simplest of dispensing equipment.

That does not mean that transmission control studies should be carried out on a self-help basis. Self-help programs introduce a number of quite separate problems involving motivation and education of the population. Demonstration of transmission control is a prior research step in which carefully controlled, imposed, but simple application techniques and schedules are evaluated precisely. If the imposed program succeeds, then the final stage of development will be the implementation of a self-help program.

Transmission control studies may be carried out with or without other control measures such as chemotherapy. Some investigators might argue that the effectiveness of the plant molluscicide can be best evaluated in the absence of any other control measures. However, in areas of high endemicity, where the infection is of public health significance, it may be desirable initially to provide treatment with a safe drug – either selectively, to persons with heavy infections or clinical signs and symptoms, or to the entire population in both the study and the comparison areas. It seems more and more unlikely in any case that mollusciciding alone will be used in serious attempts to control transmission; the availability of new drugs suitable for population-based chemotherapy makes it probable that mollusciciding will be used as a supplement to chemotherapy.

The drug treatment can serve as a point of comparison. Since it is unlikely that a single round of chemotherapy will eliminate the parasite entirely, transmission will begin again after treatment in the absence of any other control measures (Jordan et al., 1982; Machado, 1982). Such a build-up in the comparison area and its prevention or delay in the study area can provide evidence of the effectiveness of the plant molluscicide. The differentiation of light infections surviving chemotherapy from genuine reinfections might present some problems in the calculation of incidence of new infections, but these should be resolved as the study progresses.

We do not know to what level snail populations must be reduced to stop transmission in specific field conditions. Regular field snail sampling, using techniques discussed earlier, may provide information to this end. Any snails recovered can be tested for infection, but the main purpose is to monitor the effectiveness of the mollusciciding program so that changes can be made if it appears to be breaking down.

Measurement

Several methods of measuring transmission control are available, including studies of field snails, cercariometry, exposure of rodents, and studies of prevalence, intensity, and incidence in humans.

One of the simplest and most effective ways of detecting transmission is to examine field snails, either by cercarial shedding or by crushing to detect cercariae and other intramolluscan stages of the parasite (Webbe, 1965b; Chu and Dawood, 1970; Sturrock, 1973a; Sturrock et al., 1979; Chu et al., 1981). These techniques have provided valuable information on transmission and could be used in precontrol studies and to monitor chemotherapy – with the reservation that detection of cercariae in this way does not mean that they are necessarily present in the field. If a mollusciciding program is successful, few if any snails will be recovered; but those that are should be examined for infection. Even undetectably low snail populations might suffice to maintain transmission.

A second means of measuring transmission is cercariometry, the direct recovery of cercariae from water. This technique has been studied by a number of authors (Rowan, 1957, 1965; Barrett and Ellison, 1965; Sandt, 1972, 1973; Upatham, 1976; Theron, 1979; Kloos et al., 1982; M. A. Prentice, personal communication, 1982). Its attraction lies in the rapid measurement of cercarial densities in a body of water at a specific time. Of the various methods tested, those currently considered most promising involve killing the cercariae with formalin as soon as the water sample is taken, then filtering the water and straining the preparation. Coarse prefiltration removes large particles, and the use of a filter with a suitable pore size lets fine particles pass through, leaving a preparation of cercariae that can be stained and counted easily and quickly with the aid of a low-power microscope. Standardization of the water volumes used allows the techniques to be quantified (that is, expressed as number of cercariae per liter).

Recent progress with cercariometry is encouraging, but the technique is still in the developmental stage. Turbid waters, especially those containing colloidal and fine biological matter such as filamentous algae, phytoplankton, and zooplankton, can affect the recovery of cercariae by limiting the volume of water that can be filtered and by producing dirty preparations. Both affect the sensitivity of the technique in measuring field cercarial densities. The precision with which the densities can be measured varies with seasonal changes in the turbidity of field waters.

In addition, cercariometry is a subsampling technique and in many habitats it is improbable that the cercariae are randomly distributed. Little information is available with which to design a sampling schedule to provide accurate, quantitative estimates of field populations of cercariae. Other problems include the difficulty of differentiating various species from humans when more than one is present, and separating human from animal schistosome cercariae. Moreover, the infectivity of any cercariae recovered is unknown.

A third means of measuring transmission, the indirect recovery of cercariae by exposure of animals – usually rodents – has been used with some success

(Pesigan et al., 1958; Pitchford and Vissar, 1962; Webbe, 1965a, 1966; Dazo et al., 1966; Sturrock, 1973b; Barbosa and Costa, 1981). The recovery of adult worms is conclusive proof that infective cercariae were present in the site, but the absence of worms is not proof that they were not. As with cercariometry, problems arise in subsampling a large body of water in which the cercariae are not randomly distributed. Mortalities may occur in the six- to eight-week period between exposure and recovery of adult worms. Facilities to produce and maintain the animals are expensive, and results are difficult to quantify.

To detect miracidia, uninfected laboratory snails have been exposed in much the same way as rodents. The information derived by this means can be useful in studying the overall dynamics of transmission (Upatham, 1972, 1976; Shiff, 1974) and in monitoring the dramatic drop in the number of eggs passing into the environment that results from chemotherapy (Christie and Upatham, 1977). However, the longevity of schistosomes is such that significant reductions in the number of miracidia are unlikely for several years unless chemotherapy is applied in addition to mollusciciding (Sturrock et al., 1974).

Ultimately, the only way to make a true assessment of a plant molluscicide's effect on transmission is to monitor the human population. Techniques have been developed to do that with conventional synthetic molluscicides and other control methods (Farooq and Hairston, 1966; Jordan and Webbe, 1969, 1982; Ansari, 1973; Gilles et al., 1973; Jordan, 1977; Jordan et al., 1982; Scott et al., 1982). These techniques rely on the detection of eggs in the excreta, preferably by quantitative measurements such as prevalence, intensity, and incidence.

Prevalence is the proportion of a population infected at a specific time. Intensity – the number of eggs excreted – is taken to indicate the number of worms present; it may be calculated for the entire population or just those infected. For statistical purposes the geometric mean is often preferred to the arithmetic mean. If uninfected persons (with egg counts of zero) are included, the raw data will need to be transformed – usually to $\log_{10}(x+1)$ – in calculating the geometric mean.

If chemotherapy is used, the resulting substantial drop in the number of worms is reflected in a reduction in both prevalence and intensity, measured before and after control is applied. Quantitative egg-counting techniques are used to measure this drop (Jordan and Webbe, 1969, 1982; WHO, 1974, 1980a; Peters et al., 1976). Pre- and posttreatment egg counts must be estimated with the same degree of precision that is used in pre- and post-treatment snail population sampling. Examination of three separate stool samples, particularly if replicate counts are made on each, can detect egg counts of 0.5 to 1.0 per sample or more with a probability of greater than 95% (Uemura, 1973; WHO, 1974).

If chemotherapy is not used, the worm burden in man is not affected. Although transmission may be stopped completely, the prolonged life expectancy of the adult worms – half-life estimates range from three to eight years depending on the species (WHO, 1974), which means that a substantial number live considerably longer – it will be several years before a decline in prevalence or intensity in the population as a whole can be detected. This drop has been observed in heavily infected populations (Jordan, 1977; Jordan et al., 1982) but was not evident in a lightly infected community where transmission had apparently been stopped for five years (Hiatt et al., 1980).

The schistosomiasis age-prevalence curve in a community rises slowly in children under five years old, accelerates in children from five to 15, and stabilizes or declines thereafter. If, in the absence of chemotherapy, control is successful in eliminating transmission, this age-prevalence curve will shift to the right progressively as time goes by. The prevalence in a cohort of children under five years old remains low on successive reexaminations if transmission is successfully stopped, and so does that in a cohort of five- to nine-year-olds, although at a higher level. Five- to nine-year-olds examined several years after successful control has been instituted should show reduced prevalence compared with that in the same age group examined before control began. Thus, measuring prevalence and intensity in successive annual intakes of primary schoolchildren may demonstrate a declining trend (Negron and Jobin, 1979). However, a more rapid method of demonstrating control of transmission is desirable.

Incidence is a measure of the rate at which uninfected persons become infected over a specific time period (Farooq and Hairston, 1966). An initial survey at time T_1, using a sensitive and preferably quantitative egg-counting technique, identifies uninfected and infected persons. The same persons, infected and uninfected, are reexamined at time T_2. The proportion of uninfected persons who have become infected between the two surveys is the incidence (i). It is generally converted to an annual incidence (I) using the formula

$$I = 1 - q^{12/t}$$

where q is the proportion still negative at T_2 and t is the time in months between T_1 and T_2. However, determining the annual incidence is not essential. Shiff (1973) used a 120-day period, partly to detect seasonality in transmission and partly to provide a rapid check on the effectiveness of control measures. The reexamination of known infected persons at T_2 is necessary to correct I if a substantial proportion lose their infections. Such losses occur mainly among lightly infected persons whose egg counts are at the counting technique's lower limit of sensitivity and who are consequently misclassified (Shiff, 1973).

Incidence is easiest to obtain from primary schoolchildren. They are readily accessible and usually cooperative; they include enough uninfected subjects to provide a negative cohort at the start of a study; and they represent the group with maximum exposure (Dalton, 1977). Moreover, their infections are usually of recent origin, and few complications arise as a result of the possible development of immunity.

Quantitative egg counts are not imperative, since a qualitative assessment – are the subjects infected or not? – is all that is really required. The value of using quantitative methods is that problems associated with positives reverting to negatives can be related to whether or not the initial infection was light. Immunodiagnostic methods can be used (Negron and Jobin, 1979), but the tests so far developed are probably not sufficiently sensitive or specific for this purpose (Mott and Dixon, 1982).

Field Trials

The two primary considerations in selecting a study area for field trials are that the human population be large enough and sufficiently infected to permit statistically reliable estimates of changes in transmission due to control and that exposure be confined to transmission sites within the study area. It follows that all the transmission sites must be identified and treated with the molluscicide, that the area selected must be large enough to encompass all the normal activities of the human population, and that migration to and from other transmission areas must be negligible.

These constraints apply equally to the comparison area, in which no mollusciciding will be done. It should be ecologically similar if not identical to the study area.

Indiscriminate chemotherapy by government or private physicians cannot be allowed within either area if meaningful results are to be obtained. That does not preclude controlled treatment of individuals for whom it is considered necessary. In that case, however, defining the criteria to be applied is essential before deciding to give treatment, and records should be kept of persons who receive it.

Precontrol Studies. Before field trials can begin, precontrol studies must be conducted. The minimum information that must be gathered on the human population in these studies includes a demographic survey of the area and two surveys to define prevalence, intensity, and incidence in either the whole population or the segment selected for monitoring changes in transmission due to the mollusciciding program. For the snail population, the studies must identify, characterize, and map all habitats and provide information on seasonal population fluctuations and variations in transmission. Methods by which this information can be obtained have already been discussed.

The minimum period for precontrol studies is one seasonal cycle. Previous studies suggest that two or three cycles would be better (Sturrock, 1973a; Jordan, 1977).

Precontrol studies can be of use in cases where comparison areas are not available. Based on past experience (Gilles et al., 1973; Jordan, 1977), it may be impossible to find two identical sites for use as the test and comparison areas. Minor differences may have to be accepted, but significant differences may necessitate the rejection of the comparison area. In that case, the results from extended precontrol studies will have to be used as a time control to assess changes following the introduction of mollusciciding. The process will be helped if any seasonality in transmission has been related to such climatic factors as rainfall or temperature or to other cyclical changes such as water management on irrigation schemes.

Control Period Studies. The crucial assessment of reduction of transmission is best based on detecting changes in incidence, probably among schoolchildren, or, if chemotherapy is used simultaneously, on reinfection curves. Incidences are normally monitored by repeated annual surveys, but shorter periods may be used to obtain a more rapid assessment, especially if information is available on the seasonality or otherwise of transmission. Reinfection curves require more frequent examinations at three- to four-month intervals. In either case, allowance must be made for the six- to 12-week prepatent period (depending on the species) before eggs can be detected from infections acquired just before control is initiated. Too frequent sampling may antagonize human subjects and result in their refusal to provide specimens, complicating and possibly even invalidating the calculation of incidences or reinfection rates (Jordan et al., 1982).

Supplementary information on the success of the program in controlling snails and interrupting transmission can be obtained by the sampling and measurement techniques already discussed.

If an identical comparison area is used, significant reductions in transmission or differences in the rates of reinfection may be detected in the study area in a year or less. However, because the comparison area is unlikely to be identical to the study area and the amount of transmission may vary from year to year (Sturrock, 1973a; Sturrock et al., 1983; Jordan, 1977), obtaining an unequivocal result will almost certainly take much longer (Machado, 1982). A minimum of four years is recommended by some authorities (P. Jordan, personal communication, 1982).

Application of Plant Molluscicide. The form and characteristics of the plant molluscicide determine the method of application. Every effort should be made to use it in the same way that the local population is expected to use it in self-help programs. The method should already have been tested at the isolated field site.

A comprehensive application plan must be prepared on the basis of information already obtained on the molluscicide's duration of activity, which determines the frequency of application, and on where and when transmission is expected to occur. A suitable model can be developed for the rational use of a molluscicide (Jobin and Michelson, 1967; Sturrock, 1973c). If the plan fails to control the snail populations as detected by routine snail sampling, it must be modified accordingly.

Local labor may be needed to apply the molluscicide. Trained supervisors must teach these workers and make sure that the applications are made as required. Workers may also be drawn from staff members who carried out the preliminary trials and precontrol studies.

The total amount of plant molluscicide required will be determined from the information obtained on short- and long-term snail mortalities, the areas to be covered, and the frequency and duration of application. Adequate supplies must be maintained, perhaps for as long as four years. The plant may have to be specially cultivated, and doing so in the study area has obvious advantages if the local population is expected eventually to grow it themselves.

Records and Costs. The success of the trial depends on everyone concerned knowing what to do, when to do it, and how to record it. All tasks must be carried out correctly and diligently throughout the study. Everyone must be trained, and supervisors must check and double-check to see that the program is adhered to and that changes made during the program are properly effected.

Proper recording of all costs is essential to determine whether the use of the plant molluscicide is economically feasible. Because the field-testing phase is imposed on the community, costs will be incurred that are not applicable to a self-help program, but proper record keeping will allow for these expenditures. Calculation of cost-effectiveness is problematic, depending on what unit is used – cost per head of population, cost per unit of water treated, reduction of clinic or hospital attendance following successful control of transmission – but various procedures have been devised (WHO, 1980b; Jordan et al., 1982).

SELF-HELP PROGRAM

The preceding phases of the testing program should have provided information on whether the molluscicide, when applied by simple means, can kill enough snails under local field conditions to reduce transmission significantly. This next phase is concerned with how to use the molluscicide in a simple self-help program and make it cost-effective. It is assumed that any problems related to large-scale cultivation of the plant have been solved and that the safety of the molluscicide is beyond question. Methods will also have been developed to monitor transmission.

In the selection of study sites for the self-help program, the same conditions apply that applied to the selection of study areas in the field trials. Similar precontrol studies should be done in this phase. Experience gained in doing the previous precontrol studies may shorten the time required at this stage. That also applies to the design of a mollusciciding schedule.

The involvement of people in the local community makes it important to institute a health education program so that they will understand what they are supposed to do and why. They are likely to collaborate if they understand the threat posed to them by schistosomiasis and how their participation can help minimize it. The exact nature of a suitable health education program depends on local circumstances. Even if transmission is confined to a few easily circumscribed sites that are technically quite easy to treat, it may still be difficult to identify persons responsible for carrying out the treatment. If the sites are numerous and widespread, as on an irrigation network, then the collaboration of the whole community is essential. Other variables include the type of program suitable for the local population and the availability of medical services. For self-help programs, the strength of community leadership and the history of community involvement in previous health projects may be important.

Presumably, farmers will need to be taught how to grow the plant, either in or outside of the study area; some form of agricultural extension service may need to be set up. The desired level of production may not be achieved for several seasons. Siting of demonstration plots within the area may serve a dual purpose: to teach the local farmers how to grow the plant and to provide a reserve supply until local production has reached the required level.

The local population must be taught how to process the plant and how, when, and where to apply it. The simpler the instructions, the more likely they will be followed properly. Arrangements must be made to ensure that the necessary tools are available – buckets, sieves, funnels, scoops. Suitable storage space may need to be provided. The precise timing of application will be based on experience obtained during the earlier phases.

While the local population is expected to grow and apply the plant molluscicide, someone must confirm that they are doing so correctly. Supervisors or inspectors should check to see that enough plants are grown, that they are harvested correctly, and that the material is applied properly at the correct time. These inspectors should develop friendly relations with the farmers and the rest of the community and should be willing and able to solve any problems that arise. The program should not be jeopardized by shortages that could have been anticipated or mistakes that inspectors and local persons should have observed. Supervisors may also need to devise measures to encourage participation in the program.

As in the field-trial phase, the principal evaluation of the effect on transmission will be through human studies, using the methods already devised and tested. Supplementary information may be obtained from cercariometry and animal exposure. Regular snail searches should be carried out to detect any failures in the program. If additional plant material is available, additional mollusciciding may be done and suitable alterations to production and application schedules made.

At least a year will be required to carry out precontrol studies for the self-help program and to teach the local people to grow, process, and apply the plant. After application, monitoring should continue for four years to evaluate the effect of the program on transmission, even if evidence is obtained before then that transmission is being controlled. The initial novelty will wear off, with an accompanying loss of enthusiasm, so it is important to see that the program remains effective once it is established on a routine, repetitive basis. Premature evaluation may be misleading (Gilles et al., 1973; Jordan, 1977; Jordan et al., 1982), whatever the control methods employed.

As in the previous phases, the keeping of accurate records is essential to demonstrate that the program is effective, to assess its cost, and to consider the possibility of more widespread use of the technique.

CONCLUSION

Because of uncertainties about which plant molluscicide is to be used and where, this chapter can give only the barest outline of a plan of action, but it is hoped that the principles will be applicable to specific molluscicides. In any case, one point cannot be overemphasized: Field testing must be done on a sufficiently large scale and for a long enough period to obtain reliable and acceptable results that demonstrate the efficacy of the plant.

The long-term objective is to develop a cheap and effective method that people can use to protect themselves from schistosomiasis. But this development itself can be done neither cheaply nor quickly. It requires the long-term commitment of funds and personnel. That fact must be appreciated by private, national, or international groups interested in sponsoring the development of a plant molluscicide. Shortcuts may be possible by overlapping some of the stages proposed in this chapter. Nevertheless, a decade will be required from the initial trials under simulated field conditions to the final evaluation of a full trial of the plant molluscicide applied on a self-help basis by a local community.

A development program of this duration and magnitude cannot be achieved in a piecemeal fashion. Careful planning and effective administration and execution are essential for its success. In particular, the activities of the various specialists involved must be coordinated with those of government and other agencies and the local population. The establishment of a suitable organizational system is an immediate priority. Its exact nature depends on local circumstances, but it should be controlled by an overall project manager who has the appropriate authority to run the program and tenure to provide continuity throughout its existence.

Acknowledgments

We are grateful to Professor G. Webbe, Department of Medical Helminthology, London School of Hygiene and Tropical Medicine, England, and Professor C. F. A. Bruijning, Laboratorium voor Parasitologie der Rijksuniversiteit de Leiden, The Netherlands, for their helpful comments during the preparation of this paper.

REFERENCES

Adewunmi, C. O., S. K. Adesina, and V. O. Marquis. On the laboratory and field trials of *Tetrapleura tetraptera*. *Bull. Anim. Health Prod. Afr.* 30:89–94 (1982).

Ansari, N., ed. *Epidemiology and Control of Schistosomiasis.* Basel: S.Karger (1973).

Barbosa, F. S., and D. P. P. Costa. A long-term schistosomiasis control project with molluscicides in a rural area of Brazil. *Ann. Trop. Med. Parasitol.* 75:41–52 (1981).

Barrett, P. O., and P. R. Ellison. A continuous flow centrifuge for testing the presence of bilharzia cercariae in water. *Cent. Afr. J. Med.* 11:338–340 (1965).

Berrios-Duran, L. A., L. S. Ritchie, and H. A. Wessal. Field tests on molluscicides against *Biomphalaria glabrata* in flowing water. *Bull. WHO* 39:316–320 (1968).

Chernin, E., and V. L. Adler. Effects of desiccation on eggs of *Australorbis glabaratus*. *Ann. Trop. Med. Parasitol.* 61:11–14 (1967).

Christie, J., and E. S. Upatham. Control of *Schistosoma mansoni* transmission in Saint Lucia: II. Biological results. *Am. J. Trop. Med. Hyg.* 26:894–898 (1977).

Chu, K. Y. Trials of ecological and chemical measures for the control of *Schistosoma haematobium* transmission in a Volta Lake village. *Bull. WHO* 56:313–322 (1978).

Chu, K. Y., and I. K. Dawood. Cercarial transmission seasons of *Schistosoma mansoni* in the Nile Delta area. *Bull. WHO* 42:575–580 (1970).

Chu, K. Y., J. A. Vanderburg, and R. K. Klumpp. Transmission dynamics of miracidia of *Schistosoma haematobium* in the Volta Lake. *Bull. WHO* 59:555–560 (1981).

Crossland, N. O. A mud-sampling technique for the study of the ecology of aquatic snails, and its use in the evaluation of the efficacy of molluscicides in field trials. *Bull. WHO* 27:125–133 (1962).

———. Field trials to evaluate the effectiveness of the molluscicide N-tritylmorpholine in irrigated systems. *Bull. WHO* 37:23–42 (1967).

Dalton, P. A socioecological approach to the control of *Schistosoma mansoni* in Saint Lucia. *Bull. WHO* 54:587–595 (1977).

Dazo, B. C., N. S. Hairston, and I. K. Dawood. The ecology of *Bulinus truncatus* and *Biomphalaria alexandrina* and its implications for the control of bilharziasis in the Egypt-49 project. *Bull. WHO* 35:339–356 (1966).

El-Sawy, M. F., J. Duncan, T. F. de C. Marshall, H. K. Bassiouny, and M.A.-R. Shehata. The molluscicidal properties of *Ambrosia maritima* L. (Compositae):

Design for a molluscicide field trial. *Tropenmed. Parasitol.* 34:11–14 (1983).

El-Sawy, M. F., J. Duncan, T. F. deC. Marshall, M. A.-R. Shehata, and N. Brown. The molluscicidal properties of *Ambrosia maritima* L. (Compositae): 2. Results from a field trial using dried plant material. *Tropenmed. Parasitol.* 35:100–104 (1984).

Farooq, M., and N. G. Hairston. The epidemiology of *Schistosoma haematobium* and *S. mansoni* infections in the Egypt-49 project area: 4.Measurement of the incidence of bilharzia. *Bull. WHO* 35:331–338 (1966).

Gilles, H. M., A. A.-A. Zaki, M. H. Soussa, S. A. Samaan, S. S. Soliman, A. Hassan, and F. Barbosa. Results of a seven-year snail control project on the endemicity of *Schistosoma haematobium* infection in Egypt. *Ann. Trop. Med. Parasitol.* 67:45–65 (1973).

Hairston, N. G. Statistical analysis of molluscicide field trials. *Bull. WHO* 32:289–296 (1965).

Hairston, N. G., B. Hubendick, and J. Watson. An evaluation of techniques used in estimating snail populations. *Bull. WHO* 19:661–672 (1958).

Hiatt, R. A., B. L. Cline, E. Ruiz-Tiben, W. B. Knight, and L. A. Berrios-Duran. The Boqueron project after five years: A prospective community-based study of infection with *Schistosoma mansoni* in Puerto Rico. *Am. J. Trop. Med. Hyg.* 29:1228–1240 (1980).

Hopf, H. S. S., J. Duncan, J. S. S. Beesley, D. J. Webley, and R. F. Sturrock. Molluscicidal properties of organotin and organolead compounds with particular reference to triphenyllead acetate. *Bull. WHO* 36:955–961 (1967).

Jobin, W. R. Population dynamics of aquatic snails in three farm ponds of Puerto Rico. *Am. J. Trop. Med. Hyg.* 19:1038–1048 (1970).

Jobin, W. R., and E. H. Michelson. Mathematical simulation of an aquatic snail population. *Bull. WHO* 37:657–664 (1967).

Jordan, P. Schistosomiasis – Research to control. *Am. J. Trop. Med. Hyg.* 26:877–886 (1977).

Jordan, P., and G. Webbe. *Human Schistosomiasis.* London: Heinemann (1969).

——. *Schistosomiasis: Epidemiology, Treatment and Control.* London: Heinemann (1982).

Jordan, P., R. K. Bartholomew, E. Grist, and E. Auguste. Evaluation of chemotherapy in the control of *Schistosoma mansoni* in Marquis Valley, Saint Lucia. *Am. J. Trop. Med. Hyg.* 31:76–86 (1982).

Kloos, H., C. H. Gardiner, A. Selim, and G. I. Higashi. Laboratory and field evaluation of a direct filtration technique for recovery of schistosome cercariae. *Am. J. Trop. Med. Hyg.* 3l:122–127 (1982).

Klumpp, R. K., and K. Y. Chu. Ecological studies of *Bulinus rohlfsi*, the intermediate host of *Schistosoma haematobium* in the Volta lake. *Bull. WHO* 55:715–730 (1977).

Lemma, A., P. Goll, J. Duncan, and B. Mazengia. Control of schistosomiasis with the use of endod in Adwa, Ethiopia: Results of a five-year study. In *Proceedings of the International Conference on Schistosomiasis.* Vol. 1, 415–436. Cairo: Ministry of Health (1978).

Machado, P. A. The Brazilian program for schistosomiasis control 1975-1979. *Am. J. Trop. Med. Hyg.* 31:103–110 (1982).

Mott, K. E., and H. Dixon. Collaborative study on antigens for immunodiagnosis of schistosomiasis. *Bull. WHO* 60:729–753 (1982).

Negron-Aponte, H., and W. R. Jobin. Schistosomiasis control in Puerto Rico: Twenty-five years of operational experience. *Am. J. Trop. Med. Hyg.* 28:515–525 (1979).

Pesigan, T. P., N. G. Hairston, J. J. Jauregui, E. G. Garcia, A.T.Santos, B. C. Santos, and A. A. Besa. Studies on *Schistosoma japonicum* infection in the Philippines: 2. The molluscan host. *Bull. WHO* 18:481–578 (1958).

Peters, P. A., K. S. Warren, and A. A. F. Mahmoud. Rapid accurate quantification of schistosome eggs in nuclepore filters. *J. Parasitol.* 62:154–155 (1976).

Pitchford, R., and P. S. Visser. The role of naturally infected rodents in the epidemiology of schistosomiasis in the Eastern Transvaal. *Trans. R. Soc. Trop. Med. Hyg.* 56:126–135 (1962).

Rowan, W. B. A simple device for determining population density of *Schistosoma mansoni* cercariae in infected waters. *J. Parasitol.* 43:696–697 (1957).

——. The ecology of schistosome transmission foci. *Bull. WHO* 33:63-71 (1965).

Sandt, D. G. Evaluation of an overlay technique for the recovery of *Schistosoma mansoni* cercariae. *Bull. WHO* 47:125–127 (1972).

——. Direct filtration for recovery of *Schistosoma mansoni* cercariae in the field. *Bull. WHO* 48:27–34 (1973).

Scott, D., K. Senker, and E. C. England. Epidemiology of human *Schistosoma haematobium* infection around Lake Volta, Ghana. *Bull. WHO* 60:89–100 (1982).

Shiff, C. J. Studies on *Bulinus* (*Physopsis*) *globosus* in Rhodesia: III. Bionomics of a natural population existing in a temporary habitat. *Ann. Trop. Med. Parasitol.* 58:240–255 (1964).

——. The value of incidence for the assessment of schistosomiasis control: A study in Southern Rhodesia. *Bull. WHO* 48:409–414 (1973).

——. Seasonal factors influencing the location of *Bulinus* (*Physopsis*) *globosus* by miracidia of *Schistosoma haematobium* in nature. *J. Parasitol.* 60:578–583 (1974).

Sturrock. R. F. Field studies on the population dynamics of *Biomphalaria glabarata* (Say), intermediate host of *Schistosoma mansoni* on Saint Lucia, West Indies. *Int. J. Parasitol.* 3:169–174 (1973a).

——. Field studies on the transmission of *Schistosoma mansoni* and on the bionomics of its intermediate host *Biomphalaria glabrata* on Saint Lucia, West Indies. *Int. J. Parasitol.* 3:175–194 (1973b).

——. Control of *Schistosoma mansoni* transmission: Strategy for using molluscicides on Saint Lucia. *Int. J. Parasitol.* 3:795–801 (1973c).

——. Distribution of the snail *Biomphalaria glabrata*, intermediate host of *Schistosoma mansoni*, within a Saint Lucian field habitat. *Bull. WHO* 52:263–272 (1975).

Sturrock. R. F., and G. Barnish. The aerial application of molluscicides with special reference to schistosomiasis control. *Bull. WHO* 49:283–285 (1973).

Sturrock, R. F., G. Barnish, and E. S. Upatham. Snail findings from an experimental mollusciciding program to control *Schistosoma mansoni* transmission on Saint Lucia. *Int. J. Parasitol.* 4:231–240 (1974).

Sturrock, R. F., S. J. Karamsadkar, and J. Ouma. Schistosome infection rates in field snails: *Schistosoma mansoni* in *Biomphalaria pfeifferi* from Kenya. *Ann. Trop. Med. Parasitol.* 73:369–375 (1979).

Sturrock, R. F., R. Kimani, B. J. Cottrell, A. E. Butterworth, H. M. Seitz, T. K. Siongok, and V. Houba. Observations on possible immunity to reinfection among Kenyan school children after treatment for *Schistosoma mansoni*. *Trans. R. Soc. Trop. Med. Hyg.* 77:363–371 (1983).

Theron, A. A differential filtration technique for the measurement of schistosome cercarial densities in standing water. *Bull. WHO* 57:971–975 (1979).

Uemura, K. Detection of schistosome eggs in urine and stool. In *Epidemiology and Control of Schistosomiasis*, ed. N. Ansari, 729–730. Basel: S. Karger (1973).

Upatham, E. S. Exposure of caged *Biomphalaria glabrata* (Say) to investigate dispersion of miracidia of *Schistosome mansoni* Sambon in outdoor habitats in Saint

Lucia. *J. Helminthol.* 46:297–306 (1972).

——. Field studies on the bionomics of the free-living stages of Saint Lucian *Schistosoma mansoni*. *Int. J. Parasitol.* 6:239–245 (1976).

Webbe, G. The transmission of *Schistosoma haematobium* in an area of Lake Province, Tanganyika. *Bull. WHO* 27:59–85 (1962a).

——. Population studies on intermediate hosts in relation to transmission of bilharziasis in East Africa. In *Bilharziasis: Ciba Foundation Symposium*, ed. G. E. Wolstenholme and M. O'Connor, 7–22. London: Churchill (1962b).

——. Transmission of bilharziasis: 1. Some essential aspects of snail population dynamics and their study. *Bull. WHO* 33:147–153 (1965a).

——. Transmission of bilharziasis: 2. Production of cercariae. *Bull. WHO* 33:155–162 (1965b).

——. The effect of water velocities on the infection of animals exposed to *Schistosoma mansoni* cercariae. *Ann. Trop. Med. Parasitol.* 60:78–84 (1966).

WHO. Molluscicide screening and evaluation. *Bull. WHO* 33:567–581 (1965a).

——. Snail control in the prevention of bilharziasis. *WHO Monograph Series*, 50 (1965b).

——. Schistosomiasis control. *WHO Tech. Rep. Series* 515:26–34 (1973).

——. Immunology of schistosomiasis. *Bull. WHO* 51:533–595 (1974).

——. Quantitative aspects of the epidemiology of *Schistosoma japonicum* infection in a rural community of Luzon, Philippines. *Bull. WHO* 58:629–638 (1980a).

——. Environmental management for vector control. *WHO Tech. Rep. Series* 649:32–36 (1980b).

Yeo, D. A preliminary statistical analysis of snail counts. *Bull. WHO* 27:183–187 (1962).

11

PLANT MOLLUSCICIDE STUDIES IN THE PEOPLE'S REPUBLIC OF CHINA

Y. H. Kuo

Institute of Parasitic Diseases
Chinese Academy of Preventive Medicine
Shanghai, China

Schistosomiasis due to *Schistosoma japonicum* is endemic in China, and its snail host, *Oncomelania hupensis*, is widespread. Chemical control by means of synthetic compounds and plants is a necessary supplement to the large-scale environmental control methods now in use. Plant molluscicides have the advantage of being less expensive and less polluting than the synthetics; at the same time they afford clues for synthesizing new compounds once the effective ingredient is known.

More than 500 species of plants have been screened for molluscicidal effects, most of them by the Institute of Parasitic Diseases. Dried plants are used. A 1% stock solution is prepared by adding 3 g of plant to 300 mL of boiling water; this mixture is boiled for two hours in the water bath, then the solution is filtered and dechlorinated water added to 300 mL. After cooling, the solution is ready for use in screen testing.

Live adult snails are selected for the tests. Groups of 30 are exposed to various concentrations of the solutions for 24, 48, 72, 96, and 120 hours, usually at temperatures from 20° to 25° C. After exposure, the snails are washed thoroughly in running water and kept under observation for 24, 48, and 72 hours. Snail death is determined by tactile stimulation of the operculum and is followed by crushing of the snails.

Of the 500-plus species tested, 20 demonstrated high molluscicidal effect, with snail mortality reaching 90% to 100% at concentrations of 30 to 10000 ppm (Table 1) (Institute of Parasitic Diseases, 1956; Mao et al., 1963).

Table 1. Plant Species with High Molluscicidal Effect

Species	Parts Tested	Conc. Tested (ppm)	Tempera ture (°C)	Time Exposure (hours)	% Mortality
Aconitium lycoctomum	roots	5,000	25	72	100
Anemone rivularis	flowers	1,000	20–25	48	90
Areca catechu	fruits	10,000	25	72	100
Astragalus sinicus	leaves	1,000	20–25	48	90
Belamcanda chinensis	stems, roots	1,000	20–25	24	100
Berberidaceae sp.	leaves, stems	10,000	25	72	90
Camellia oleosa	seeds	2,000	20–25	24	100
		1,000	20–25	24	99
Croton tiglium	seeds	100	25	24	100
	(alcohol extract)	30	21–29	24	93
Dryobalanops aromatica	stems	10,000	25	72	100
Euphorbia helioscopia	juice of leaves, stems	1,000	25	48	100
Gleditsia sinensis	seeds	10,000	25	72	100
Kochia scoparia	seeds	10,000	25	72	90
Melia azedarach	roots, leaves	1,000	20–25	48	90
Rhododendron molle	flowers	5,000	25	48	100
		2,500	25	120	100
	leaves	1,000	33	72	90
Rhus toxicodendron	leaves	1,000	25	120	95
Sabia japonica	leaves, stems	10,000	25	72	90
Sapindus mukorossi	roots	1,000	20–25	48	90
Schima argentea	stems	5,000	25	72	97.8
Strychnos ignatii	seeds	100	25	72	93.3
Strychnos nux vomica	seeds	1,000	25	72	100

Table 2. Plant Species with Moderate Molluscicidal Effect

Anemarrhana asphodeloides	*Gardenia jasminoides*
Angelica pubescens	*Inula britannica*
Aralia chinensis	*Justicia procumbens*
Artemisia apiacea	*Luffa cylindrica*
Aster tataricus	*Magnolia liliflora*
Buddledia officinalis	*Momordica cochinchinensis*
Cannabis sativa	*Morus alba*
Cimicifuga foctida	*Perilla frutescens*
Citrus medica	*Piper nigrum*
Coptis chinensis	*Polygala tenuifolia*
Dichroa febrifuga	*Saposhnikovia divaricata*
Dioscorea hypoglauca	*Scutellaria baicalensis*
Euphorbia sieboldiana	*Toona sinensis*
Evodia rutaecarpa	*Veratrum nigrum*

Twenty-eight species were shown to have moderate effect, with mortality of 51% to 89% at concentrations of 10000 ppm (Table 2) (Institute of Parasitic Diseases, 1956). Another 114 species had only slight effect, killing 1% to 50% of snails at concentrations of 10000 ppm (Table 3) (Institute of Parasitic Diseases, 1956). The rest were not toxic at all to *Oncomelania* snails (Table 4) (Institute of Parasitic Diseases, 1956).

Camellia oleosa, which has 100% mortality at 2000 ppm (24 hours), grows in ten provinces of South China. Its seeds are oleaginous; after extraction of the oil, the dregs are made into tea cake, which is used by villagers as a base fertilizer and as a soap for washing clothes. It kills leeches, *Oncomelania* and *Lymnaea* snails, and schistosome cercariae at various concentrations, but has a very low toxicity to mammals. The cost of tea cake is only US$0.008 per kg. *Camellia oleosa* could become an effective and economic molluscicide when used in combination with fertilization.

Rhododendron molle contains ericolin and andromedotoxin. Distributed over vast areas of China, it has been used as an insecticide against *Lygus lucorum* and *Adelphocoris taeniophorus* but is not harmful to plants. Its flowers produced 100% snail mortality at 2500 ppm (120 hours) and 5000 ppm (48 hours).

Croton tiglium is found in the southern provinces and in Taiwan. Its seeds contain saponin, which, after extraction by alcohol, has high molluscicidal activity: 100% mortality at 100 ppm (24 hours). The plant is also applied as an insecticide.

Belamcanda chinensis occurs widely in China. It contains essential oil and is used as a traditional medicine for the treatment of sore throat. Its stems and roots showed 100% snail mortality at 1000 ppm (24 hours). Its great availability would allow it to be used as a molluscicide in many areas.

Table 3. Plant Species with Low Molluscicidal Effect

Acanthopanax gracilistylus
Achyranthes bidentata
Acorus gramineus
Adnophora trachelioides
Akebia quinata
Aloe vera
Alpinia galanga
Alpinia officinarum
Amomum cardamomum
Amomum villosum
Arctium lappa
Aristolochia debilis
Artemisia anomala
Asarum heterotropoides
Atractylodes lancea
Biota orientalis
Bletilla striata
Boehmeria nivea
Boswellia carterii
Brassica alba
Carthamus tinctorius
Cassia tora
Celosia argentea
Celosia cristata
Citrus aurantium
Clematis chinensis
Commiphora myrrha
Corydalis yanhusuo
Crataegus pinnatifida
Cremastra variabilis
Curculigo orchioides
Curcuma aromatica
Curcuma longa
Cynanchum inamoenum
Cyperus rotundus
Daphne genkwa
Dictamnus dasycarpus
Dioscorea bulbifera
Dipsacus asper
Dolichos lablab
Ephedra sinica
Equisetum hiemale
Equisetum palustre
Eriobotrya japonica
Eriocaulon buergerianum
Erythrina variegata
Euonymus alatus
Eupatorium fortunei
Euphorbia lathyris

Euphorbia pekinensis
Foeniculum vulgare
Forsythia suspensa
Fortunella margarita
Gentiana macrophylla
Gentianopsis barbata
Ginkgo biloba
Glycyrrhiza uralensis
Howttuynia cordata
Hydnocarpus anthelmintica
Hydrangea strigosa
Impatiens balsamina
Juncus effusus
Laminaria japonica
Ligusticum wallichii
Ligustrum lucidum
Lonicera japonica
Lycium chinensis
Magnolia officinalis
Malva verticillata
Nardostachys chinensis
Nicotiana tabacum
Notopterygium incisum
Paeonia lactiflora
Paeonia suffruticosa
Phaseolus angularis
Phellodendron amurense
Photinia serrulata
Polygonum multiflorum
Polyporus mylittae
Poncirus trifoliata
Prinsepia uniflora
Prunus armeniaca
Pueraria lobata
Pulsatilla chinensis
Punica granatum
Quercus infectoria
Quisqualis indica
Rehmannia glutinosa
Rhodiola henryi
Rosa rugosa
Rubia codifolia
Salvia miltiorrhiza
Sanguisorba officinalis
Santalum album
Saussurea lappa
Schisandra chinensis
Schizonepeta tenuifolia
Scirpus validus

Table 3 *cont'd.*

Semiaquilegia adoxoides	*Trigonella foenumgraecum*
Sophor japonica	*Typha angustata*
Spodiopogon sagitti	*Uncaria rhynchophylla*
Stellaria dichotoma	*Vaccaria segelalis*
Sterculia scaphigera	*Vaccinium bracteatum*
Taraxacum mongolicum	*Vitex rotundifolia*
Terminalia chebula	*Xanthium sibiricum*
Tribulus terrestris	*Zingiber officinale*

Table 4. Plant Species Without Molluscicidal Effect

Acacia catechu	*A. coeruleum*
Achyranthes longifolia	*A. macrostemon*
Aconitum artemisaefolium	*A. m. uratense*
A. balfourii	*A. sativum*
A. carmichaeli	*A. tuberosum*
A. c. fortunei	*Alpinia katsumada*
A. c. hwangshanicum	*A. oxyphylla*
A. c. pubescens	*Ammannia baccifera*
A. delavayi	*Amomum globosum*
A. hemsleyanum	*A. tsaoko*
A. karakolicum	*A. xanthioides*
A. kusenezoffii	*Ampelopsis aconitifolia*
A. paniculigerum	*A. japonica*
A. richardsonianum	*Anemone hupehensis*
A. soongaricum	*A. raddeana*
A. stylosum	*Angelica acutiloba*
A. sungpanense	*A. sinensis*
A. taipeicum	*Arisaema amurense*
A. transsectum	*A. consanguineum*
Adenophora axilliflora	*A. heterophyllum*
A. capillaris	*Aristolochia fangchi*
A. lilifolioides	*A. hetrophylla*
A. polyantha	*A. mollissima*
A. potaninii	*Artemisia argyi*
A. tetraphylla	*A. capillaris*
Aesculus chinensis	*A. japonica*
Agastache rugosa	*A. nova*
Agrimonia pilosa	*A. tridentata*
A. p. viscidula	*Arundina chinensis*
Akebia trifoliata	*Astrasalus chrysopterus*
A. t. australis	*A. complanatus*
Albizzia julibrissin	*A. floridus*
Aleurites fordii	*A. membranaceus*
Alisma ptantago-aquatica orientale	*A. mongholicus*
Allium chinense	*A. tibetanus*

Cont'd.

Table 4 *cont'd.*

A. tongolensis	Cuscuta chinensis
Bambusa textilis	Cynanchum atratum
Baphicacanthus cusia	C. glaucescens
Benincasa cerifera	C. stauntoni
B. hispida	Cynomorium songaricum
Blechnum orientale	Daemonorops draco
Boschniakia rossica	Dalbergia odorifera
Brainia insignis	Dendrobium nobile
Broussonetia papyrifera	Desmodium pulchellum
Caesalpinia bonducella	D. styracifolium
C. sappan	Dichrostachys glomerata
Campanumoea lancifolia	Dioscorea collettii
Campsis grandiflora	D. futschanensis
Canarium album	D. gracillima
Canavalia gladiata	D. tokoro
Capsella bursa-pastoris	Dipsacus japonicus
Cassia acutifolia	Dracaema ombet
C. angustifolia	Drynaria fortunei
C. occidentalis	Dryoteris crassirhizoma
Centipeda minima	Echinops grijisii
Centranthera cochinchinensis	E. latifolius
Cephalanoplos segetum	E. ritro
Chenopodium hybridum	Elsholtzia splendens
Chrysanthemum boreale	Epimedium brevicornum
C. indicum	E. grandiflorum
C. lavandulaefolium	E. sagittatum
C. morifolium	Erodium stephanianum
Cibotium barometz	Eucommia ulmoides
Cinnamomum cassia	Eupatorium chinensis
Cirsium japonicum	E. japonicum
Citrullus vulgaris	Euphorbia humifusa
Citrus erythrosa	Euryale ferox
C. tangerina	Ficus carica
C. wilsonii	F. simplicissima
Clerondendron cyrtophyllum	Firmiana simplex
Cocculus trilobus	Fritillaria cirrhosa
Codonopsis pilosula	Gastrodia elata
Coix lachryma	Gentiana loureiri
Coptis chinensis brivesepala	G. scabra
C. deltoidea	G. triflora
C. omeiensis	Glehnia littoralis
C. tectoides	Gloiopeltis furcata
Cordyceps sinensis	Glycine max
Cornus officinalis	G. soja
Corylus cuneata	Gueldenstaedtia multiflora
Crylus hetrophylla	G. pauciflora
Cucumis melo	Hedysarum polybotrys
Cupressus funebris	Helicteres angustifolia
Curcuma zedoaria	Hibiscus syriacus

Table 4 *cont'd.*

Hierochloe odorata	*Monochoria vaginalis*
Homalomena occulta	*Morinda officinalis*
Hordeum vulgare	*Myristica fragans*
Hydrocharis asiatica	*Nandina domestica*
Hygrophila salicifolia	*Nelumbo nucifera*
Ilex cornuta	*Ophiopogon japonicus*
I. latifolia	*Oryza sativa*
Imperata cylindrica	*Osbeckia crinita*
Indigofera tinctoria	*Osmund japonica*
Inula linearifolia	*Paeonia emodi*
Iris pallasii	*P. longum*
Isatis indigotica	*P. officinalis*
I. tinctoria	*Paris chinensis*
Jasminum sambac	*P. delavayi*
Juglans regia	*P. polyphylla*
Lactuca sativa	*P. p. pubescens*
Leonurus heterophyllus	*P. p. stenophylla*
Leptochloa chinensis	*P. p. yunnanensis*
Ligusticum glaucescens	*P. quadrifolia*
Lindera strychnifolia	*Parnassia palustris*
Liquidambar orientalis	*Patrinia hetrophylla*
Liriope kansuensis	*P. scabiosacfolia*
L. minor	*Paulownia fortunei*
L. platyphylla	*P. tomentosa*
L. spicata	*Peucedanum decursivum*
Litchi chinensis	*P. terebinthaceum*
Lithospermum erythrorhizon	*Phaseolus calcaratus*
Litsea cubela	*P. radiatus*
Lonicera confusa	*Phragmites communis*
L. dasystyla	*Physalis alkekengi franchetii*
L. fuchsioides	*Phragmites communis*
L. henryi	*Phyllostachys nigra*
L. hypoglauca	*P. n. henonis*
L. lanceolata	*Phytolacca acinosa*
L. similis	*Pinellia pedatisecta*
Lophatherum gracile	*P. ternata*
Loranthus gracilifolius	*Pinus massoniana*
L. parasiticus	*P. tabulaeformis*
Lunathyrium acrostichoides	*P. yunnanensis*
Lypodicum japonicum	*Piper cubeba*
Mahonia bealei	*P. kadsura*
M. fortani	*P. longum*
M. japonicum	*Plantago asiatica*
Marsilea quadrifolia	*P. depressa*
Matteuccia struthiopteris	*P. major*
Melica scabrosa	*Pleomele cambodiana*
M. Spicata	*Pogostemon cabin*
Mesona chinensis	*Polygala telephioides*
Monascus purpureus	*Polygonatum cirrhifolium*

Cont'd.

Table 4 *cont'd.*

P. cytonema	*Sapium sebiferum*
P. involucratum	*Sargassum fusiforme*
P. kingianum	*Saussurea japonica*
P. macropodium	*Scrophularia ningpoensis*
P. odoratum	*Selaginella involvens*
P. roseum	*Sesamum indicum*
P. sibiricum	*Shiraia bambusicola*
P. verticilla	*Siegesbeckia orientalis*
Polygonutum curvistylum	*S. o. glabrescens*
P. erythrocarpum	*S. o. pubescens*
P. filipes	*Smilax glabra*
P. lasianthum	*Solanum melongena*
Polygonum orientale	*Sophora flavescens*
P. perfoliatum	*Sparganium stoloniferum*
P. tinctorium	*Spatholobus suberectus*
Polyonatum inflatum	*Speranskia tuberculata*
Polyporus umbellatus	*Stachyurus himalaicus*
Populus pseudosimonii	*Stemona japonica*
Poria cocos	*S. parviflora*
Potentilla fulgens	*S. sessilifolia*
Portulaca grandiflora	*S. tuberosa*
P. oleracea	*Stephania tetrandra*
Prunella hispida	*Sterculia nobilis*
P. vulgaris	*Tamarix chinensis*
Prunus davidiana	*T. juniperina*
P. humilis	*T. ramosissima*
P. japonica	*Tetrapanax papyriferus*
P. j. nakaii	*Trachelospernum jasminoides*
P. mandshurica	*Trachycarpus wagnerianus*
P. mume	*Trichosanthes kirilowii*
P. persica	*Triticum aestivum*
P. pseudocerasus	*Tropaeolum majus*
P. tomentosa	*Tussilago farfara*
P. triloba	*Ulmus marcocarpa*
Psoralea corylifolia	*Umbilicaria esculenta*
Pteris multifida	*Uncaria gambier*
Pyrola atropurpurea	*Verbana officinalis*
P. decorata	*Veronica anagallisaquatica*
P. rotundifolia	*V. peregrina*
Pyrrosia lingua	*V. serpyllifolia*
Raphanus sativus	*Viola japonica*
Rhaponticum uniflorum	*V. yedoensis*
Rheun officinale	*Viscum album*
R. palmatum	*V. coloratum*
R. tanguticum	*V. c. lutescens*
Rosa chinensis	*V. c. rubroaurantiacum*
R. laevigata	*Vitex negundo*
Rubus chingii	*Woodwardia japonica*
Salvia chinensis	*Ziziphus jujuba*

Based on our investigations, many other wild plants and medicinal herbs in various localities in China are potential molluscicides and worthy of further study.

REFERENCES

Institute of Parasitic Diseases, Chinese Academy of Medical Sciences. *Collection of the Studies on Schistosomiasis Japonica*, 451–456 (1956).

Mao, S. P., Y. H. Kuo, H. Q. Tan, S. Z. Yu, T. Y. Lu, and S. L. Hu. *Studies on the Eradication of Oncomelania Snail*, 2nd ed., 122–125. Shanghai: Shanghai Science and Technology Publishing House (1963).

12

PLANT MOLLUSCICIDE RESEARCH – AN UPDATE

K. Hostettmann and A. Marston

Institute of Pharmacognosy and Phytochemistry
School of Pharmacy of the University
Lausanne, Switzerland

Since the 1983 meeting of the Scientific Working Group on Plant Molluscicides in Geneva, the number of reports on the use of plant molluscicides has been augmented considerably. Reviews covering various aspects of the field have appeared (Hostettmann, 1985; Marston and Hostettmann, 1985; Duncan, 1985), and increasing numbers of research teams around the world have become involved.

Various classes of compounds have been found responsible for the molluscicidal activities of plants, but saponins still seem to hold the greatest promise for control of the snail vector. Not only are some saponins as strongly molluscicidal as synthetic agents, but also the content of saponins in vegetable material is often high; some plant parts contain as much as 30% saponin. Earlier field studies on the applicability of *Phytolacca dodecandra* (Phytolaccaceae) were reported by Lugt (1981); now, the saponin-containing fruits and seed pods of *Swartzia madagascariensis* (Leguminosae) have also been subjected to field trials as molluscicidal agents (Suter et al., 1986).

Although data on the activities of pure compounds, such as saponins, is being thoroughly investigated, the tendency is to report molluscicidal activities of plant extracts and not those of the corresponding constituents of the plants. That is unfortunate, because identification of the substances responsible for the activity is of inestimable value in the search for other sources of plant-derived molluscicides, in investigations of toxicity, and so forth (Marston and Hostettmann, 1985). In fact, of more than 1000 plant

299

species tested for molluscicidal activity, only about 100 natural products with recognized molluscicidal activity have been isolated.

This chapter covers some of the recent reports on plant molluscicides, with emphasis on two areas: saponin-containing and tannin-containing vegetable material.

Expression of snail kill results for pure compounds varies according to literature sources. Some authors prefer LC_{100} values (the dose of molluscicide required to kill 100% of all snails), while previously established WHO standards are LC_{90} and/or LC_{50} values (see chapter 9).

CLASSES OF MOLLUSCICIDES FROM PLANT SOURCES

Triterpene Saponins

In addition to the triterpene glycosides *3*, *4*, and *5*, further saponins (*7*, *17*, *20-22*) have been isolated from *P. dodecandra* (Tables 1 to 3). The monodesmosidic saponins *17* and *21* (glycoside chain in position 3) are active against *Biomphalaria glabrata* snails, whereas the bidesmosidic saponins *7*, *20*, and *22* (glycoside chain in positions 3 and 28) are inactive (Domon and Hostettmann, 1984). This mono-/bidesmosidic difference seems to be a general feature of the triterpene saponins tested so far for molluscicidal activity. When the *P. dodecandra* plant material is extracted with water, predominantly monodesmosidic saponins are obtained; when extracted with methanol, more bidesmosidic saponins are obtained (Domon and Hostettmann, 1984). The reason is probably that fermentation occurs during the water-extraction process and the ester-linked glycosides are hydrolyzed, leaving monodesmosidic saponins. A similar phenomenon occurs with saponins from the tubers of *Talinum tenuissimum* (Portulacaceae). Extraction of the tubers with methanol gave mainly the inactive saponin *9* with small amounts of the monodesmosidic saponin *8*, whereas an aqueous extraction gave predominantly the strongly molluscicidal saponin *8* and only traces of *9* (Gafner et al., 1985).

Besides the mono-/bidesmosidic differences, the molluscicidal activities of saponins also vary with the nature of the sugar chains, the sequence of the sugars, the interglycosidic linkages, and the substitution patterns of the aglycone. The aglycones oleanolic acid and hederagenin, on the other hand, are inactive (Domon and Hostettmann, 1983).

The leguminous tree *S. madagascariensis*, found widely distributed in Africa, has been known for a long time as a fish poison (Haerdi, 1964). The saponin-containing fruits also have molluscicidal properties (Mozley, 1952).

Since many of the criteria for plant molluscicides (see the Annex to this book) are fulfilled by *S. madagascariensis*, it was selected for limited trials in the field (Suter et al., 1986). Large quantities (up to 40 kg) of fruits can be obtained from each tree, and a 100 mgL^{-1} water extract of dry seed pods kills over 90% of *Bulinus globosus* snails exposed for 24 hours to the molluscicidal extract. The major saponin responsible for the molluscicidal activity was shown to be a monodesmoside of oleanolic acid with structure *10* and LC$_{100}$ toxicity to *B. globosus* and *Biomphalaria glabrata* snails of 3 mgL^{-1} (Suter et al., 1986). The area chosen for study, in southeast Tanzania, has a high incidence of urinary schistosomiasis, with *B. globosus* as the intermediate host responsible for the spread of the disease. Treatment of ponds containing the host snails at a concentration of 100 mgL^{-1} of *S. madagascariensis* crushed fruits showed a dramatic decrease in *B. globosus* density, giving a virtual extinction of snails after one week. The concentration of saponins in the pond water was controlled by TLC and hemolysis experiments. In addition, the mortality of caged snails was investigated. Eggs of *B. globosus* were not affected by the treatment with *S. madagascariensis* extracts, meaning that a single application of molluscicide is not feasible; at least one subsequent application is necessary to kill remaining snails. Although the half-life of saponins in the pond environment was short, it is known that they are piscicidal and there is always the possibility of other adverse influences on the local ecology. The toxicity and mutagenicity of *S. madagascariensis* fruit extracts is under test, however, and the local environment of the ponds is constantly being monitored (Suter et al., 1986).

Spirostanol Saponins

A number of spirostanol glycosides are now known to be molluscicides (Table 4). As in the case of the triterpene glycosides, the bidesmosidic glycosides *30* and *31* are not noticeably active against *B. glabrata* snails.

Sesquiterpenes

Although frequently possessing cytotoxic and allergenic properties (Towers, 1979), sesquiterpenes have been widely investigated for their molluscicidal activities. Most of them have been isolated from members of the family Compositae. Some of the recent discoveries are listed in Table 5. Helenalin (*34*), for example, is both a strong cytotoxic and molluscicidal compound. However, the α-methylene-γ-lactone moiety, important for cytotoxic activity, is not essential for molluscicidal activity, and bipinnatin (*37*) was inactive against *Biomphalaria havanensis* snails (Marchant et al., 1984).

Oleanolic acid:
$R^1 = R^2 = H$

Table 1. Molluscicidal Activities of Oleanolic Acid Glycosides

Compound	R^1	R^2	Plant	Molluscicidal Activity (mgL^{-1})	Reference
1	GlcA-	H	Lonicera nigra	2 (LC_{100})	Domon and Hostettmann, 1983
2	Glc-^{2}Ara-	H	Lonicera nigra	2 (LC_{100})	Domon and Hostettmann, 1983
3 (Lemmatoxin C)	Rha-^{2}Glc-^{2}Glc-	H	Phytolacca dodecandra	3 (LC_{90})	Parkhurst et al., 1973a

4 (Oleanoglycotoxin-A)	Glc-^{4}Glc-
	  |^2
	  Glc
		H	*Phytolacca dodecandra*	6 (LC$_{100}$)	Domon and Hostettmann, 1984

				3 (LC$_{90}$)	Parkhurst et al., 1973b

5 (Lemmatoxin)	Glc-^{4}Glc-
	  |^3
	  Gal
		H	*Phytolacca dodecandra*	1.5 (LC$_{90}$)	Parkhurst et al., 1974

6	Glc-^{2}Ara-	Glc-^{6}Glc-	*Lonicera nigra*	n.a.	Domon and Hostettmann, 1983

7	Glc-^{4}Glc-
	  |^2
	  Glc
		Glc	*Phytolacca dodecandra*	n.a.	Domon and Hostettmann, 1984

8	Xyl-^{3}GlcA-	H	*Talinum tenuissimum*	1.5 (LC$_{100}$)	Gafner et al., 1985
9	Xyl-^{3}GlcA-	Glc-	*Talinum tenuissimum*	n.a.	Gafner et al., 1985
10	Rha-^{3}GlcA-	H	*Swartzia madagascariensis*	3 (LC$_{100}$)	Suter et al., 1986

n.a.: not active; LC$_{100}$: 100% lethal concentration to *Biomphalaria glabrata* snails; LC$_{90}$: 90% lethal concentration to *Biomphalaria glabrata* snails (LC values after 24 hr); Ara: α-L-arabinopyranosyl; Rha: α-L-rhamnopyranosyl; Xyl: β-D-xylopyranosyl; Glc: β-D-glucopyranosyl; GlcA: β-D-glucuronopyranosyl; Gal: β-D-galactopyranosyl.

Adapted with permission from A. Marston and K. Hostettmann, *Phytochemistry*, **24**(4): 639–652, copyright 1985, Pergamon Journals Ltd.

Hederagenin:
R¹=R²=H

Table 2. Molluscicidal Activities of Hederagenin Glycosides

Compound	R¹	R²	Plant	Molluscicidal Activity (mgL^{-1})	Reference
11	Glc-	H	*Hedera helix*	15 (LC_{100})	Hostettmann, 1980
12	Ara-	H	*Hedera helix*	3 (LC_{100})	Hostettmann, 1980
			Lonicera nigra		Domon and Hostettmann, 1983
13	GlcA-	H	*Lonicera nigra*	16 (LC_{100})	Domon and Hostettmann, 1983

14	Glc-^{2}Glc-	H	*Hedera helix*	12 (LC$_{100}$)	Hostettmann, 1980
15	Rha-^{2}Ara-	H	*Hedera helix*	8 (LC$_{100}$)	Hostettmann, 1980
16	Glc-^{2}Ara-	H	*Lonicera nigra*	8 (LC$_{100}$)	Domon and Hostettman, 1983
17	Glc-^{4}Glc-$\mid^2$Glc	H	*Phytolacca dodecandra*	12 (LC$_{100}$)	Domon and Hostettmann, 1984
18	Ara-	Glc-^{6}Glc-	*Lonicera nigra*	n.a.	Domon and Hostettmann, 1983
19	Glc-^{2}Ara-	Glc-^{6}Glc-	*Lonicera nigra*	n.a.	Domon and Hostettmann, 1983
20	Glc-^{4}Glc-$\mid^2$Glc	Glc-	*Phytolacca dodecandra*	n.a.	Domon and Hostettmann, 1984

Abbreviations: see Table 1.

Bayogenin:
$R^1 = R^3 = H$, $R^2 = CH_3$
Phytolaccagenin:
$R^1 = R^3 = H$, $R^2 = COOCH_3$

Table 3. Molluscicidal Activity of Bayogenin and Phytolaccagenin Glycosides

Compound	R^1	R^2	R^3	Plant	Molluscicidal Activity (mgL^{-1})	Reference
21	Glc-^{4}Glc-	CH_3	H	*Phytolacca dodecandra*	12 (LC_{100})	Domon and Hostettmann, 1984
22	Glc-^{4}Glc-	CH_3	Glc-	*Phytolacca dodecandra*	n.a.	Domon and Hostettmann, 1984
23	Xyl-	$COOCH_3$	H	*Phytolacca americana*	60	Johnson and Shimizu, 1974
24	Glc-^{2}Xyl-	$COOCH_3$	H	*Phytolacca americana*	80	Johnson and Shimizu, 1974

Abbreviations: see Table 1.
Adapted with permission from A. Marston and K. Hostettmann, *Phytochemistry*, **24**(4): 639–652, copyright 1985, Pergamon Journals Ltd.

37

Similarly, damsin from *Ambrosia maritima* was not toxic to the snails (Marchant et al., 1984), thus contradicting reports by Shoeb and El-Aman (1978). One reason for inactivity might be the limited solubility of these sesquiterpene lactones in water.

A schistosomicidal heliangolide has been isolated from *Eremanthus goyazensis* (Vichnewski et al., 1976) and may be worth investigating as a molluscicide, especially since a number of sesquiterpene lactones have been characterized from the molluscicidal extract of *Eremanthus glomerulatus* (Barros et al., 1985).

Monoterpenes

A brief report has appeared on the toxicity toward *B. glabrata* snails of thymol, carvacrol (not strictly terpenes but often included in this class of natural products), and limonene from plants of the genus *Lippia* (Bezerra et al., 1981).

Iridoids

The iridoid glycosides ligstroside and oleuropein from the fruits of *Olea europaea* (Oleaceae) have been claimed to possess activity against *B. glabrata* snails (100 mgL^{-1} and 250 mgL^{-1}, respectively) (Kubo and Matsumoto, 1984b), but the activities of the pure compounds are so weak that they do not even meet the minimum level required for the molluscicidal activity of plant extracts. However, oruwacin (*38*), an iridoid ferulate from *Morinda lucida* leaves, was 100% active against *Bulinus rohlfsii*, *B. globosus*, and *Biomphalaria pfeifferi* snails at concentrations varying between 2.5 and 5 mgL^{-1} (Adewunmi and Adesogan, 1984).

Sarsapogenin: R=H

Yamogenin: R=H

C   22-Methoxy (25S),5β-furostan-3β,26-diol: R¹=R²=H

D   22-Hydroxy-(25S),5β-furostan-3β,26-diol: R¹=R²=H

Table 4. Molluscicidal Activities of Spirostanol Saponins

Compound	Parent Triterpene	R	Plant	Molluscicidal Activity mgL⁻¹ (LC₁₀₀)	Reference
25	A	Rha-⁴Glc- |² Glc	*Asparagus curillus*	20	Sati et al., 1984
26	A	Ara-⁴Glc- |² Rha	*Asparagus curillus*	5	Sati et al., 1984

		Structure	Plant		Reference
27	A	$\text{Rha-}^4\text{Glc-}\overset{2}{\underset{\text{Glc}}{\mid}}$ (with $\overset{\text{Ara}}{\underset{\mid}{6}}$)	*Asparagus curillus*	5	Sati et al., 1984
28	B	$\text{Rha-}^2\text{Glc-}$	*Asparagus plumosus*	25	Sati et al., 1984
29	B	$\text{Rha-}^3\text{Glc-}\overset{2}{\underset{\text{Rha}}{\mid}}$	*Asparagus plumosus*	20	Sati et al., 1984
30	C	$\text{Rha-}^4\text{Glc-}\overset{2}{\underset{\text{Glc}}{\mid}}$ (with $\overset{\text{Ara}}{\underset{\mid}{6}}$)	*Asparagus curillus*	n.a.	Sati et al., 1984
31	D	$\text{Rha-}^4\text{Glc-}\overset{2}{\underset{\text{Glc}}{\mid}}$ (with $\overset{\text{Ara}}{\underset{\mid}{6}}$)	*Asparagus curillus*	n.a.	Sati et al., 1984

Abbreviations: see Table 1.

32   R=OH
33   R=H

34

35

36

Table 5. Molluscicidal Activities of Sesquiterpenes

Compound	Plant	Molluscicidal Activity		Reference
		mgL^{-1}	Snails Tested	
32 (7 -Hydroxy-3-deoxyzaluzanin)	Podachaenium eminens	1 (LC$_{100}$)	Biomphalaria glabrata	Fronczek et al., 1984
33 (3-Deoxyzaluzanin)	Podachaenium eminens	n.a.	Biomphalaria glabrata	Fronczek et al., 1984
34 (Helenalin)	Helenium	10 (LC$_{100}$)	Biomphalaria havanensis	Marchant et al., 1984
35 (Pyrethrosin)	Chrysanthemum	15 (LC$_{100}$)	Biomphalaria havanensis	Marchant et al., 1984
36 (Desacetylisotenulin)	Psathyrotes ramosissima	20 (LC$_{50}$)	Biomphalaria glabrata	Kubo & Matsumoto, 1984a

Abbreviations: see Table 1; LC$_{50}$: 50% lethal concentration.

38

Quinones

The molluscicidal activities of members of this class of natural products have recently been reported for the first time, with details on the effects of naphthoquinones against bilharzia-transmitting snails (Marston et al., 1984a). Simple naphthoquinones are very effective at snail killing (Table 6), whereas prenylated (Vitamin K_1) and dimeric naphthoquinones (*48, 49*) are inactive. Introduction of a hydroxy substituent into the quinonoid ring (isojuglone, lapachol) causes a significant decrease in activity, observed also for 3-methoxy-7-methyljuglone, an artifact obtained during the extraction of *Diospyros usambarensis* (Ebenaceae) with methanol (Marston et al., 1984a). The synthetic prothrombogenic drug menadione (Vitamin K_3) shows marked molluscicidal activity.

Naphthoquinones are frequently found in species of African and Oriental plants (Thomson, 1971) and may be utilizable for local control of bilharzia-transmitting snails. However, the naphthoquinones from *D. usambarensis* were isolated from the roots; to avoid destroying the parent trees, it would be more suitable to use other naphthoquinone-containing plant parts. Berries or fruits are the obvious preference.

Anthraquinones and an anthraquinol methyl-ether from the leaves of *Morinda lucida* have also been shown to possess molluscicidal activity (Adewunmi and Adesogan, 1984). Rubiadin (*50*), rubiadin-1-methyl ether (*51*), 2-methylanthraquinone (*52*), damnacanthal (*53*), soranjidiol diacetate (*54*), and oruwal (*55*) all show 100% toxicity to *B. rohlfsii* and *B. globosus* snails at 10 mgL^{-1}.

Table 6. Molluscicidal Activities of Naphthoquinones

Compound	R^1	R^2	R^3	R^4	R^5	Plant	Molluscicidal Activity mgL^{-1} (LC$_{100}$)	Reference
39 (Juglone)	H	H	OH	H	H	(*Juglans regia*) (Juglandaceae)	10	Marston et al., 1984a
40 (Isojuglone)	OH	H	H	H	H	(*Lawsonia inermis*) (Lythraceae)	50	Marston et al., 1984a
41 (7-Methyljuglone)	H	H	OH	CH$_3$	H	*Diospyros usambarensis*	5	Marston et al., 1984a

Compound						Source		Reference
42 (Plumbagin)	CH$_3$	H	OH	H	H	(*Drosera rotundifolia*) (Droseraceae) artifact	2	Marston et al., 1984a
43 (3-Methoxy-7-methyljuglone)	H	OCH$_3$	OH	CH$_3$	H		50	Marston et al., 1984a
44 (Vitamin K$_3$; menadione)	CH$_3$	H	H	H	H	commercial chemical	3	Marston et al., 1984a
45 (Naphthazarin)	H	H	OH	H	OH	commercial chemical	50	Marston et al., 1984a
46 (Lapachol)						commercial chemical	50	Marston et al., 1984a
47 (Vitamin K$_1$)						commercial chemical	50	Marston et al., 1984a
48 (Isodiospyrin)						*Diospyros usambarensis*	50	Marston et al., 1984a
49 (Mamegakinone)						*Diospyros usambarensis*	50	Marston et al., 1984a

Abbreviations: see Table 1.
Adapted with permission from A. Marston and K. Hostettmann, *Phytochemistry*, **24**(4): 639–652, copyright 1985, Pergamon Journals Ltd.

Before wider applications of quinones in bilharzia control are considered, careful investigations of their toxicities to nontarget organisms must be carried out.

50   R = H
51   R = CH₃

52

53

54

55

Flavonoids and Rotenoids

The flavonol glycoside quercetin 3-(2'-galloylglucoside) from *Polygonum senegalense* has been reported to kill *B. glabrata* snails at concentrations as low as 10 mgL^{-1} (Dossaji and Kubo, 1980). The closely related compound quercetin 3-(2'-galloylgalactoside), however, has shown no activity on the same snails in our laboratory, even up to concentrations of 70 mgL^{-1}.

In view of the abundance of flavonoids in plants, molluscicidal activities of at least some members of this class of compounds would be very desirable for the control of vector snails.

Following reports that *Tephrosia vogelii* (Leguminosae) leaves are active against snails (Ransford, 1948), the rotenoids of this plant were investigated for molluscicidal properties (Marston et al., 1984b). Although a petroleum ether extract of the leaves gave some activity (400 mgL^{-1}), the major rotenones, deguelin and tephrosin, proved to be inactive. The commercially available insecticide rotenone was similarly inactive in concentrations up to 100 mgL^{-1} (Marston and Hostettmann, 1985).

Coumarins

The recent discovery that roots of the Nigerian medicinal plant *Clausena anisata* (Rutaceae) have molluscicidal properties (LC_{90} of the 80% methanol extract is 80 mgL^{-1} and the powdered crude drug kills 60% of *B. globosus* snails at 100 mgL^{-1}) led to the investigation of the molluscicidal activity of the constituent coumarins. Although these constituents are only sparingly water soluble and need to be dissolved in propylene glycol before application, their activities approach relatively low LC_{100} values. For example, both heliettin and imperatorin from *C. anisata* have LC_{100} values of 8 mgL^{-1} against *B. globosus*. Structure-activity relationships with other coumarins are being studied in the same laboratory (Adesina and Adewunmi, 1985).

Tannins

A number of plants rich in tannins have strong snail-killing activity (Schaufelberger and Hostettmann, 1983). Typical sources of tannins, such as *Krameria triandra*, *Hamamelis virginiana*, and *Quercus*, give extracts that are active at 50, 100, and 200 mgL^{-1}, respectively. Even a methanol extract of Japanese green tea leaves (*Camellia* sp.) kills *B. glabrata* snails at 200 mgL^{-1} (Schaufelberger and Hostettmann, 1983). Commercial tannic acid (tanninum) is active against *B. glabrata* snails at 50 ppm. Gallic acid, ellagic acid, and (+)-catechin are, however, inactive up to 100 mgL^{-1} (Schaufelberger and Hostettmann, 1983).

Following reports that fruits of *Acacia nilotica* (Leguminosae) possessed molluscicidal properties, it was discovered that aqueous extracts were active against *B. pfeifferi* and *Bulinus truncatus* snails at 200 mgL^{-1} (Ayoub, 1982). Furthermore, a special preparation called TAN, which is a spray-dried powder of aqueous extracts of the fruits, was active against *B. truncatus* at 50 mgL^{-1} and *B. pfeifferi* at 75 mgL^{-1} (Ayoub, 1982). It was then reported that ethyl acetate extracts of the fruits were also active (Ayoub and Yankov, 1984) – a somewhat retrograde step, in view of the economics of organic solvent extractions.

The fruits of *A. nilotica* contain up to 23% tannins, and it is reasonable to assume that they are the molluscicidal agents; indeed, it has been claimed that (-)-epigallocatechin-7-gallate and (-)-epigallocatechin-5,7-digallate are responsible for the action of *A. nilotica* extracts (Ayoub, 1984). The activities of these two flavonol derivatives are only as high as those of the extracts, however, implying that other molluscicidal compounds must also be present.

Twenty or so other species of *Acacia* have also been tested for their snail-killing abilities; activity would appear to correlate with tannin content (Ayoub, 1985).

Because tannins may prove to be less toxic to nontarget organisms than saponins, they appear to be a class of natural products worthy of further investigation – although it must be emphasized that their structure elucidation is not straightforward and their activities are weak.

Isobutylamides

Three isobutylamides have been shown to kill snails but at rather high concentrations in water (Table 7).

Alkaloids

The only example of a molluscicidal alkaloid with any activity to date is the quinolizine alkaloid 2,3-dehydro-0-(2-pyrrolylcarbonyl)-virgiline (*59*) from the leaves of *Calpurnia aurea* (Leguminosae) (Kubo et al., 1984b). This compound kills *B. glabrata* snails at 130 mgL^{-1} within 48 hours but will never have any practical use.

59

CONCLUSION AND OUTLOOK

Despite the number of new plant molluscicides that have been documented since the 1983 meeting of the Scientific Working Group concerned with this aspect of schistosomiasis, very few actually satisfy the criteria for effective large-scale application (see Annex).

Many simply do not have sufficient activity. The LC$_{90}$ should be less than 10 mgL^{-1} to be competitive with synthetic molluscicides and to avoid the use of prohibitively large amounts of plant material containing the active compounds for the treatment of infected sites. Extracts of plants for application to infected sites should originate from regenerating parts, such as fruits and leaves, so that the plant itself is not actually destroyed while collecting the vegetable material. Therefore, extracts of roots are impractical for mollusciciding. The size of the plant is also important.

The size factor and other criteria for effective plant molluscicides are well illustrated by *S. madagascariensis*. Its saponin-containing fruits are large, and the yield per tree per year can reach 40 kg. The amount of raw material available and the widespread occurrence of this tree in areas of Africa with

56

57

58

Table 7. Molluscicidal Activities of Isobutylamides

| Compound | Plant | Molluscicidal Activity | | Reference |
		mgL^{-1}	Snails Tested	
56 (Affinin)	*Heliopsis longipes*	50 (LC$_{100}$)	*Physa occidentalis*	Johns et al., 1982
(N-Isobutyl-2,6,8-decatrienamide)	*Spilanthes oleraceae* (Compositae) *Wedelia parviceps* (Compositae)			
57 (N-Isobutyl-2E,4E-octadienamide)	*Fagara macrophylla*	200 (LC$_{50}$)	*Biomphalaria glabrata*	Kubo et al., 1984a
58 (Fagaramide)	*Fagara macrophylla*	200 (LC$_{50}$)	*Biomphalaria glabrata*	Kubo et al., 1984a

Abbreviations: see Table 1.
Adapted with permission from A. Marston and K. Hostettmann, *Phytochemistry*, **24**(4): 639–652, copyright 1985, Pergamon Journals Ltd.

a high incidence of schistosomiasis give it advantages over the other promising molluscicidal plant, *P. dodecandra*.

Saponins from *S. madagascariensis* and *P. dodecandra* are among the most powerful pure plant-derived molluscicides and have activities of the same order of magnitude as synthetic molluscicides. The chemistry and structure-activity relationships of the saponins from both plants are under investigation in the Lausanne laboratories.

Moreover, a series of field trials with *S. madagascariensis* to evaluate its applicability as a large-scale plant molluscicide in focal areas is under way in Southeast Tanzania (Suter et al., 1986), and the first results look very promising.

Among the other classes of compounds with some potential as molluscicides are the tannins. The activities of pure tannins have not been systematically investigated, but the high concentrations of tannins in plant extracts and their widespread occurrence in nature mean that they may find some practical use in the control of schistosomiasis. The one restriction on the few molluscicidal tannins so far structurally characterized is their weak activity when compared, for example, with saponins.

A number of groups worldwide are actively searching for plant-derived molluscicides, and new sources and perhaps new classes of plant molluscicides will be discovered in the future. It remains an exciting question as to how they will measure up to the possibilities offered by the molluscicides currently available.

Acknowledgments

Financial support from the UNDP/WORLD BANK/WHO Special Programme for Research and Training in Tropical Diseases and the Swiss National Science Foundation is gratefully acknowledged.

REFERENCES

Adesina, S. K., and C. O. Adewunmi. Molluscicidal agents from the root of *Clausena anisata*. *Fitoterapia* 56:289–292 (1985).

Adewunmi, C. O., and E. K. Adesogan. Anthraquinones and oruwacin from *Morinda lucida* as possible agents in fascioliasis and schistosomiasis control. *Fitoterapia* 55:259–263 (1984).

Ayoub, S. M. H. Molluscicidal properties of *Acacia nilotica*. *Planta Med.* 46:181–183 (1982).

——. Effect of the galloyl group on the molluscicidal activity of tannins. *Fitoterapia* 55:343–345 (1984).

——. Flavonol molluscicides from the Sudan acacias. *Int. J. Crude Drug Res.* 23:87–90 (1985).

Ayoub, S. M. H., and L. K. Yankov. Field trials for the evaluation of the molluscicidal activity of *Acacia nilotica*. *Fitoterapia* 55:305–307 (1984).

Barros, D. A. D., J. L. C. Lopes, W. Vichnewski, J. N. C. Lopes, P. Kulanthaivel, and W. Herz. Sesquiterpene lactones in the molluscicidal extract of *Eremanthus glomerulatus*. *Planta Med*. 38–39 (1985).

Bezerra, P., A. G. Fernandes, A. A. Craveiro, C. H. S. Andrade, F. J. A. Matos, J. W. Alencar, M. I. L. Machado, G. S. B. Viana, F. F. Matos, and M. Z. Rouquayrol. Chemical composition and biological activity of essential oils of plants from Northeast Brazil – genus *Lippia*. *Cienc. Cult*. 33 (Supl., Simp. Plant. Med. Bras., 6th, 1980):1–14 (1981).'

Domon, B., and K. Hostettmann. Saponins with molluscicidal properties from *Lonicera nigra* L. *Helv. Chim. Acta* 66:422–428 (1983).

——. New saponins from *Phytolacca dodecandra* l'Herit. *Helv. Chim. Acta* 67:1310–1315 (1984).

Dossaji, S., and I. Kubo. Quercetin 3-(2'-galloylglucoside), a molluscicidal flavonoid from *Polygonum senegalense*. *Phytochemistry* 19:482(1980).

Duncan, J. The toxicology of plant molluscicides. *Pharmacol. Ther*. 27:243–264 (1985).

Fronczek, F. R., D. Vargas, N. H. Fischer, and K. Hostettmann. The molecular structure of 7α-hydroxy-3-desoxyzaluzanin C, a molluscicidal sequiterpene lactone. *J. Nat. Prod*. 47:1036–1039 (1984).

Gafner, F., J. D. Msonthi, and K. Hostettmann. Phytochemistry of African medicinal plants: Part 3. Molluscicidal saponins from *Talinum tenuissimum* Dinter. *Helv. Chim. Acta* 68:555–558 (1985).

Haerdi, F. Afrikanische Heilpflanzen. *Acta Tropica* Suppl. 8:1 (1964).

Hostettmann, K. Saponins with molluscicidal properties from *Hedera helix* L. *Helv. Chim. Acta* 63:606–609 (1980).

——. On the use of plants and plant-derived compounds for the control of schistosomiasis. *Naturwissenschaften* 71:247–251 (1985).

Johns, T., K. Graham, and G. H. N. Towers. Molluscicidal activity of affinin and other isobutylamides from the Asteraceae. *Phytochemistry* 21:2737–2738 (1982).

Johnson, A. L., and Y. Shimizu. Phytolaccinic acid, a new triterpene from *Phytolacca americana*. *Tetrahedron* 30:2033–2036 (1974).

Kubo, I., and T. Matsumoto. Desacetylisotenulin, a molluscicide from the desert plant *Psathyrotes ramosissima*. *Agric. Biol. Chem*. 48:3147–3149 (1984a).

——. Molluscicides from *Olea europaea* and their efficient isolation by countercurrent chromatographies. *J. Agric. Food Chem*. 32:637–638 (1984b).

Kubo, I., T. Matsumoto, J. A. Klocke, and T. Kamikawa. Molluscicidal and insecticidal activities of isobutylamides isolated from *Fagara macrophylla*. *Experientia* 40:340–341 (1984a).

Kubo, I., T. Matsumoto, M. Kozuka, A. Chapya, and H. Naoki. Quinolizine alkaloids from the African medicinal plant *Calpurnia aurea*: Molluscicidal activity and structrual study by 2D-NMR. *Agric. Biol. Chem*. 48:2839–2841 (1984b).

Lugt, C. B. *Phytolacca Dodecandra Berries as a Means of Controlling Bilharzia-Transmitting Snails*. Addis Ababa: Litho Printers (1981).

Marchant, Y. Y., F. Balza, B. F. Abeysekera, and G. H. N. Towers. Molluscicidal activity of sesquiterpene lactones. *Biochem. Systematics Ecol*. 12:285–286 (1984).

Marston, A., and K. Hostettmann. Plant molluscicides. *Phytochemistry* 24:639–652 (1985).

Marston, A., J. D. Msonthi, and K. Hostettmann. Naphthoquinones of *Diospyros usambarensis*: Their molluscicidal and fungicidal activities. *Planta Med*. 50:279–280 (1984a).

——. On the reported molluscicidal activity from *Tephrosia vogelii* leaves. *Phytochemistry* 23:1824–1825 (1984b).

Mozley, A. *Molluscicides*. London: Lewis (1952).

Parkhurst, R. M., D. W. Thomas, W. A. Skinner, and L. W. Cary. Molluscicidal saponins of *Phytolacca dodecandra*: Lemmatoxin C. *Indian J. Chem.* 11:1192–1195 (1973a).

——. Molluscicidal saponins of *Phytolacca dodecandra*: Oleanoglycotoxin-A. *Phytochemistry* 12:1437–1442 (1973b).

——. Molluscicidal saponins of *Phytolacca dodecandra*: Lemmatoxin. *Can. J. Chem.* 52:702–705 (1974).

Ransford, O. N. Schistosomiasis in the Kota Kota district of Nyasaland. *Trans. R. Soc. Trop. Med. Hyg.* 41:617–625 (1948).

Sati, O. P., G. Paul, and K. Hostettmann. Potent molluscicides from asparagus. *Pharmazie* 39:581 (1984).

Schaufelberger, D., and K. Hostettmann. On the molluscicidal activity of tannin-containing plants. *Planta Med.* 48:105–107 (1983).

Shoeb H. A., and M. A. El-Emam. The molluscicidal properties of substances gained from *Ambrosia maritima*. In *Proceedings of the International Conference on Schistosomiasis*, 487–494. Cairo: S.O.P. Press (1978).

Suter, R., M. Tanner, C. Borel, K. Hostettmann, and T. A. Freyvogel. Laboratory and field trials at Ifakara (Kilombero District, Tanzania) on the plant molluscicide *Swartzia madagascariensis*. *Acta Tropica* 43:69–83 (1986).

Thomson, R. H. *Naturally Occurring Quinones*. London: Academic Press (1971).

Towers, G. H. N. Contact hypersensitivity and photodermatitis evoked by Compositae. In *Toxic Plants*, ed. A. D. Kinghorn, 171–193. New York: Columbia University Press (1979).

Vichnewski, W., S. J. Sarti, B. Gilbert, and W. Herz. Goyazensolide, a schistosomicidal heliangolide from *Eremanthus goyazensis*. *Phytochemistry* 15:191–193 (1976).

ANNEX

GUIDELINES FOR EVALUATION OF PLANT MOLLUSCICIDES

These guidelines are intended to provide a general background for scientists involved in research on molluscicides of plant origin. The interpretation of these guides is meant to be flexible to accommodate local needs.

1. The identification of the plant must be confirmed by a botanist and specimens submitted to a national herbarium and at least one international herbarium. The voucher specimen should include the following information:
   a) name of collector, collector number, herbarium number;
   b) date and season of collection;
   c) geographic location of collection (longitude and latitude);
   d) altitude of collection and habitat;
   e) local name of the plant;
   f) the specimen submitted should be fully flowered and/or with ripe fruit as appropriate;
   g) uses and reason for collection; and
   h) type of plant (tree, shrub, or herb).

2. All subsequent communications, particularly scientific publications, concerning the plant should include:
   a) the names of the herbaria where voucher specimens are deposited;
   b) genus, species, and authority;
   c) season in which the plant was collected;
   d) whether wild or cultivated tree, shrub, or herb; and
   e) part of the plant tested.

3. The molluscicidal activity must be evaluated and include the following minimal results from laboratory screening tests, if a candidate plant is to be considered for field trials. The species and origin of the laboratory snail should be fully described. The exact methods and conditions of experimentation should be carefully documented. All laboratory, toxicological and field tests should be carried out with a standard preparation* of plant material

---

* Interested scientists should write to the Office of the Director of the Special Programme, World Health Organization, 1211 Geneva 27, Switzerland, to ascertain if descriptions of specific standard preparations are available.

and the procedure, as well as the biological and chemical assays used to verify the constituents described.

3.1 Aquatic Snails

 a) The crude plant material (recorded as dry weight) should be active at equal to or less than 100 mg/liter (ppm) to kill 90% of snails exposed for 24 hours at constant water temperature.

 b) A cold and/or hot water extract of the plant material should be active at equal to or less than 20 mg/liter to kill 90% of snails exposed for 24 hours at defined water temperature.

 c) Optionally, an alcoholic solvent (methanol) extract of the plant material should be active at equal to or less than 20 mg/liter to kill 90% of snails exposed for 24 hours at defined water temperature.

 d) Optionally, a lipophilic solvent extract of the plant material should be active at equal to or less than 20 mg/liter to kill 90% of snails exposed for 24 hours at defined water temperature.

3.2 Amphibious Snails

The molluscicidal activity on amphibious snails may be further evaluated when a concentration at 100–200 mg/liter or less kills 90% or more snails exposed within three days in a preliminary screening test.

3.3 If the molluscicidal activity is found only in the alcoholic and/or lipophilic solvent fractions, the potential for further industrial development may be limited by the processing required for use in endemic areas.

4. The level of safety testing should parallel developmental stages of a plant molluscicide. Before it is used operationally, all safety testing requirements should be fulfilled, taking into account the degree of exposure of operational personnel, the general population, and the environment.

 a) The plant should not cause dermal toxic effects upon exposure to those involved in handling and processing it.

 b) Toxicity to nontarget organisms must be at acceptable levels.

5. When research on a plant product has:

 a) defined molluscicidal activity;

 b) shown it to be acceptably safe; and

 c) reached the point of showing feasibility of cultivation,

the research group is encouraged to collaborate with laboratories having competence for determination of chemical structures. It is concluded that a full description of the chemical characteristics of active molluscicide constituents is desirable, but is not a prerequisite, for field evaluation and development of the plant molluscicide.

6. The agricultural aspects to be considered include:

 a) cultivation of the plant under a wide range of local environmental conditions while maintaining high, consistent molluscicidal activity;

b) possibility of propagation from seeds or cuttings;
c) a high yield of molluscicidal activity per plant or unit area cultivated;
d) resistance to diseases and pests; and
e) maintenance – in storage conditions for at least one year – of molluscicidal activity of the plant material, which is produced seasonally.

7. Other General Considerations
   a) If preparation of the plant material is necessary, ideally it should require only crushing or grinding by appropriate local technology.
   b) If extraction is necessary, the active chemical principle should be extractable by a simple apparatus and commonly available solvents, preferably water.
   c) Knowledge on the part of the local population of growing habits and requirements, toxicity, and any medicinal properties of plants is an asset.
   d) Absence of cultural uses of candidate plants and aversions based on folklore and magic, which might interfere with their use for snail control, is desirable.
   e) Suitability of the plants for other uses, such as medicines, pesticides, and other manufactured goods, food and other domestic uses, erosion control and reforestation, is highly desirable.
   f) Appropriate environmental and plant conservation should be observed in collection of the candidate plant so as to maintain satisfactory ecological balance.

8. Field Evaluation
When the above criteria have been met and approval has been obtained from national authorities, the following sequence of field evaluations is suggested:
   a) Application of plant molluscicide by one or several techniques in defined natural snail habitats according to a protocol similar to that described in this document. After successful completion of this first evaluation, operational evaluation becomes appropriate as follows:
   b) Use of the plant molluscicide in a defined endemic area in combination with chemotherapy and in comparison with a similar area where chemotherapy alone is used; and/or an acceptable synthetic molluscicide (niclosamide) is used in combination with chemotherapy (this trial would be optional, according to requirements of the specific endemic countries).

PARTICIPANTS

Prof. C. O. Adewunmi, Drug Research and Production Unit, Faculty of Pharmacy, University of Ife, Ile-Ife, Nigeria.

Dr. A. A. Daffalla, Gezira Schisotosomiasis Project, Medical Research Council, P. O. Box 1304, Khartoum, Sudan.

Dr. J. Duncan, Center for Overseas Pest Research, 56 Gray's Inn Road, London WC1X 8LU, UK.

Prof. N. R. Farnsworth, Program for Collaborative Research in the Pharmaceutical Sciences, University of Illinois at Chicago, Box 6998, Chicago, IL 60680, USA.

Dr. B. Gilbert, Instituto de Ciencias Biomedicas, Universidade Federal do Rio de Janeiro, Rio de Janeiro, Brazil.

Prof. T. O. Henderson, Department of Biological Chemistry, M/C 536, University of Illinois College of Medicine at Chicago, 1853 West Polk Street, Chicago, IL 60612, USA.

Prof. K. Hostettmann, Institute of Pharmacognosy and Phytochemistry, School of Pharmacy of the University, Rue Vuillermet 2, CH-1005 Lausanne, Switzerland.

Dr. H. Kloos, 2307 North Backer, Fresno, CA 93703, USA. (Present address: Department of Geography, Addis Ababa University, P. O. Box 31609, Addis Ababa, Ethiopia.)

Prof. J. H. Koeman, Department of Toxicology, Agricultural University, Biotechnion, De Dreijen 12, NL-6703 BC Wageningen, The Netherlands.

Dr. Y. H. Kuo, Institute of Parasitic Diseases, Chinese Academy of Preventive Medicine, 207 Rui Jin Er Lu, Shanghai, China.

Dr. J. D. H. Lambert, Department of Biology, Carleton University, Ottawa K1S 5B6, Canada.

Dr. C. B. Lugt, c/o Royal Netherlands Embassy, 18 Hassan Sabri, Zamalek Street, Cairo, Egypt. (Present address: Borneostraat 124, NL-2585 TW The Hague, The Netherlands.)

Prof. G. E. H. Mahran, Pharmacognosy and Medicinal Plants Department, Faculty of Pharmacy, University of Cairo, Cairo, Egypt.

Dr. E. Malek, Department of Tropical Medicine, School of Medicine, Tulane University, 1430 Tulane Avenue, New Orleans, LA 70112, USA.

Dr. L. Rey, c/o OMS, Caixa Postal 377, Maputo, Mozambique. (Present address: Superintendente INCQS, Fundação Oswaldo Cruz, Caixa Postal 926, CEP 20 000 Rio de Janeiro RJ, Brazil.)

Dr. R. F. Sturrock, London School of Hygiene and Tropical Medicine, Winches Farm Field Station, 395 Hatfield Road, St. Albans, Herts. AL4 OXQ, UK.

Prof. G. Webbe, Department of Medical Helminthology, London School of Hygiene and Tropical Medicine, Winches Farm Field Station, 395 Hatfield Road, St. Albans, Herts. AL4 OXQ, UK. (Chairman)

Dr. G. E. Wickens, Economic and Conservation Section, Royal Botanic Gardens, Kew, Richmond, Surrey TW9 3AB, UK.

WHO Secretariat

Dr. C. O. Akerle, Traditional Medicine, Division of Diagnostic, Therapeutic and Rehabilitative Technology.

Dr. K. Y. Chu, Special Programme for Research and Training in Tropical Diseases, WHO Regional Office for the Western Pacific.

Dr. A. Davis, Director, Parasitic Diseases Programme.

Mr. H. Dixon, Epidemiological and Statistical Methodology, Division of Epidemiological Surveillance and Health Situation and Trend Assessment.

Dr. F. S. McCullough, Ecology and Control of Vectors, Division of Vector Biology and Control.

Dr. K. E. Mott, Chief, Schistosomiasis and Other Trematode Infections, Parasitic Diseases Programme.

Dr. M. Vandekar, Pesticides Development and Safe Use, Division of Vector Biology and Control.

Miss A. Wehrli, Pharmaceuticals, Division of Diagnostic, Therapeutic and Rehabilitative Technology.